Molecular Materials
Preparation, Characterization, and Applications

Molecular Materials

Preparation, Characterization, and Applications

Sanjay V. Malhotra, B. L. V. Prasad, and
Jordi Fraxedas

CRC Press is an imprint of the
Taylor & Francis Group, an **informa** business

CRC Press
Taylor & Francis Group
6000 Broken Sound Parkway NW, Suite 300
Boca Raton, FL 33487-2742

First issued in paperback 2022

ISBN-13: 978-1-482-24595-0 (hbk)
ISBN-13: 978-1-03-233969-6 (pbk)
DOI: 10.1201/9781315118697

Library of Congress Cataloging-in-Publication Data

Names: Malhotra, Sanjay V., editor. | Prasad, B. L. V., editor. | Fraxedas, Jordi, editor.
Title: Molecular materials: preparation, characterization, and applications/
edited by Sanjay Malhotra, B. L. V. Prasad, and Jordi Fraxedas.
Description: Boca Raton, FL : CRC Press, [2017] | Includes bibliographical
references and index.
Identifiers: LCCN 2016053998| ISBN 9781482245950 (hardback : alk. paper) |
ISBN 9781482245967 (ebook)
Subjects: LCSH: Thin films. | Nanostructured materials. | Layer structure (Solids)
Classification: LCC TA418.9.T45 M65 2017 | DDC 530.4/275–dc23
LC record available at https://lccn.loc.gov/2016053998

Visit the Taylor & Francis Web site at
http://www.taylorandfrancis.com

and the CRC Press Web site at
http://www.crcpress.com

Contents

Preface

The field of molecular materials, that is, those materials essentially made out of molecules, is extremely broad and dynamic, covering a long list of diverse disciplines such as biotechnology, biomedicine, microelectronics, materials science, and chemistry, to mention but a few. The activity related to this field has experienced a burst with the advent of nanotechnology in the present millennium since strategies have been developed with the quest to engineer materials with predefined functionalities. Here, the advantage of molecular materials over their inorganic counterparts is overwhelming, since molecular materials are intrinsically flexible in character, a property rarely found in inorganic materials. The complexity and variety of structures that can be obtained with molecular materials go well beyond what can be achieved with inorganic materials. The flexibility arises mainly from the abundant number of carbon–carbon, carbon–hydrogen, and carbon–oxygen bonding in the molecules that enable a host of adaptable and robust structures, since, for example, carbon–carbon bonding is one of the strongest bonds in nature but at the same time rotation pivoting on this bond is possible. Flexibility is a necessary condition for adaptability, which in the case of biomolecules is mandatory in order to sustain favorable architectures for self-sustained processes that enable life. Thus, molecular materials can be conceived as the interface connecting the pure inorganic and bioorganic realms. However, this mediation role is non-exclusive for biochemistry and can be extended to other non-biological environments.

Given the plethora of available molecular materials, with their particular functionalities, it would be a titanic task to try to summarize all the reported materials and applications in a single, manageable book. Such a book would end up in an encyclopedic format that would become out of date in a short time period. From this perspective, we have structured this book in two main areas, preparation and applications, and organized it into four sections: polymers, self-assembly and supramolecular architectures, nanostructures/metal oxides, and devices, with the aim to cover myriad of highly relevant topics. The first three sections contain the synthesis and preparation of molecular materials in different structures and architectures, that is, thin films, nanoparticles, supramolecular structures, and assemblies on surfaces, while the fourth deals with some applications.

Chapters 1 through 3 (polymers) are devoted to Langmuir–Blodgett films, porphyrin-based nanomaterials, and conjugated polymers, respectively. Chapters 4 through 6 (self-assembly and supramolecular architectures) introduce new tools for hierarchical assembly, the use of solid surfaces to guide assembly, and the synthesis and physical properties of organic conductors and superconductors, respectively. Chapters 7 through 9 (nanostructures/metal oxides) focus on silicate and phosphate nanomaterials, the control of nanoparticle size and morphologies, and the use of metal-organic precursors, respectively. Finally, Chapters 10 through 12 (devices) explore selected examples of applications in microelectronics, biomedicine, and drug delivery systems.

Sanjay V. Malhotra

B. L. V. Prasad

Jordi Fraxedas

Editors

Sanjay V. Malhotra earned a PhD in chemistry at Seton Hall University, New Jersey, in 1995, under the supervision of Prof. Robert L. Augustine. Subsequently, he did postdoctoral research at Purdue University under the supervision of Prof. Herbert C. Brown and started his independent career as assistant professor at the New Jersey Institute of Technology. In 2006, he moved to the Frederick National Laboratory for Cancer Research in Maryland and established the Laboratory of Synthetic Chemistry to support the drug discovery and development programs of the National Cancer Institute. In 2015, he joined the faculty at Stanford University, Palo Alto, California. His research interests are in the design and discovery of synthetic and natural product–inspired small molecules that can be used as probes for developing the understanding of biological phenomena and translational research in drug discovery, development, imaging, and radiation. He has coauthored more than 100 peer-reviewed scientific articles and edited 4 books.

B. L. V. Prasad earned a degree in chemistry at the University of Hyderabad (India) in 1989 and earned a PhD in 1997 at the same university under the supervision of Prof. T. P. Radhakrishnan. After two postdoctoral stints—one at Tokyo Institute of Technology (2-year JSPS fellowship and 1-year research associateship) and the other at Kansas State University (2.5 years)—he joined NCL in 2003 and is continuing there as a principal scientist. His group is actively working in the general area of material synthesis and in particular on nanoparticles and nanoscale materials. The areas his group is currently working on include the identification of novel nanoparticle synthetic routes, preparation of biomolecule–nanoparticle conjugates and their applications in drug delivery systems, bio-implants, and diagnostics. He has published more than 100 papers in peer-reviewed international journals and filed 15 patent applications. Eleven students have obtained PhDs under his supervision so far. For his achievements, Dr. Prasad has received medals from the Materials Research Society of India and the Chemical Research Society of India and also a Young Career Award from the Department of Science and Technology (India)—Nano Mission. Recently he was elected fellow of the Indian Academy of Sciences, Bangalore (INDIA).

Jordi Fraxedas earned a degree in physics at the University of Zaragoza (Spain) in 1985 and earned a PhD (Dr. rer. nat.) in 1990 at the University of Stuttgart (Germany). His thesis work was performed at the Max Planck Institut für Festkörperforschung and at the Berliner Speicherring für Synchrotronstrahlung (BESSY) under the supervision of Professor M. Cardona. After a postdoctoral position at the European Synchrotron Radiation Facility (ESRF) in Grenoble (France) and an established researcher position at the European Laboratory for Particle Physics (CERN) in Geneva (Switzerland), he joined the Solid State Research Institute of Barcelona (ICMAB) of the Spanish Research Council (CSIC) in 1995 and worked as a research associate at the Centre National de la Recherche Scientifique (CNRS) in 2002. Since 2007, he has led the Force Probe Microscopy and Surface Nanoengineering Group at the Catalan Institute of Nanoscience and Nanotechnology (ICN2). His research activity is focused on interfacial phenomena and surface science. He has coauthored more than 120 peer-reviewed scientific articles and published the books *Molecular Organic Materials: From Molecules to Crystalline Solids* (Cambridge University Press, 2006) and *Water at Interfaces: A Molecular Approach* (Taylor & Francis CRC, 2014).

Contributors

Daniel A. Bernards
Department of Bioengineering and Therapeutic Sciences
University of California, San Francisco
San Francisco, California

Mario Cabodi
Department of Biomedical Engineering
Boston University
Boston, Massachusetts

Dominique de Caro
Coordination Chemistry Laboratory
National Center for Scientific Research
Toulouse, France

K. Chakrapani
Department of Inorganic and Physical Chemistry
Indian Institute of Science
Bangalore, Karnataka, India

Tejal A. Desai
Department of Bioengineering and Therapeutic Sciences
University of California, San Francisco
San Francisco, California

and

Joint Graduate Group in Bioengineering
University of California, Berkeley
Berkeley, California

Christophe Faulmann
Coordination Chemistry Laboratory
National Center for Scientific Research
Toulouse, France

Rachel Gamson
Miramonte High School
Orinda, California

Timothy S. Gehan
Department of Chemistry
University of Massachusetts, Amherst
Amherst, Massachusetts

Andrew C. Jamison
Department of Chemistry and the Texas Center for Superconductivity
University of Houston
Houston, Texas

S. Kiruthika
Chemistry and Physics of Materials Unit and Thematic Unit of Excellence in Nanochemistry
Jawaharlal Nehru Centre for Advanced Scientific Research
Bangalore, Karnataka, India

Catherine M. Klapperich
Department of Biomedical Engineering
Boston University
Boston, Massachusetts

Giridhar U. Kulkarni
Centre for Nano and Soft Matter Sciences
Bangalore, Karnataka, India

T. Randall Lee
Department of Chemistry and the Texas Center for Superconductivity
University of Houston
Houston, Texas

Lasya Maganti
School of Chemistry
University of Hyderabad
Hyderabad, Telangana, India

Sadananda Mandal
Department of Materials Science
Indian Association for the Cultivation of Science
Kolkata, West Bengal, India

Gangaiah Mettela
Chemistry and Physics of Materials Unit and Thematic Unit of Excellence in Nanochemistry
Jawaharlal Nehru Centre for Advanced Scientific Research
Bangalore, Karnataka, India

Takehiko Mori
Department of Materials Science and Engineering
Tokyo Institute of Technology
Tokyo, Japan

Ramaswamy Murugavel
Department of Chemistry
Indian Institute of Technology Bombay
Mumbai, Maharashtra, India

Amitava Patra
Department of Materials Science
Indian Association for the Cultivation of Science
Kolkata, West Bengal, India

B. L. V. Prasad
Physical/Materials Chemistry Division
CSIR-National Chemical Laboratory
Pune, Maharashtra, India

T. P. Radhakrishnan
School of Chemistry
University of Hyderabad
Hyderabad, Telangana, India

Puspanjali Sahu
Physical/Materials Chemistry Division
CSIR-National Chemical Laboratory
Pune, Maharashtra, India

S. Sampath
Department of Inorganic and Physical Chemistry
Indian Institute of Science
Bangalore, India

Erica Schlesinger
Department of Bioengineering and Therapeutic Sciences
University of California, San Francisco
San Francisco, California

and

Joint Graduate Group in Bioengineering
University of California, Berkeley
Berkeley, California

Sonika Sharma
Department of Chemistry
Loyola Academy Degree and PG College
Secunderabad, Telangana, India

Jayesh Shimpi
Physical/Materials Chemistry Division
CSIR-National Chemical Laboratory
Pune, Maharashtra, India

Lydie Valade
Coordination Chemistry Laboratory
National Center for Scientific Research
Toulouse, France

D. Venkataraman
Department of Chemistry
University of Massachusetts, Amherst
Amherst, Massachusetts

Pratap Vishnoi
Department of Chemistry
Indian Institute of Technology Bombay
Mumbai, Maharashtra, India

Sharon Y. Wong
Department of Biomedical Engineering
Boston University
Boston, Massachusetts

Crystal A. Young
Department of Chemistry and the Texas Center for Superconductivity
University of Houston
Houston, Texas

Section I

Polymers (Organic/Inorganic)

1 Polyelectrolyte-Templated Langmuir/Langmuir–Blodgett Films

Lasya Maganti, Sonika Sharma, and T. P. Radhakrishnan

CONTENTS

1.1 INTRODUCTION

The significant mutualistic impacts of molecular chemistry and nanoscience are the development of capabilities for the observation of single molecules, deployment of molecules as devices, and the growing prominence of molecular-level control of chemical processes and assembly. Single molecules can be observed by their spectroscopic signatures or even directly by microscopy.[1,2] There is a burgeoning interest in developing molecular-scale electronic, photonic, and spintronic devices.[3–6] Reactions have been carried out on selected individual molecules,[7] and the reaction-induced changes in the covalent structure of single molecules during the course of a reaction can be monitored.[8,9] Molecular assemblies can be constructed with a high degree of control on the congregation of the building blocks. Self-assembly approaches for the generation of supramolecular structures are quite popular.[10,11] Even covalently bound architectures have been fabricated through tailored assembly of molecular legos.[12] The subtle, yet crucial influence that the environment exerts on the attributes and functions of molecules highlights the significant role of assembly in developing novel molecular materials and devices.

Molecular nanomaterials occupy the critical space between bulk materials on one hand and the single molecule limit on the other.[13] Their fabrication and functions rely heavily on the molecular level control of the assembly. Methodologies for the fabrication of nanomaterials follow break-down (top-down) or build-up (bottom-up) protocols. Molecular nanomaterials are generally constructed through soft chemical approaches that lead to the organized build-up of molecules. The structural attributes of molecules that give rise to a variety of noncovalent interactions such as ionic, H-bonding, dispersion, and π-stacking guide their self-assembly[14] into supramolecular entities. Even though such a spontaneous process is extensively exploited in the fabrication of molecular materials, steered assembly realized through external controls such as electric or magnetic fields[15–17] and mechanical forces add a new dimension to it. The celebrated Langmuir–Blodgett (LB) technique[18,19] is of special interest in the context of using mechanical control that steers the assembly of molecules.

Amphiphilic molecules spread at the air–water interface can be packed into monolayer structures called Langmuir films and subsequently transferred to suitable substrates to form LB films.[20] The LB technique is a unique and elegant method to assemble molecules into mono- or multilayer structures, controlling systematically their packing density and hence the extent of aggregation. This method facilitates the fabrication of ultrathin films with tailored dimensions and molecular organization, serving as a valuable tool for the bottom-up synthesis of molecular nanomaterials. An inherent problem, however, is the metastability of LB films transferred to a substrate, as the molecules were originally assembled at the air–water interface under a mechanical constraint that does not exist any longer. Among the various approaches that have been explored to address this problem and enhance the stability of the LB films, a particularly attractive approach is to utilize a template, preferably a polymeric one with specific and strong interactions with the amphiphiles, to arrest the structural transformations of the assembly. This fruitful interplay of self-assembly and steered assembly that harnesses the benefits and efficacy of the LB technique and simultaneously realizes the stability required in a functional material forms the basic theme of this chapter.

We provide an overview of the LB technique and note in particular the issue of instability of the ultrathin films. A brief review of selected studies on the interaction of polymers with surfactants at the air–water interface, particularly protein–lipid systems, provides the background to the development of the idea of polyelectrolyte templating to stabilize Langmuir and LB films. The basic methodology involved in this approach and its relevance is described, followed by a listing of case studies that demonstrate the utility of the technique in suppressing molecular aggregation, enhancing materials responses and controlling molecular assembly and their reactions. Finally, possible extensions of the use of polyelectrolyte-templating in novel applications of LB films are outlined.

1.2 LANGMUIR–BLODGETTRY

Molecules possessing a hydrophilic head group and a hydrophobic tail, commonly referred to as amphiphiles, tend to disperse at the air–water interface and reduce the interfacial (surface) tension. The common head groups of these surfactant molecules include ionic, dipolar, or H-bonding functionalities, and the tails are typically hydrocarbon chains with at least 14–16 carbons; fatty acids such as stearic and arachidic acids are classic examples. The dependence of the surface tension (γ) on the surfactant concentration (c) at temperature T is given by the Gibb's adsorption isotherm

$$\left(\frac{d\gamma}{d\ln c}\right)_T = -\Gamma RT$$

where

Γ is the surface excess (moles per unit area)
R is the universal gas constant

A typical isotherm is shown in Figure 1.1a, and the distribution of the amphiphiles at the air–water interface at the various stages is shown schematically in Figure 1.1b.

1.2.1 Surface Pressure–Area Isotherm

Decrease in the surface tension from that of pure water is denoted as the surface pressure, π ($=\gamma_0 - \gamma$). Isothermal plot of the surface pressure against the area of the monolayer film, the pressure–area (π–A) isotherm, is a fundamental characterization of the monolayer at the air–water interface. A simple π–A isotherm is shown in Figure 1.1c; the regimes I, II, and III are described as the "gaseous," "liquid," and "solid" states and correspond, respectively, to a low density of amphiphiles with no contact between them, increased density leading to initiation of contact and enhancement of the surface pressure, and finally a close-packed two-dimensional lattice with low compressibility. Isotherms reflecting different states of the "liquid" phase as well as phase equilibria (biphasic regimes) of the amphiphile monolayers are often encountered.

The surface pressure is usually determined by measuring the force acting on a filter paper (Wilhelmy plate) contacting the water surface. A typical experiment to record the π–A isotherm involves the following steps. High-purity water is taken as the subphase in a shallow trough with large surface area, mostly constructed with a hydrophobic material such as Teflon®; in principle, any nonvolatile liquid can be taken as the subphase, but water is the most common choice. A dilute solution of the amphiphile in a volatile organic solvent, such as chloroform, is spread on the subphase surface; the concentration and volume of the solution are adjusted so that the number of molecules spread is less than that required to form a close-packed monolayer at the air–water interface. The amphiphile molecules are compacted by compression using a mechanical barrier, again usually fabricated using Teflon®, sliding on the subphase surface. As the total area of the monolayer and hence the area per molecule decrease, the surface pressure increases; opening the barrier reverses the process. Hysteresis in the isotherm is quite common and indicative of the different kinetics involved in the packing and unpacking of the amphiphilic molecules. Compression of the monolayer to very high pressures (above π_c in Figure 1.1c) can lead to collapse of the Langmuir film through forced dissolution into the subphase, formation

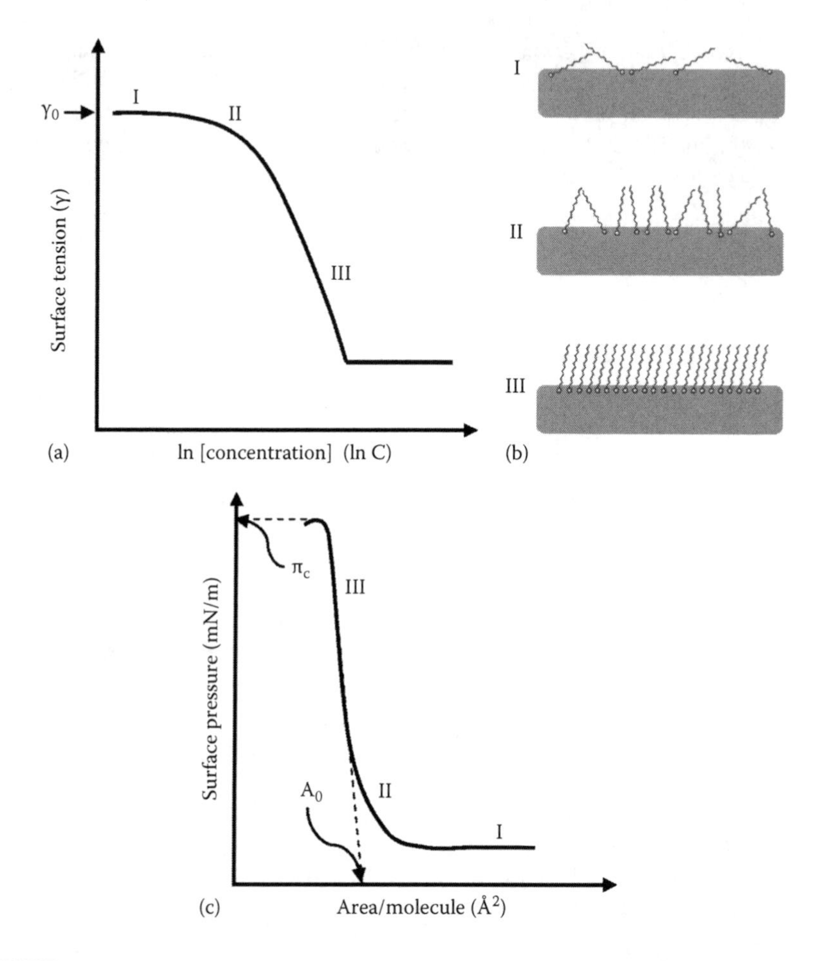

FIGURE 1.1 (a) Schematic diagram of the Gibbs isotherm; γ_0 is the surface tension of pure water. (b) Regions I, II, and III represent the surfactant monolayer at the air–water interface in the "gas," "liquid," and "solid" state, respectively; the water subphase (gray box) and the amphiphile with head group (open circle) and tail (wiggly line) are shown. (c) Schematic plot of a simple surface pressure–area isotherm; surface pressure $\pi = \gamma_0 - \gamma$, π_c is the collapse pressure and A_0 is the limiting molecular area of the monolayer.

of multilayers, or accidental loss under the mechanical barrier. Repeated cycles of compression–expansion (isocycling) are useful to obtain a reproducible isotherm and ensure uniform and efficient compaction of the film. The limiting area extrapolated from the "solid" region of the isotherm, A_0 in Figure 1.1c, gives an estimate of the molecular area representing the projection of the amphiphile on the surface.

1.2.2 Langmuir Film

The π–A isotherm can be translated into plots of "inverse compressibility" that describes the stiffness of the Langmuir film ($1/\kappa = -d\pi/dA$) versus the area/molecule.

The stability of the Langmuir film at the air–water interface is assessed by recording the area of a film maintained at a constant surface pressure as a function of time. If these "creep plots" show negligible changes in the area over extended periods of time, typically a few hours, the film is deemed to be mechanically robust. Brewster angle microscopy (BAM) based on the local variations in the refractive index at the air–water interface provides a powerful tool to image Langmuir films. Changes in the refractive index at the interface due to the formation of molecular assemblies lead to partial reflection of p-polarized light incident on the surface at the Brewster angle corresponding to pure water. Images with a spatial resolution of a few μm can be recorded using BAM, providing a map of possible domain formation and merger in the Langmuir film that occurs during the mechanical compression. Surface potential measurements using electrodes placed above and below the monolayer provide useful information about the dipole orientation of molecules at the air–water interface.

1.2.3 Langmuir–Blodgett Film

The monolayer film formed at the air–water interface can be transferred on to a substrate by either sliding the latter vertically through the film (vertical dipping) or gently contacting it with the film by placing the plane surface parallel to the interface (horizontal dipping) (Figure 1.2a and b); typically, the film is maintained at a constant surface pressure during the transfer. The former is called the Langmuir–Blodgett (LB) film, and the latter, the Langmuir–Schaefer (LS) film. Glass or quartz plates with the surface made hydrophilic or hydrophobic using appropriate treatment

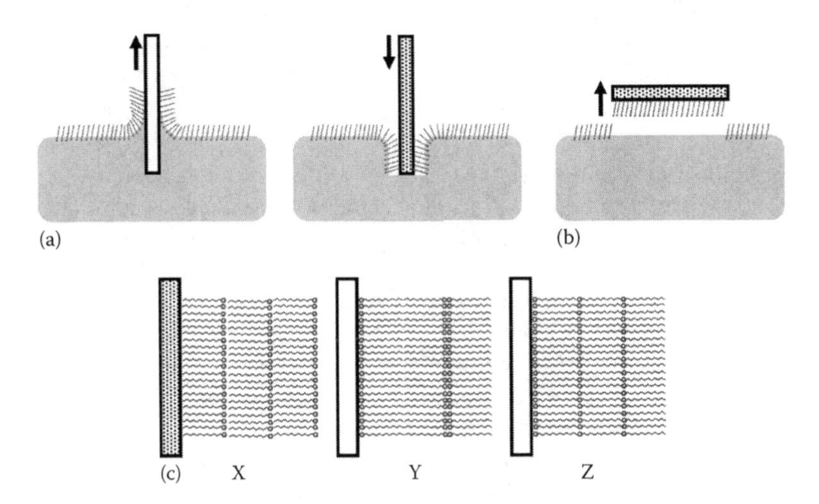

FIGURE 1.2 (a) Transfer of a monolayer during vertical dipping—upward for a hydrophilic substrate and downward for a hydrophobic substrate. (b) Transfer of a monolayer during horizontal dipping of a hydrophobic substrate. (c) Structure of X-, Y-, and Z-type multilayer LB films. Open and filled rectangles represent hydrophilic and hydrophobic substrates, respectively; the water subphase (gray box) and the amphiphile with head group (open circle) and tail (wiggly line) are shown.

are commonly used for the deposition of the LB or LS films. Mica, silicon, and highly oriented pyrolytic graphite (HOPG) substrates are also employed for specific applications. The orientation in which the amphiphilic molecule is transferred does indeed depend on the nature of the substrate surface and the direction of dipping, as shown in Figure 1.2a. The efficiency of transfer is denoted by the "transfer ratio," ratio of the area of film transferred from the air–water interface (as estimated from the movement of the mechanical barrier while maintaining the surface pressure constant) to the area of substrate dipped through the monolayer. One of the major advantages of the LB technique is that it allows facile and controlled transfer of mono- or multilayer molecular films (Figure 1.2c) onto the preferred substrate. The surface pressure at which the transfer is carried out provides an important handle to tune the density of molecules in the transferred films. Depending on the attachment of the amphiphile on the substrate and their relative orientation in successive layers, the LB films are classified as X, Y, and Z type.

A wide range of spectroscopy, diffraction, and microscopy tools can be employed to characterize LB films and investigate the material attributes of the ultrathin films. The substrates are chosen based on the specific techniques being employed; for example, the films are deposited directly on polymer-coated copper grids for transmission electron microscope (TEM) imaging and often on mica surface for atomic force microscopy (AFM).

1.2.4 Stability of Langmuir–Blodgett Film

There is a significant change in the environment and status of the molecular assembly as it moves from the Langmuir film to the LB film. In the Langmuir film, the molecules exist at the air–water interface and are under mechanical stress. For a monolayer that is typically 2 nm thick, a surface pressure of 30 mN/m translates to a bulk pressure of ~150 atm. Once transferred onto a substrate, the mechanical stress no longer exists and there are new adhesive forces with the surface. The structure of the LB film may resemble that of the Langmuir film due to kinetic factors but may be subject to transformations so as to attain a thermodynamically stable state.[21] The LB film is known to undergo molecular-level structural changes spontaneously or under external stimuli, leading to significant effects on their functionality. LB films of amphiphilic diacetylene mixed with polyallylamine have been shown to undergo morphological changes over time as a result of self-organization of the neutral state of the amphiphiles.[22] Gradual decay and disappearance of the optical absorption of LB films of a strongly zwitterionic diaminodicyanoquinodimethane–based amphiphile has been attributed to the spontaneous realignment of the chromophore groups and suppression of the intramolecular charge transfer.[23] Reorientation of chromophores in LB films of amphiphilic azobenzene dyes upon photoirradiation[24] and thermally induced order–disorder transition in molecular orientation and structure in domains of amphiphilic TCNQ-based LB films[25] have been reported. Laser irradiation has been found to induce molecular aggregation and consequent decay of optical second harmonic generation response of LB films of a hemicyanine-based amphiphile.[26] All these examples point to the inherent instability of LB films that significantly impairs the potential functionality of these materials.

1.3 POLYMER–SURFACTANT INTERACTIONS

The importance of interactions between polymers and surfactants, primarily in the context of surfactants assembled at the air–water interface, has been recognized for a long time. An early focus was on the feasibility of exploiting such interactions to engage a wide range of polymers to prepare LB film–based perm selective membranes, without actually having to synthesize many amphiphilic polymers. A typical example is the study of the gegenion complex LB films composed of poly(ethyleneimine) and arachidic acid and the effect of pH on the polymer–surfactant interaction.[27] Related studies have addressed the effect of temperature, surfactant chain length, and polymer concentration on the complex.[28–30] Figure 1.3 shows quite vividly the impact of varying the concentration of the polymer in the subphase on the domain structure of arachidic acid monolayer.[28]

A major part of the investigations related to polymer–surfactant interactions has been centered on biologically relevant problems. A topic of special interest is protein–lipid interactions and their important role in pharmaceutical research, biosensing, and membrane science. A review of the early significant advances in the general area of nanobiosciences[31] highlighted the relevance of oriented attachment of hydrophilic enzymes to glycolipidic monolayers achieved through the recognition of a monoclonal antibody

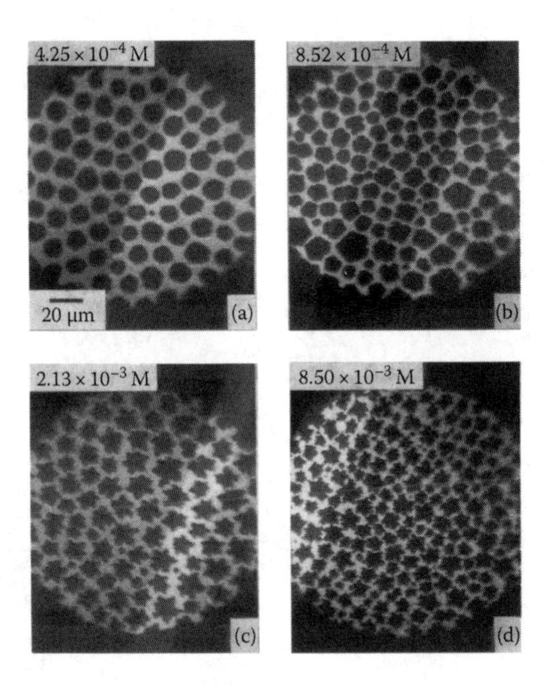

FIGURE 1.3 Fluorescence micrographs of arachidic acid monolayers for various concentrations, (a) 4.25×10^{-4} M, (b) 8.52×10^{-4} M, (c) 2.13×10^{-3} M, and (d) 8.50×10^{-3} M, of poly(ethyeleneimine) in the aqueous subphase. (Reprinted with permission from Chi, L.F., Johnston, R.R., and Ringsdorf, H., Fluorescence microscopy investigations of the domain formation of fatty acid monolayers induced by polymeric gegenions, *Langmuir*, 7, 2323–2329. Copyright 1991 American Chemical Society.)

embedded in the lipid layer. The proteo-lipidic nanostructures are useful in explorations of biocatalysis by immobilized enzymes in a biomimetic environment. Further, these systems interfaced with appropriate luminescent devices have great utility in biosensor applications. Langmuir films at the air–water interface facilitate detailed investigation of the interactions between proteins and lipids and the structure of the resulting complexes.

Brewster angle microscopy investigation of the interaction of hemoglobin with lipid monolayers at the air–water interface showed that the protein enters and remains in the lipid layer for extended periods of time.[32] Study of a model protein β-lactoglobulin (the major whey protein of bovine milk that gets adsorbed into the interface of lipid fat globules and stabilizes the oil-in-water emulsion) with lipid films at the air–water interface showed that the interactions between the two are driven by hydrophobic as well as electrostatic interactions, the phospholipids in an unfolded state being preferred.[33] The high encapsulation efficiency of the cytoplasmic protein, superoxide dismutase, in liposomes is likely to be due to specific protein–lipid interactions. Monitoring the relaxation of monolayers of 1,2-dipalmitoyl-*sn*-glycero-3-phosphocholine, 1,2-distearoyl-*sn*-glycero-3-phosphocholine, and cholesterol at the air–water interface, with and without the protein in the aqueous subphase, showed that the lipid length and mole percent of cholesterol determined the protein–lipid interactions; Figure 1.4 summarizes the main observation in this study.[34] A detailed investigation of the interaction of ovalbumin with cationic (octadecylammonium), zwitterionic (1,2-dipalmitoyl-*sn*-glycero-3-phosphocholine), and anionic (stearate) monolayers at the air–water interface showed that the incorporation of the protein in the cationic layer is most efficient, whereas the unfolding of the protein is least with the zwitterionic one;[35,36] the protein also tends to get squeezed out of the monolayer at high surface pressures. A Langmuir film study of the interaction of the

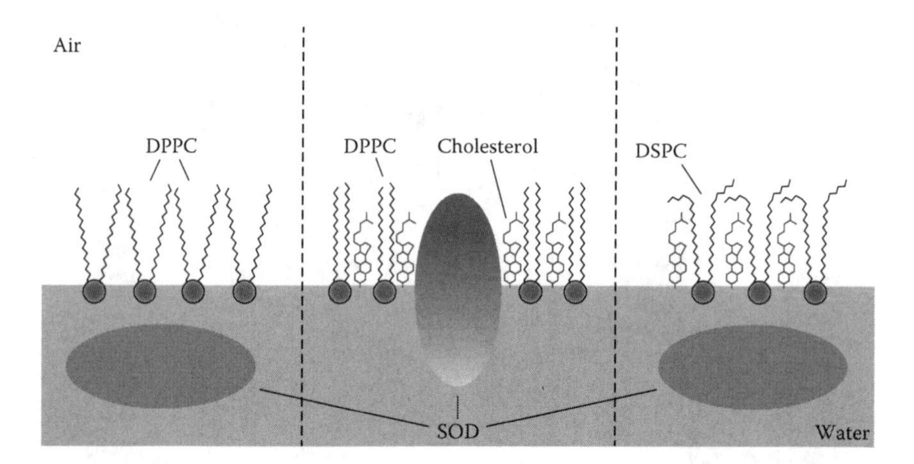

FIGURE 1.4 Schematic representation showing the conditions favoring encapsulation of superoxide dismutase (SOD) in lipid monolayers; DPPC : 1,2-dipalmitoyl-*sn*-glycero-3-phosphocholine, DSPC: 1,2-distearoyl-*sn*-glycero-3-phosphocholine. (Reprinted with permission from Costa, A.P., Xu, X., and Burgess, D.J., Langmuir balance investigation of superoxide dismutase interactions with mixed-lipid monolayers, *Langmuir*, 28, 10050–10056. Copyright 2012 American Chemical Society.)

filamentous protein dystrophin with anionic/zwitterionic membrane lipids showed that the anchorage of the protein to the membrane is promoted by the cholesterol-rich domains in the inner leaflet of the sarcolemma membrane.[37]

The LB technique allows the fabrication of simplified model structures that can be investigated as useful membrane mimics. Protein–lipid interactions form the basis of many of these studies. Monolayers of distearoyl-phosphatidylcholine-containing lipid molecules derivatized with iminodiacetate groups and bound Cu(II) ions were shown to bind myoglobin strongly, via surface-accessible histidine residues.[38] This provides a model to understand specific protein attachment to artificial membranes through metal ion mediation. Myelin model membranes on solid substrates were constructed using multilayer LB films of myelin basic protein with lipid molecules such as dimyristoyl-Lα-phosphatidic acid.[39,40] Neutron reflection, FTIR, CD, and x-ray reflectivity studies revealed the formation of ordered lipid layers with the protein stacked in between; zinc ions were found to aid the assembly of ordered stacks, and water molecules to affect structural rearrangements. The specificity of lipid–protein interaction has been demonstrated by studying LB films prepared from different lipids, cholesterol, and the integral membrane protein, glycophorin A;[41] the inner and outer leaflets of cellular membrane were mimicked by films containing dipalmitoylphosphatidylcholine and dipalmitoylphosphatidylethanolamine, respectively, and imaging based on time of flight secondary ion mass spectrometry showed the former to have a homogeneous phase while the latter displayed heterogeneity with domain separation. The effect of sulfated glycosaminoglycans on cell membranes, relevant to the treatment of diseases related to blood clotting and degenerative joint diseases, has been modeled by investigating the molecular-level interaction of chondroitin sulfate with phospholipid monolayers.[42] Interfacial studies of the interaction of human tear–lipid films with model-tear proteins using Langmuir film and sessile-bubble methods have been critically reviewed recently.[43]

An aspect of major interest in studying protein–lipid interactions at the air–water interface is to assess the impact of the interaction on the enzymatic activity of the protein. For example, evaluation of the interactions between the proteins such as cytochrome C, insulin and concanavalin A, and several types of lipid films revealed that the bound proteins retain their enzymatic activity or specific binding capability, providing models for membrane-bound receptors of importance in biochemistry and medicine.[44] Investigations of the Langmuir and LB films of various surfactants, including fatty acids and phospholipids complexed, respectively, with proteins such as glutathione S-transferase and enzymes like penicillinase showed that the enzymatic activities are largely retained by careful choice of the fabrication conditions.[45,46] Langmuir film studies of the complexation and incorporation of cytochrome C in monolayers of the phospholipid cardiolipin suggest concomitant conformational transitions and formation of catalytically reactive bubbles or hydrophilic pores.[47]

1.4 POLYELECTROLYTE TEMPLATING OF LANGMUIR AND LB FILMS

As discussed earlier, the LB film on a chosen substrate is in a different state compared to the corresponding Langmuir film at the air–water interface. As there are no strong (covalent) interactions that tie the individual molecules together, the LB films in their new environment are susceptible to spontaneous structural transformations as well as reorganization in the presence of external stimuli, such as radiations. A common approach to solve this problem is to admix the surfactant molecules of interest with amphiphiles such as fatty acids and alcohols that are known to form very stable Langmuir and LB films. This technique, however, leads to the reduction in the concentration of the active amphiphiles in the film, the molecules of interest in a specific application. The discussion in the previous section, regarding the extensively studied phenomenon of polymer–surfactant interaction, suggests the feasibility of exploiting the stabilizing influence of a polymer on the amphiphile film. This would be particularly effective, if strong electrostatic interactions could be exploited by the choice of appropriate ionic head group on the amphiphile and the counter charge on the polymer chain. In other words, an optimal choice of stabilizing partner for LB films of ionic surfactant molecules is polyelectrolytes.

A schematic representation of a typical instability of domain segregation that can occur in monolayer LB films of some amphiphiles is shown in Figure 1.5a. The traditional approach to stabilizing such films by admixing of surfactants that form stable LB

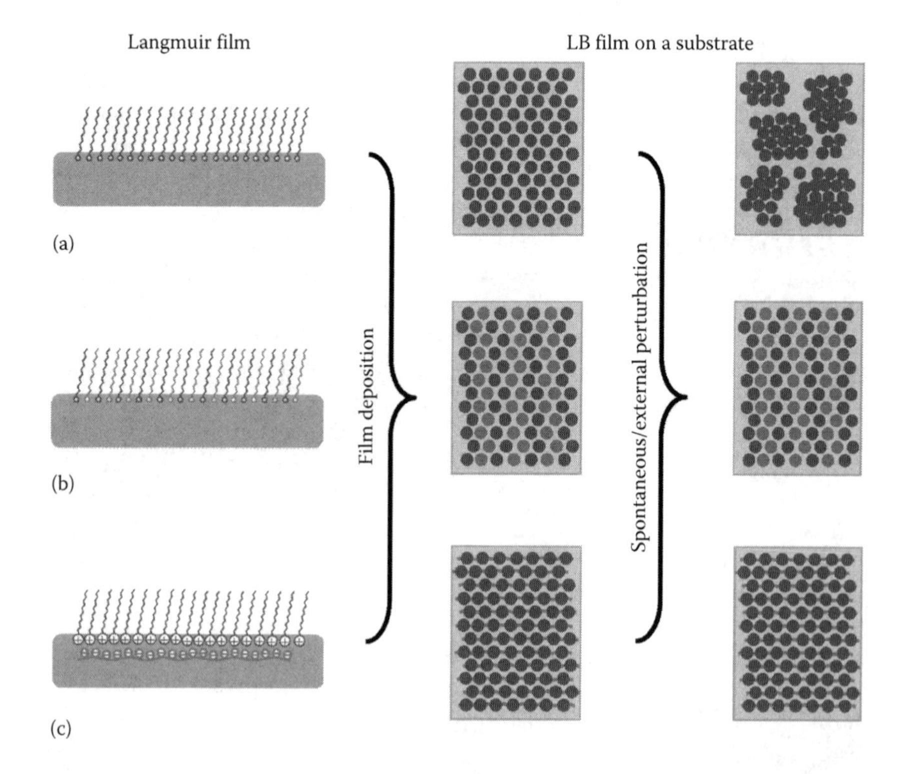

FIGURE 1.5 Schematic diagram illustrating the transfer of a Langmuir film as LB film (schematic of the top view) and its potential reorganization (spontaneous or under an external perturbation) or lack of it: (a) only amphiphiles of interest, (b) amphiphiles of interest mixed with another one that forms stable LB films, and (c) amphiphiles with cationic head groups and polyanion introduced in the subphase.

films, indicated in Figure 1.5b, shows how the two-dimensional density of the active amphiphile in the resulting LB film is significantly reduced; the reduction is 50% when the stabilizing surfactant is added in equivalent quantity. The concept of polyelectrolyte templating of the Langmuir film leading to enhanced stability of the resulting LB film made up of the ionic amphiphile–polyelectrolyte complex, without compromising significantly the density of the active amphiphile, is shown schematically in Figure 1.5c.

The polyelectrolyte templating approach is a versatile strategy to stabilize LB films of specific amphiphiles of interest, when either by themselves they do not form stable films at the air–water interface or the fabricated films tend to undergo reorganization spontaneously or under an external stimulus. No synthetic modifications of the amphiphile of interest are required except the introduction of a charge if not present already, and a wide range of polyelectrolytes can be explored to identify the optimal templating agent. Examples of polyanions and polycations that can be introduced in the aqueous subphase are shown in Figure 1.6. The average separation of the ionic

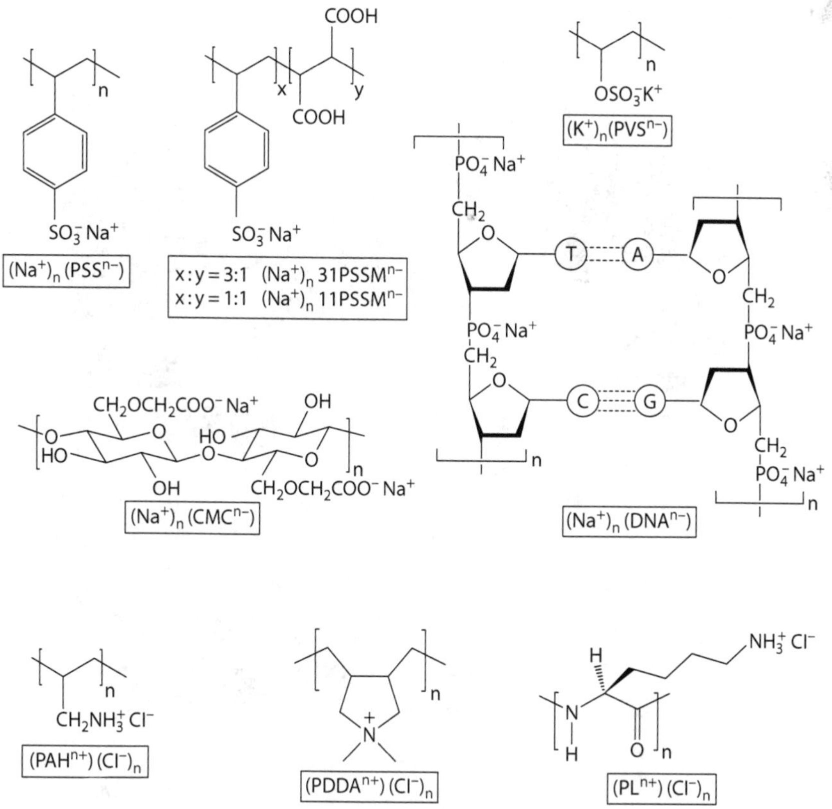

FIGURE 1.6 Examples of polyanions [PSS^{n-} = poly(styrene sulfonate); $PSSM^{n-}$ = poly(styrene sulfonate-*co*-maleic acid); PVS^{n-} = poly(vinyl sulfate); CMC^{n-} = carboxymethylcellulose ester; DNA^{n-} = deoxyribonucleic acid] and polycations [PAH^{n+} = poly(allylammonium); $PDDA^{n+}$ = poly(diallyldiamethylammonium); PL^{n+} = poly-*L*-lysinium] that can be employed as templating agents for LB films.

groups on the polyelectrolyte can be used as a controlling factor to effectively tune the area occupied by the complexed amphiphile and hence its packing density in the LB film. The mol ratio between the amphiphiles and the monomer units in the polyelectrolyte is an important factor that needs to be optimized to achieve efficient complexation at the air–water interface and stability of the resulting LB film. In the actual experiment, either the subphase can be prepared with the polyelectrolyte already dissolved in it before spreading the amphiphile monolayer or the polyelectrolyte can be injected into the subphase after forming the amphiphile monolayer at the air–water interface.

DNA has been an important choice of polyanion to study the interaction with cationic surfactants assembled in Langmuir films. Many of these studies were motivated by the need to understand the basic phenomena involved in processes such as DNA transfection, gene delivery and DNA sensing, and the fabrication of novel biomaterials. Basic insight into the organization of DNA–surfactant monolayers has been gained through cryo-TEM-based tomography;[48] Figure 1.7 shows the cryo-TEM images of DNA, DNA-surfactant monolayer, and the three-dimensional reconstruction of

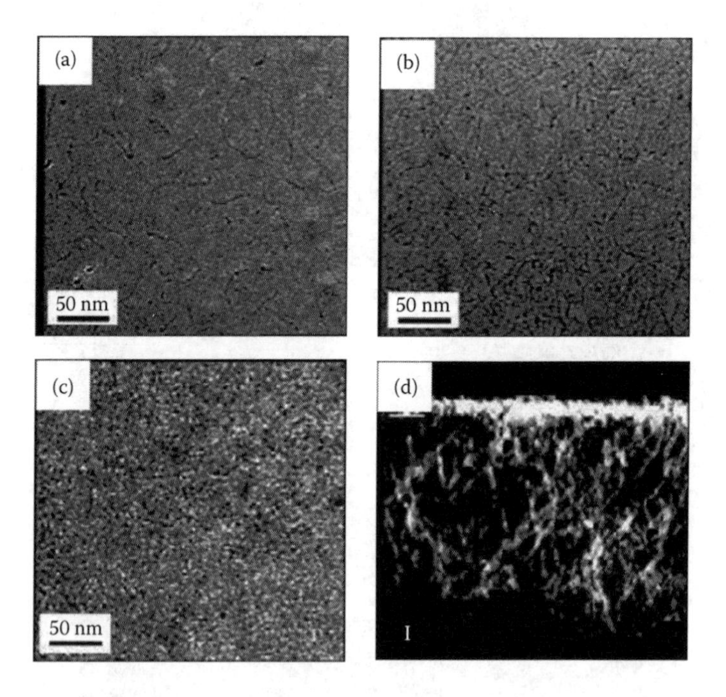

FIGURE 1.7 Cryo-TEM images of (a) a DNA solution, (b) a DNA-surfactant monolayer with surfactant molecules spread on a subphase containing DNA, and (c) same as (b) with DNA injected underneath the preformed monolayer. (d) Projection in the xz plane of the three-dimensional reconstructed volume of the surfactant-DNA complex shown in (c); white bar represents 10 nm. (Reprinted with permission from Vos, M.R.J., Bomans, P.H.H., de Haas, F., Frederik, P.M., Jansen, J.A., Nolte, R.J.M., and Sommerdijk, N.A.J.M., Insights in the organization of DNA-surfactant monolayers using Cryo-electron tomography, *J. Am. Chem. Soc.*, 129, 11894–11895. Copyright 2007 American Chemical Society.)

the complex. Study of the adsorption of DNA on monolayers of the cationic lipid, 1,2-dioleoyl-3-trimethylammonium propane and cholesterol or heptafluorocholesterol, demonstrated the possibility of using the fluorinated cholesterol as a helper lipid in DNA transfection vectors.[49] The relevance of the interaction of DNA with gemini surfactants in gene therapy–related applications has been reviewed recently.[50]

A concept closely related to the general idea of polyelectrolyte templating described earlier and demonstrated by the DNA examples has been described as "glued LB films."[51,52] Polyanions such as PSS^{n-} (Figure 1.6) have been used to stabilize LB films of singly and multiply charged cationic surfactants. The ionic cross-linking demonstrated by the increased surface viscosity of the monolayers resulted in LB films with improved stability and enhanced gas permeation selectivity. In the next section, we describe several examples of the application of polyelectrolyte templating in functional materials applications.

1.5 MATERIALS BASED ON POLYELECTROLYTE-TEMPLATED LB FILMS

An early experiment based on a simple pyridinium-based amphiphile and a collection of polyelectrolytes demonstrated unambiguously the concept of polyelectrolyte templating.[53] When N-octadecyl-4-dimethylaminopyridinium bromide (ODP^+Br^-) was spread on a pure water subphase at 25°C in an LB trough and compressed, it was found to collapse readily (Figure 1.8a), producing no stable Langmuir film. However, the introduction of polyanions such as PVS^{n-}, PSS^{n-}, $31PSSM^{n-}$, and $11PSSM^{n-}$ (Figure 1.6) in the subphase (typically in a 1:8 mole ratio between the amphiphile and the monomer unit of the polyanion) led to varying degrees of stabilization of the monolayer as seen from the corresponding π–A isotherms (Figure 1.8a).

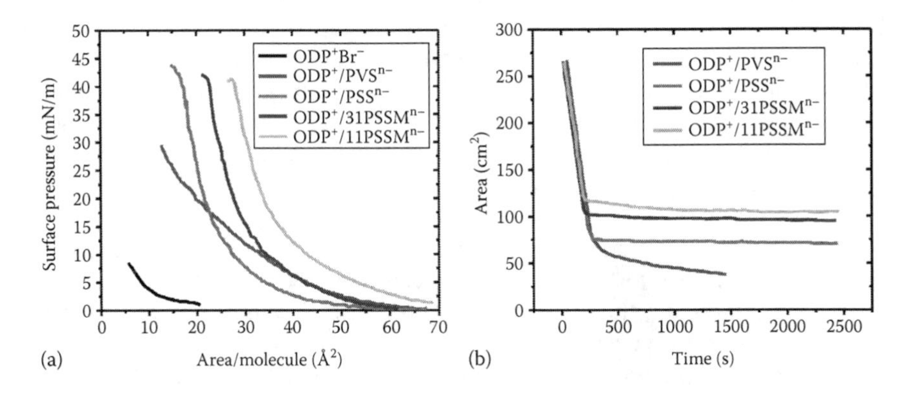

FIGURE 1.8 (a) π–A isotherms and (b) constant pressure area–time plots of ODP^+Br^- spread on a subphase of pure water and subphases containing different polyelectrolytes. The x-axis in (a) refers to area/molecule of ODP^+; the constant pressure in (b) is 18 mN/m for PVS^{n-} and 30 mN/m for the other polyanions. (Adapted with permission from Sharma, S., Chandra, M.S., and Radhakrishnan, T.P., Stabilization of a cationic amphiphile monolayer by polyanions in the subphase and computational modeling of the complex at the air-water interface, *Langmuir*, 17, 8118–8124. Copyright 2001 American Chemical Society.)

This experiment was prompted by the observation that even the introduction of dianions such as succinate provided limited stability for the ODP^+ monolayer at the air–water interface. Area–time creep plots (Figure 1.8b) with the different polyelectrolytes in the subphase indicated clearly that PSS^{n-} is the best suited polyanion to impart high stability to the ODP^+ monolayer. Figure 1.8a clearly shows that the area/molecule occupied by the same ODP^+ amphiphile in the monolayer increases when the polyelectrolyte changes from PSS^{n-} to $31PSSM^{n-}$ to $11PSSM^{n-}$. This is related to the increasing average separation between the sulfonate groups in the polyanion chains and points to the effective complexation between these anionic moieties and the cationic head group of the amphiphile. Computational modeling studies provided insight into the likely structure of the ionic complex formed; it was concluded that the compact packing of the complex leads to the enhanced stability of the ODP^+ monolayer on the PSS^{n-} template.

The example described here illustrates that the organization of amphiphiles is strongly influenced by the templating effect of the polyelectrolyte; significantly, the effective area occupied by an amphiphile molecule and hence the physical separation between them in the monolayer can be controlled. Aggregation of head groups of amphiphiles in Langmuir and LB films can have deleterious impact on some of their optical and electronic properties. A particularly sensitive effect that has been extensively explored is the optical second harmonic generation (SHG) response of LB films based on the hemicyanine dye.[54–60] Aggregation of the nonlinear optical chromophore leads to structures with diminished SHG response. Various approaches, including the formation of composites with insoluble polymers[54] and fatty acids,[55–57,60] were attempted to overcome this problem; while reducing the aggregation, these techniques lead also to the dilution of the chromophore density. Studies on the crystals of a hemicyanine salt and semiempirical quantum chemical analysis of molecular clusters in these crystals[61] showed clearly that the electronic spectra of LB films of N-*n*-octadecyl-4-[2-(4-dimethylaminophenyl)ethenyl]pyridinium bromide ($ODEP^+Br^-$) fabricated without and with polyanions in the subphase could be attributed to aggregated and deaggregated head groups, respectively. Based on these findings, a detailed study of the fabrication and SHG response of mono- and multilayer LB films of $ODEP^+Br^-$ spread on pure water as well as on aqueous solutions of the polyelectrolytes $(Na^+)_nPSS^{n-}$, $(Na^+)_n31PSSM^{n-}$, and $(Na^+)_n11PSSM^{n-}$ as the subphase was carried out.[62,63] The electronic spectra were consistent with the deaggregation of the $ODEP^+$ chromophore when the polyelectrolytes were present. The deaggregation also led to a significant enhancement of the SHG response; LB films of $ODEP^+$ complexed with PSS^{n-} showed approximately a fourfold increase over that without polyelectrolyte templating (Figure 1.9).

A subsequent study explored the impact of three polyelectrolytes—$(Na^+)_nPSS^{n-}$, $(Na^+)_nDNA^{n-}$, and $(Na^+)_nCMC^{n-}$ (Figure 1.6)—on the mode of formation of multilayer LB films of $ODEP^+Br^-$ and their SHG response.[64] The impact of polyelectrolyte complexation on the Langmuir films was clearly revealed in the π–A isotherms and BAM images. Transfer ratios observed during film deposition together with the electronic absorption spectra and AFM images of the LB films clearly showed that the polyanions exert significant influence on the mode of deposition of the multilayers. A model that describes the multilayer deposition sequence, consistent with

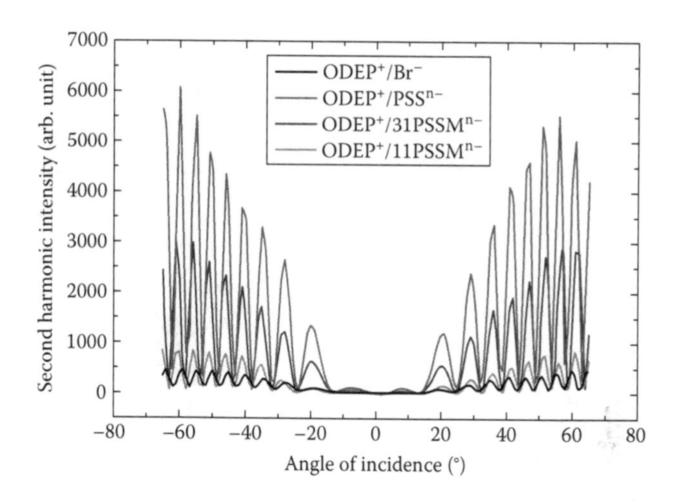

FIGURE 1.9 SHG fringes recorded for single-layer LB films of $ODEP^+$ fabricated without and with different polyelectrolytes in the subphase. (Reprinted with permission from Chandra, M.S., Ogata, Y., Kawamata, J., and Radhakrishnan, T.P., Polyelectrolyte-assisted deaggregation and second harmonic generation enhancement in hemicyanine Langmuir-Blodgett film, *Langmuir*, 19, 10124–10127. Copyright 2003 American Chemical Society.)

these information and the observed SHG responses of the LB films, is shown in Figure 1.10. Sequence A is applicable to $ODEP^+Br^-$ deposited from pure water subphase, Sequence B represents the case where PSS^{n-} and DNA^{n-} are present as templates in the subphase, and Sequence C describes the ideal one obtained with CMC^{n-} in the subphase. The variations of the SHG response in the different cases are collected in Figure 1.11a; the steady increase of the SHG response in the case of CMC^{n-} template LB films is shown in Figure 1.11b. The impact of polyelectrolyte templating on multilayer formation and the identification of CMC^{n-} as the optimal choice to realize favorable Z-type deposition with strong SHG response are demonstrated by this study. It was proposed that the structural adjustments of the sandwiched polymer layer lead to the observed deposition sequence and SHG responses.

The cellulose-based polyanion, CMC^{n-}, was found to be a versatile template to enhance the linear and nonlinear optical response of LB films of another hemicyanine-based amphiphile, N-*n*-octadecyl-4-[2-(4-(N,N-ethyloctadecylamino) phenyl)ethenyl]pyridinium bromide ($OEOEP^+Br^-$) possessing a "tail-head-tail" structure.[65] BAM images of the Langmuir film of $OEOEP^+Br^-$ on pure water and an aqueous solution of $(Na^+)_nCMC^{n-}$ clearly demonstrate the impact of the polyanion on the homogenization of the film (Figure 1.12). The electronic absorption spectra of the LB films are consistent with the reduction of aggregation of the head groups upon polyanion templating; polarized electronic absorption spectra showed further that the orientation of the monomer chromophores is significantly altered by the template. The suppression of aggregation leads to an enhancement in the fluorescence of the $OEOEP^+$-based LB film; the SHG is also enhanced considerably for films deposited at higher pressures (Table 1.1).

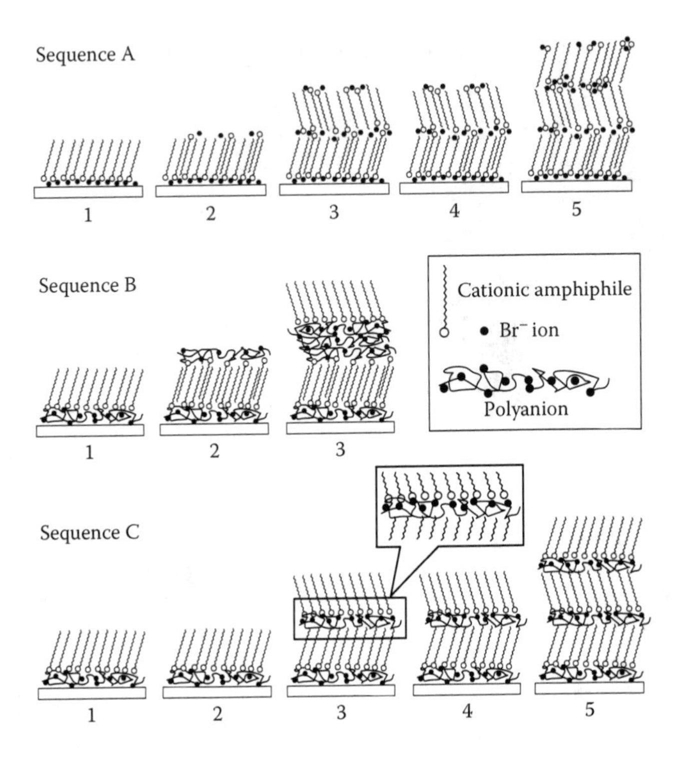

FIGURE 1.10 Deposition sequences proposed for the formation of multilayer LB films of $ODEP^+$ in the absence (Sequence A) and presence (Sequences B and C) of polyelectrolyte templates. Expanded view of the sandwiched polyelectrolyte during Sequence C is shown. Numbers refer to the deposition strokes, and the symbols used are clarified in the legend; open and filled circles indicate cationic and anionic groups, respectively. (Reprinted with permission from Rajesh, K., Chandra, M. S., Hirakawa, S., Kawamata, J., and Radhakrishnan, T. P., Polyelectrolyte templating strategy for the fabrication of multilayer hemicyanine Langmuir-Blodgett films showing enhanced and stable second harmonic generation. *Langmuir*, 23, 8560–8568. Copyright 2007 American Chemical Society.)

As noted earlier, the instability of LB films may manifest in the form of structural reorganization induced by external stimuli. A vivid demonstration of this is seen in the SHG experiments on the simple hemicyanine-based LB films in some of the studies discussed earlier. If the SHG experiment involving high-power laser irradiation is repeated on a given sample of the $ODEP^+Br^-$ LB film, the SHG response is found to decrease each time (Figure 1.13).[26] The underlying reason for this was found to be the induction of molecular aggregation, as revealed by the electronic absorption spectra of the films observed after each experiment; the monomer peak was found to decrease along with an increase in the peak due to aggregates (Figure 1.14a). Templating $ODEP^+$ amphiphiles by polyanions not only enhances the SHG response, as stated earlier, but also arrests the decay upon repeated laser irradiation (Figure 1.13). The effect is most prominent in the case of PSS^{n-} templating. The stabilization of the head groups against molecular aggregation that leads to the SHG response

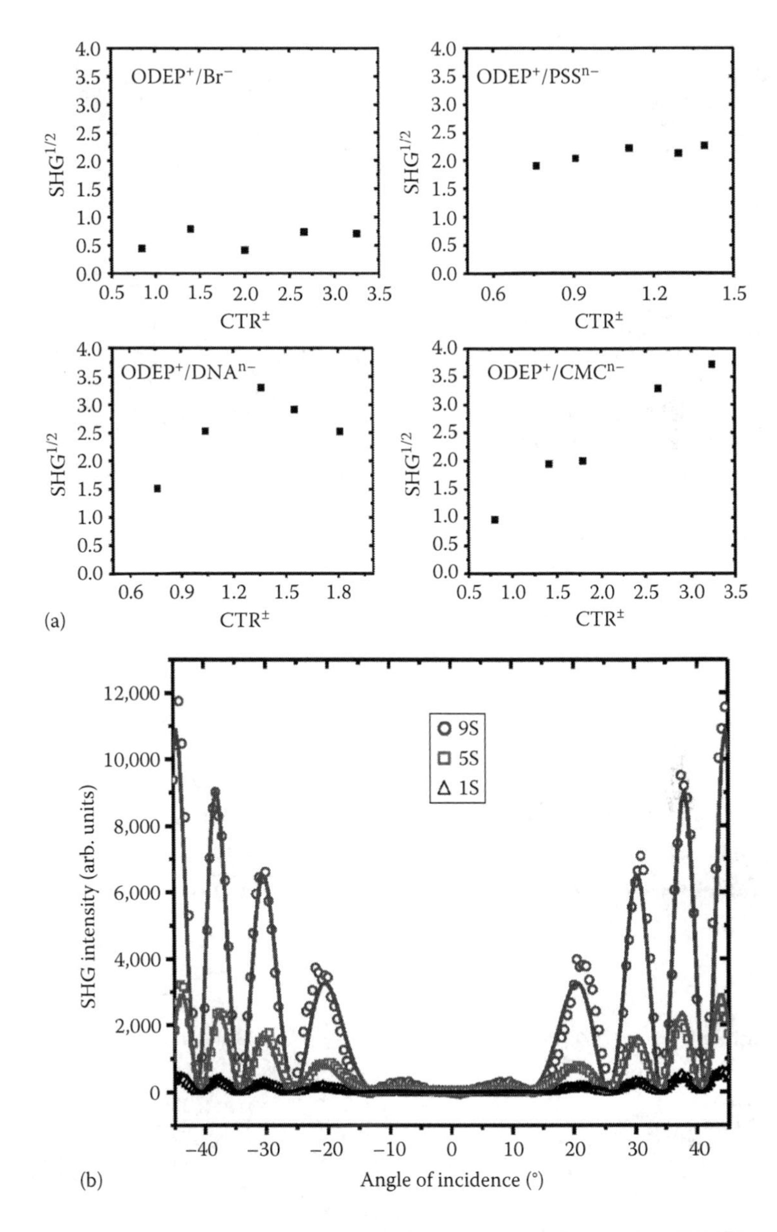

FIGURE 1.11 (a) Plot of the square root of normalized SHG intensity ($SHG^{1/2}$) versus the cumulative transfer ratio, CTR± (defined in Reference 64), for multilayer LB films of $ODEP^+Br^-$ on pure water and $(Na^+)_nPSS^{n-}$, $(Na^+)_nDNA^{n-}$, and $(Na^+)_nCMC^{n-}$ solutions as the subphase; (b) SHG fringe patterns for the 1-stroke, 5-stroke, and 9-stroke multilayer LB films deposited from the subphase of $(Na^+)_nCMC^{n-}$ solution; experimental points and the corresponding theoretical fit are indicated. (Adapted with permission from Rajesh, K., Chandra, M. S., Hirakawa, S., Kawamata, J., and Radhakrishnan, T. P., Polyelectrolyte templating strategy for the fabrication of multilayer hemicyanine Langmuir-Blodgett films showing enhanced and stable second harmonic generation. *Langmuir*, 23, 8560–8568. Copyright 2007 American Chemical Society.)

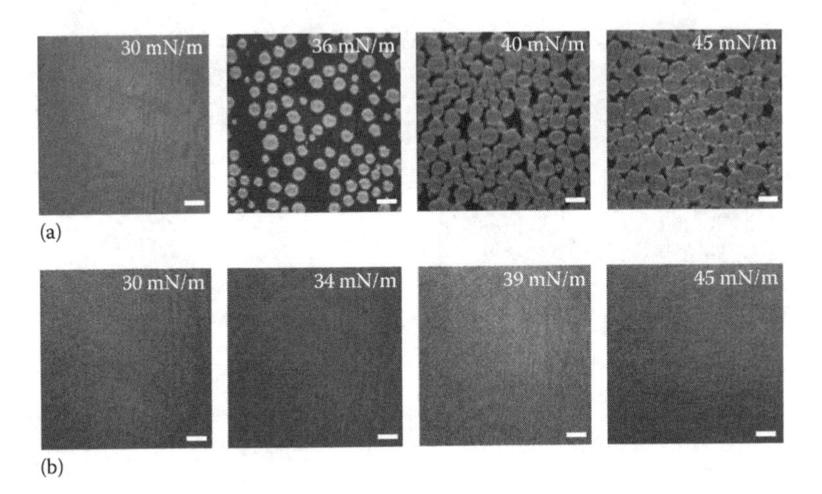

FIGURE 1.12 BAM images of Langmuir films of $OEOEP^+Br^-$ on (a) pure water and (b) solution of $(Na^+)_nCMC^{n-}$ as the subphase at 25°C at different surface pressures (scale bar = 25 μm). (Adapted with permission from Rajesh, K., Balaswamy, B., Yamamoto, K., Yamaki, H., Kawamata, J., and Radhakrishnan, T.P., Enhanced optical and nonlinear optical responses in a polyelectrolyte templated Langmuir-Blodgett film, *Langmuir*, 27, 1064–1069. Copyright 2011 American Chemical Society.)

TABLE 1.1
Normalized Fluorescence and SHG Intensities (Relative Values) of $OEOEP^+$-Based Monolayer LB Films Deposited at Two Different Surface Pressures from Pure Water ($OEOEP^+ Br^-$) and a Solution of $(Na^+)_nCMC^{n-}$ ($OEOEP^+/CMC^{n-}$) as the Subphase

		Normalized Intensity (arb. units)	
Deposition Pressure (mN/m)	**LB Film**	**Fluorescence**	**SHG**
30	$OEOEP^+ Br^-$	1.00	1.35
	$OEOEP^+/CMC^{n-}$	2.28	1.15
45	$OEOEP^+/Br^-$	1.00	0.80
	$OEOEP^+/CMC^{n-}$	2.35	1.75

Source: Rajesh, K. et al., 2011. *Langmuir* 27: 1064.

stability is clearly illustrated in the corresponding electronic absorption spectra (Figure 1.14b). Similar stabilization of electronic absorption and SHG response have been observed in the case of $ODEP^+$-based multilayer LB films templated by CMC^{n-}.[64] A recent study has illustrated the utility of polyelectrolyte–ionic surfactant complex in sequestering nonamphiphilic dye molecules in LB films and tuning their fluorescence response.[66] Mixtures of poly(N-vinylbenzyl-N,N,N-trimethylammonium)

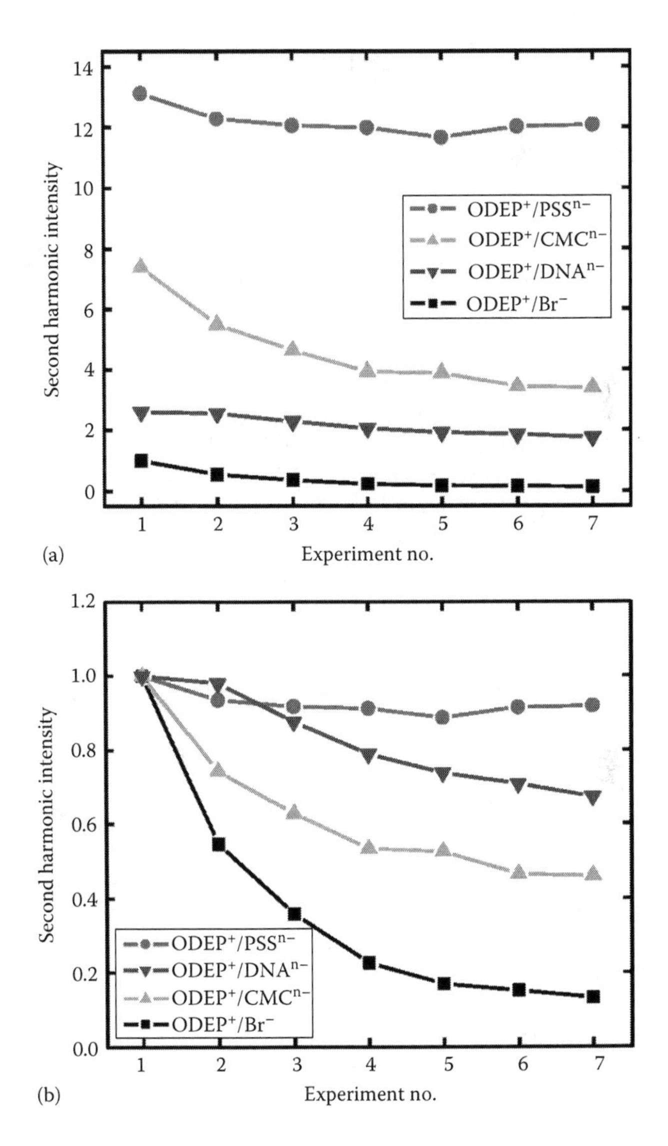

FIGURE 1.13 SHG intensity of $ODEP^+$-based LB films (deposited from pure water as well as solutions of the polyanions as the subphase) from consecutive experiments. (a) Values are scaled with respect to the initial value in each case; the relative magnitude of different films is highlighted by scaling all the initial values with respect to that of $ODEP^+/Br^-$. (b) Values are scaled with respect to 1.0 for the initial value in each case to highlight the difference in the variations. The lines connecting points are only a guide for the eye. (Chandra, M.S., Krishna, M.G., Mimata, H., Kawamata, J., Nakamura, T., and Radhakrishnan, T.P.: Laser induced second harmonic generation decay in a Langmuir-Blodgett film: Arresting by polyelectrolyte templating. *Adv. Mater.* 2005. 17. 1937–1941. Copyright Wiley-VCH Verlag GmbH & Co. KGaA. Reproduced with permission.)

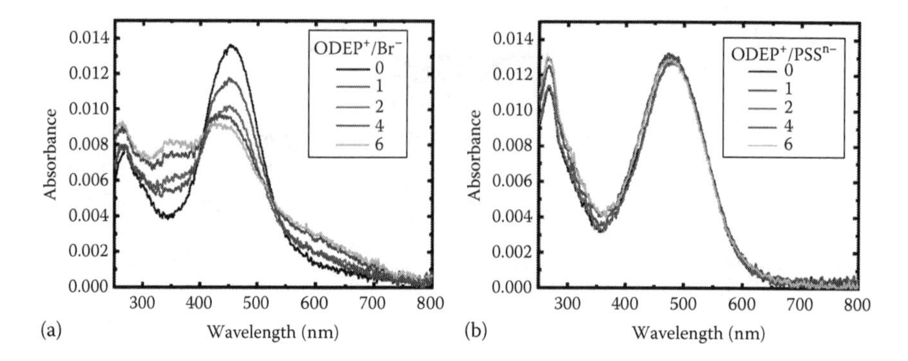

FIGURE 1.14 Electronic absorption spectra of the ODEP+-based LB films, deposited from (a) pure water and (b) solution of PSSn− as the subphase, before the first SHG experiment (0) and after SHG experiments (1, 2, 4, 6). (Chandra, M.S., Krishna, M.G., Mimata, H., Kawamata, J., Nakamura, T., and Radhakrishnan, T.P.: Laser induced second harmonic generation decay in a Langmuir-Blodgett film: Arresting by polyelectrolyte templating. *Adv. Mater.* 2005. 17. 1937–1941. Copyright Wiley-VCH Verlag GmbH & Co. KGaA. Reproduced with permission.)

dodecylbenzenesulfonate complex and Nile Red dye were found to form stable monolayers and LB films. Optical spectroscopy showed that the presence of the dye monomer and dimer in the fluorescent LB films could be tuned by varying the dye concentration in the monolayers.

An application of the polyelectrolyte templating approach, very different from the various examples discussed earlier, was demonstrated by its use in the controlled polymerization of aniline at the air–water interface.[67] Polyelectrolytes such as PVS^{n-} and CMC^{n-} introduced in the aqueous subphase were shown to have a profound influence on the kinetics of oxidative polymerization of N-octadecylamine at the air–water interface. More significantly, the polyelectrolyte templates led to the realization of enhanced alignment of the polyaniline chains and their bundles in the resulting LB and LS films. Another study of related interest is the electropolymerization of water-soluble phenosafranine dye adsorbed on arachidic acid monolayers and transferred as LB films, resulting in two-dimensional films of poly(phenosafranine) sandwiched between the surfactant layers.[68]

1.6 CONCLUSIONS

The polyelectrolyte-templating approach to Langmuir–Blodgettry essentially involves the supramolecular interaction between amphiphilic surfactant molecules in Langmuir films at the air–water interface and polymers introduced in the aqueous subphase. The strength of the interaction arises from the electrostatic attraction between the ionic head group of the surfactant molecule and the counterion moieties on the polymer. This paradigm effectively imparts the robustness and stability of the macromolecular structure to the otherwise metastable surfactant monolayer. The practical advantage stems from the fact that elaborate covalent synthesis of polymeric amphiphilic structures is circumvented while realizing high stability in the resulting LB films.

Starting with fatty acids, a wide range of amphiphilic molecules have been investigated in the form of Langmuir and LB films. Several novel amphiphiles with specific head groups have been synthesized for targeted applications, model studies, and device development. Many of these systems are easily amenable to the polyelectrolyte templating strategy in order to achieve enhanced stability of the ultrathin films and the desired density of amphiphilic molecules. The enormous developments that have taken place in the field of macromolecular chemistry and polyelectrolytes in particular can be exploited in this pursuit.

Several interesting problems remain to be explored in this area. Utility of specific polymer architectures, including polymers with tailored tacticity and block copolymers, need to be investigated. Organization of nanostructures, including metal and semiconductor nanoparticles, can benefit by the introduction of polyelectrolyte–surfactant combinations and the inherent asymmetry at the interface. Even though the focus of much of the discussion in this chapter was on the impact of polyelectrolyte templating on the structure and attributes of the amphiphile monolayer, it would be equally interesting to probe the role of the amphiphile monolayer on the conformational structure and materials characteristics of the polyelectrolyte; an interesting example of this direction is the study of aggregation of DNA chemisorbed on comb copolymers at the air–water interface.[69] This can open up new possibilities and designs in the field of macromolecule-based ultrathin film materials.

ACKNOWLEDGMENTS

We thank Drs. M. Sharath Chandra and K. Rajesh, B. Balaswamy, and Professors M. Ghanashyam Krishna, J. Kawamata, and T. Nakamura for their valuable contribution to various aspects of the work from our group described in this chapter. We also thank Professors H. Ringsdorf, D. J. Burgess, and N. A. J. M. Sommerdijk for their gracious permission to reproduce a figure from their articles in this chapter, and Professor M. J. Swamy for some useful discussions. Financial support from the Department of Science and Technology, New Delhi, and infrastructure support from the University of Hyderabad (School of Chemistry, Central Facility for Nanotechnology and Centre for Modeling, Simulation and Design) are gratefully acknowledged.

REFERENCES

1. Moerner, W. E. and Fromm, D. P. 2003. Methods of single-molecule fluorescence spectroscopy and microscopy. *Rev. Sci. Instrum.* 74: 3597–3619.
2. Gross, L., Mohn, F., Moll, N., Liljeroth, P., and Meyer, G. 2009. The chemical structure of a molecule resolved by atomic force microscopy. *Science* 325: 1110–1114.
3. Heath, J. R. 2009. Molecular electronics. *Annu. Rev. Mater. Res.* 39: 1–23.
4. Metzger, R. M. 2008. Unimolecular electronics. *J. Mater. Chem.* 18: 4364–4396.
5. Holten, D., Bocian, D. F., and Lindsey, J. S. 2002. Probing electronic communication in covalently linked multiporphyrin arrays. A guide to the rational design of molecular photonic devices. *Acc. Chem. Res.* 35: 57–69.
6. Kim, W. Y. and Kim, K. S. 2010. Tuning molecular orbitals in molecular electronics and spintronics. *Acc. Chem. Res.* 43: 111–120.

7. Katano, S., Kim, Y., Trenary, M., and Kawai, M. 2013. Orbital-selective single molecule reactions on a metal surface studied using low-temperature scanning tunneling microscopy. *Chem. Commun.* 49: 4679–4681.
8. de Oteyza, D. G., Gorman, P., Chen, Y., Wickenburg, S., Riss, A., Mowbray, D. J., Etkin, G. et al., 2013. Direct imaging of covalent bond structure in single-molecule chemical reactions. *Science* 340: 1434–1437.
9. Kim, Y., Motobayashi, K., Frederiksen, T., Ueba, H., and Kawai, M. 2015. Action spectroscopy for single-molecule reactions—Experiments and theory. *Prog. Surf. Sci.* 90: 85–143.
10. Whitesides, G. M., Mathias, J. P., and Seto, C. T. 1991. Molecular self-assembly and nanochemistry: A chemical strategy for the synthesis of nanostructures. *Science* 25: 1312–1319.
11. Stupp, S. I. and Palmer, L. C. 2014. Supramolecular chemistry and self-assembly in organic materials design. *Chem. Mater.* 26: 507–518.
12. Grill, L., Dyer, M., Lafferentz, L., Persson, M., Peters, M. V., and Hecht, S. 2007. Nano-architectures by covalent assembly of molecular building blocks. *Nat. Nanotech.* 2: 687–691.
13. Patra, A., Chandaluri, C. G., and Radhakrishnan, T. P. 2012. Optical materials based on molecular nanoparticles. *Nanoscale* 4: 343–359.
14. Wang, A., Huang, J., and Yan, Y. 2014. Hierarchical molecular self-assemblies: Construction and advantages. *Soft Matter* 10: 3362–3373.
15. Balzer, F. and Rubahn, H. 2002. Laser-controlled growth of needle-shaped organic nanoaggregates. *Nano Lett.* 2: 747–750.
16. Sakai, M., Iizuka, M., Nakamura, M., and Kudo, K. 2005. Organic nano-transister fabricated by co-evaporation method under alternating electric field. *Synth. Metals* 153: 293–296.
17. Dey, S. and Pal, A. J. 2011. Layer-by-layer electrostatic assembly with a control over orientation of molecules: Anisotropy of electrical conductivity and dielectric properties. *Langmuir* 27: 8687–8693.
18. Langmuir, I. 1917. The constitution and fundamental properties of solids and liquids. II Liquids. *J. Am. Chem. Soc.* 39: 1848–1906.
19. Blodgett, K. B. 1935. Films built by depositing successive monomolecular layers on a solid surface. *J. Am. Chem. Soc.* 57: 1007–1022.
20. Petty, M. C. 1996. *Langmuir–Blodgett Films: An Introduction*. Cambridge University Press, New York.
21. Schwartz, D. K. 1997. Langmuir–Blodgett film structure. *Surf. Sci. Rep.* 27: 241–334.
22. Tachibana, H., Yamanaka, Y., Sakai, H., Abe, M., and Matsumoto, M. 2001. Self-organization of amphiphilic diacetylenes in Langmuir–Blodgett films. *Stud. Surf. Sci. Catal.* 132: 461–464.
23. Rajesh, K. and Radhakrishnan, T. P. 2009. Optical response sensitive to the assembly in a molecular material: Ultrathin film with a vanishing electronic absorption. *Chem. Eur. J.*, 15: 2801–2809.
24. Schönhoff, M., Chi, L. F., Fuchs, H., and Lösche, M. 1995. Structural rearrangements upon photoreorientation of amphiphilic azobenzene dyes organized in ultrathin films on solid surfaces. *Langmuir* 11: 163–168.
25. Wang, Y., Nichogi, K., Terashita, S., Iriyama, K., and Ozaki, Y. 1996. Dependence of thermal behaviors on the number of layers in Langmuir–Blodgett films of 2-octadecyl-7,7,8,8-tetracyanoquinodimethane. 1. Order-disorder transitions in the films investigated by ultraviolet-visible and infrared spectroscopies. *J. Phys. Chem.* 100: 368–373.

26. Chandra, M. S., Krishna, M. G., Mimata, H., Kawamata, J., Nakamura, T., and Radhakrishnan, T. P. 2005. Laser induced second harmonic generation decay in a Langmuir–Blodgett films: Arresting by polyelectrolyte templating. *Adv. Mater.* 17: 1937–1941.
27. Stroeve, P. and Hwa, M. J. 1996. Effects of pH on gegenion complex Langmuir–Blodgett films. *Thin Solid Films* 284–285: 561–563.
28. Chi, L. F., Johnston, R. R., and Ringsdorf, H. 1991. Fluorescence microscopy investigations of the domain formation of fatty acid monolayers induced by polymeric gegenions. *Langmuir* 7: 2323–2329.
29. Kobayashi, K., Takasago, M., Taru, Y., and Takaoka, K. 1994. Structural investigations of polyion complex Langmuir–Blodgett films containing fluorocarbon chains. *Thin Solid Films* 247: 248–251.
30. Lee, B. J. and Kunitake, T. 1994. Two-dimensional polymer networks of maleic acid co polymers and Poly(allylamine) by the Langmuir–Blodgett technique. *Langmuir* 10: 557–562.
31. Girard-Egrot, A. P., Godoy, S., and Blum, L. J. 2005. Enzyme association with lipidic Langmuir-Blodgett films: Interests and applications in nanobioscience. *Adv. Colloid Interf. Sci.* 116: 205–225.
32. Kafi, A. K. M. and Kwon, Y. 2008. Brewster angle microscopic study of mixed lipid-protein monolayer at the air-water interface and its application in biosensing. *Talanta* 76: 1029–1034.
33. Junghans, A., Champagne, C., Cayot, P., Loupiac, C., and Koper, I. 2010. Protein-lipid interactions at the air-water interface. *Langmuir* 26: 12049–12053.
34. Costa, A. P., Xu, X., and Burgess, D. J. 2012. Langmuir balance investigation of superoxide dismutase interactions with mixed-lipid monolayers. *Langmuir* 28: 10050–10056.
35. Kamilya, T., Pal, P., and Talapatra, G. B. 2007. Incorporation of ovalbumin within cationic octadecylamine monolayer and a comparative study with zwitterionic DPPC and anionic Stearic acid monolayer. *J. Colloid Interface Sci.* 315: 464–474.
36. Kamilya, T., Pal, P., and Talapatra, G. B. 2007. Interaction of ovalbumin with phospholipids Langmuir–Blodgett film. *J. Phys. Chem. B* 111: 1199–1205.
37. Hir, S. A., Raguénès-Nicol, C., Paboeuf, G., Nicolas, A., Le Rumeur, E., and Vié, V. 2014. Cholesterol favors the anchorage of human dystrophin repeats 16 to 21 in membrane at physiological surface pressure. *Biochim. Biophys. Acta* 1838: 1266–1273.
38. Shnek, D. R., Pack, D. W., Sasaki, D. Y., and Arnold, F. H. 1994. Specific protein attachment to artificial membranes via coordination to lipid-bound Copper(II). *Langmuir* 10: 2382–2388.
39. Haas, H., Torrielli, M., Steitz, R., Cavatorta, P., Sorbi, R., Fasano, A., Riccio, P., and Gliozzi, A. 1998. Myelin model membranes on solid substrates. *Thin Solid Films* 327-329: 627–631.
40. Haas, H., Steitz, R., Fasano, A., Liuzzi, G. M., Polverini, E., Cavatorta, P., and Riccio, P. 2007. Laminar order within Langmuir–Blodgett multilayers from Phospholipid and Myelin basic protein: A neutron reflectivity study. *Langmuir* 23: 8491–8496.
41. Baker, M. J., Zheng, L., Winograd, N., Lockyer, N. P., and Vickerman, J. C. 2008. Mass spectral imaging of glycophospholipids, cholesterol, and glycophorin A in model cell membranes. *Langmuir* 24: 11803–11810.
42. Ceridório, F. L., Caseli, L., and Oliveira, Jr, O. N. 2016. Chondroitin sulfate interacts mainly with headgroups in phospholipid monolayers. *Colloids Surf. B: Biointerfaces* 141: 595–601.
43. Svitova, T. F. and Lin, M. C. 2016. Dynamic interfacial properties of human tear-lipid films and their interactions with model-tear proteins in vitro. *Adv. Colloid Interface. Sci.* 233: 4–24.

44. Ahlers, M., Blankenburg, R., Haas, H., Möbius, D., Möhwald, H., Müller, W., Ringsdorf, H., and Siegmund, H. U. 1991. Protein interactions with ordered lipid films: Specific and unspecific binding. *Adv. Mater.* 3: 39–46.
45. Berzina, T. S., Troitsky, V. I., Petrigliano, A., Alliata, D., Tronin, A. Yu., and Nicolini, C. 1996. Langmuir–Blodgett films composed of monolayers of amphiphilic molecules and adsorbed soluble proteins. *Thin Solid Films* 284–285: 757–761.
46. Scholl, F. A. and Caseli, L. 2015. Langmuir and Langmuir–Blodgett films of lipids and penicillinase: Studies on adsorption and enzymatic activity. *Colloids Surf. B. Biointerfaces* 126: 232–236.
47. Marchenkova, M. A., Dyakova, Y. A., Tereschenko, E. Y., Kovalchuk, M. V., and Vladimirov, Y. A. 2015. Cytochrome c complexes with cardiolipin monolayer formed under different surface pressure. *Langmuir* 31: 12426–12436.
48. Vos, M. R. J., Bomans, P. H. H., de Haas, F., Frederik, P. M., Jansen, J. A., Nolte, R. J. M., and Sommerdijk, N. A. J. M. 2007. Insights in the organization of DNA-surfactant monolayers using Cryo-electron tomography. *J. Am. Chem. Soc.* 129: 11894–11895.
49. Paiva, D., Brezesinski, G., Pereira, M. C., and Rocha, S. 2013. Langmuir monolayers of monocationic lipid mixed with Cholesterol or Fluorocholesterol: DNA adsorption studies. *Langmuir* 29: 1920–1925.
50. Ahmed, T., Kamel, A. O., and Wettig, S. D. 2016. Interactions between DNA and gemini surfactant: Impact on gene therapy: Part II. *Nanomedicine* 11: 403–420.
51. Yan, X., Janout, V., Hsu, J. T., and Regen, S. L. 2003. The gluing of a Langmuir-Blodgett bilayer. *J. Am. Chem. Soc.* 125: 8094–8095.
52. McCullough, D. H. and Regen, S. L. 2004. Don't forget Langmuir-Blodgett films. *Chem. Commun.* 2787–2791.
53. Sharma, S., Chandra, M. S., and Radhakrishnan, T. P. 2001. Stabilization of a cationic amphiphile monolayer by polyanions in the subphase and computational modeling of the complex at the air-water interface. *Langmuir* 17: 8118–8124.
54. Hayden, L. M., Kowel, S. T. and Srinivasan, M. P. 1987. Enhanced second harmonic generation from multilayered Langmuir/Blodgett films of dye. *Opt. Commun.* 61: 351–356.
55. Girling, I. R., Cade, N. A., Kolinsky, P. V., Jones, R., Peterson, I. R., Ahmed, M. M., Neal, D. B., Petty, M. C., Roberts, G. G., and Feast, W. J. 1987. Second-harmonic generation in mixed hemicyanine: Fatty-acid Langmuir-Blodgett monolayers. *J. Opt. Soc. Am. B.* 4: 950–954.
56. Schildkraut, J. S., Penner, T. L., Willand, C. S., and Ulman, A. 1988. Absorption and second-harmonic generation of monomer and aggregate hemicyanine dye in Langmuir–Blodgett films. *Opt. Lett.* 13: 134–136.
57. Carpenter, M. A., Willand, C. S., Penner, T. L., Williams, D. J., and Mukamel, S. 1992. Aggregation in hemicyanine dye Langmuir-Blodgett films: Ultraviolet-visible absorption and second harmonic generation studies. *J. Phys. Chem.* 96: 2801–2804.
58. Evans, C. E. and Bohn, P. W. 1993. Characterization of an aggregate-sensitive single-component energy-transfer system. *J. Am. Chem. Soc.* 115: 3306–3311.
59. Evans, C. E., Song, Q., and Bohn, P. W. 1993. Influence of molecular orientation and proximity on spectroscopic line shape in organic monolayers. *J. Phys. Chem.* 97: 12302–12308.
60. Wang, W., Lu, X., Xu, J., Jiang, Y., Liu, X., and Wang, G. 2000. Influence of temperature on hemicyanine Langmuir-Blodgett multilayer films. *Physica B* 293: 6–10.
61. Chandra, M. S. and Radhakrishnan, T. P. 2003. Molecular aggregation in a hemicyanine dye: Modeling by a combined crystallographic and computational approach. *Mol. Cryst. Liq. Cryst.* 403: 77–89.
62. Chandra, M. S., Ogata, Y., Kawamata, J., and Radhakrishnan, T. P. 2003. Polyelectrolyte-assisted deaggregation and second harmonic generation enhancement in hemicyanine Langmuir-Blodgett film. *Langmuir* 19: 10124–10127.

63. Chandra, M. S., Ogata,Y., Kawamata, J., and Radhakrishnan, T. P. 2004. Enhanced SHG in poly electrolyte complexed hemicyanine dye Langmuir-Blodgett films. *J. Nonlin. Opt. Phys. Mater.* 13: 347–353.
64. Rajesh, K., Chandra, M. S., Hirakawa, S., Kawamata, J., and Radhakrishnan, T. P. 2007. Polyelectrolyte templating strategy for the fabrication of multilayer hemicyanine Langmuir-Blodgett films showing enhanced and stable second harmonic generation. *Langmuir* 23: 8560–8568.
65. Rajesh, K., Balaswamy, B., Yamamoto, K., Yamaki, H., Kawamata, J., and Radhakrishnan, T. P. 2011. Enhanced optical and nonlinear optical responses in a polyelectrolyte templated Langmuir-Blodgett film. *Langmuir* 27: 1064–1069.
66. Seliverstova, E., Ibrayev, N., and Kudaibergenov, S. 2013. Fluorescencing behavior of thin solid films based on polyelectrolyte-surfactant complex and dye molecules. *J. Appl. Polym. Sci.* 129: 289–295.
67. Chandra, M. S. and Radhakrishnan, T. P. 2006. Polyelectrolyte templated polymerization in Langmuir films: Nanoscopic control of polymer chain organization. *Chem. Eur. J.* 12: 2982–2986.
68. Sawant, S. N., Doble, M., Yakhmi, J. V., Kulshreshtha, S. K., Miyazaki, A., and Enoki, T. 2006. Polymer-surfactant layered heterostructures by electropolymerization of Phenosafranine in Langmuir-Blodgett films. *J. Phys. Chem. B* 110: 24530–24540.
69. Fujimori, A., Taguchi, M., and Arai, S. 2014. Formation of aggregates of DNA molecules chemisorbed on the organized films of comb polymers containing s-triazine *Colloids Surf. A* 443: 432–438.

2 Porphyrin-Based Functional Nanomaterials and Their Functionalities

Sadananda Mandal and Amitava Patra

CONTENTS

2.1 INTRODUCTION

Significant attention has been paid to inorganic and organic-based self-assemblies owing to their unique optical and electronic properties for potential applications like catalysis, light energy conversion, optoelectronic devices, and biological sensors.[1–9] Organic-based assemblies have received special attention due to their several advantages.[10–14] Among the self-assemblies, porphyrin-based assemblies are of great interest in supramolecular chemistry, light harvesting, photosensitization, photonics, photocatalysis, and various sensor devices because of their excellent properties.[15–21] In the last few years, research on porphyrins, metalloporphyrins, and their assemblies has provided a large number of nanostructured materials, namely, particles, rods, tubes, fibers, rectangular nanotube, sheets and flowers, and so on, with unique structural and electronic properties.[22–24] Drain and coworkers have demonstrated the formation of porphyrin nanoparticles from a wide variety of *meso*-arylporphyrins.[10] It reveals that π–π stacking of porphyrin molecules produces porphyrin nanoparticles with a size range between 20 and 200 nm. Schwab et al. have reported the formation of porphyrin nanorods from aqueous acid solution of tetrakis(4-sulfonatophenyl)porphine (TPPS), and the dimension of nanorods is found to be varied with changing of the ionic strength of aqueous solutions.[16,25] Porphyrin nanotubes are synthesized by ionic assembly of two oppositely charged porphyrins ($H_4TPPS_4^{2-}$, and Sn(IV) tetrakis(4-pyridyl) porphyrin dichloride) in aqueous solution.[26] Square porphyrin nanosheets are prepared from Sn(IV) 5-(4-pyridyl)-10,15,20-triphenyl-porphyrin dichloride, which are found to be promising for electronics, photonics, and photocatalysts[27]. Highly crystalline rectangular nanotube structures are synthesized from metal-free 5, 10, 15, 20-tetra(4-pyridyl)porphyrin (H_2TPyP) by a vaporization–condensation and recrystallization process.[28] It is reported that the assembly of H_2TPyP molecules is due to hydrogen bonding, hydrogen–π interactions, and π–π interactions where rod structures are formed at an early stage, and rectangular tubular structures are formed at later growth stages. Furthermore, clover-like nanostructure was synthesized by ionic self-assembly of $Zn^{II}T(N\text{-}EtOH\text{-}4\text{-}Py)P^{4+}$ and the tin(IV) complex of $TPPS^{4-}$ ($Sn^{IV}TPPS^{4-}$).[22] Patra et al. have demonstrated the synthesis of surfactant-assisted various morphologies (sphere, rod, flake, and flower) of tetra-carboxyphenyl porphyrin nanoaggregates, and the influence of these structures on photocatalytic properties has been investigated.[19] The fabrication of ionic liquid–assisted ZnOEP nanoaggregate for singlet oxygen generation has been reported recently.[29] Various noncovalent interactions,[30–36] such as hydrogen bonding, π–π stacking, hydrophobic interactions, electrostatic interactions, and van der Waals forces, are responsible for the formation of these porphyrin-based nanostructures. The porphyrin nanostructure materials are formed by self-assembly of porphyrins either J-type (edge-to-edge arrangement) or H-type (face-to-face arrangement) aggregates.[37] In general, porphyrins and metalloporphyrins exhibit two well-defined absorption regions, a so-called high-energy B (also known as Soret) and low-energy Q bands. Soret absorption band corresponds to $S_0 \rightarrow S_2$ electronic transitions, while Q bands correspond to $S_0 \rightarrow S_1$ electronic transitions. Both the B and Q bands arise from π–π^* electronic transitions.[38] Both S_1 and S_2 excitation states are spitted into the lower- and higher-level energy excited states when (supra) molecular aggregations (H- or J-type) occur.[39] According to the exciton coupling model, the higher excited states are transitionally allowed in the case of H-aggregation, whereas lower excited states are transitionally allowed for J-aggregation (Scheme 2.1).[40] As a result, hypsochromic and bathochromic shifts are observed for H- and J-aggregates, respectively (with respect to the monomer absorption).

Porphyrin-doped conjugated polymer nanoparticles with special functionalities are found to be a promising field of research for potential applications. It is

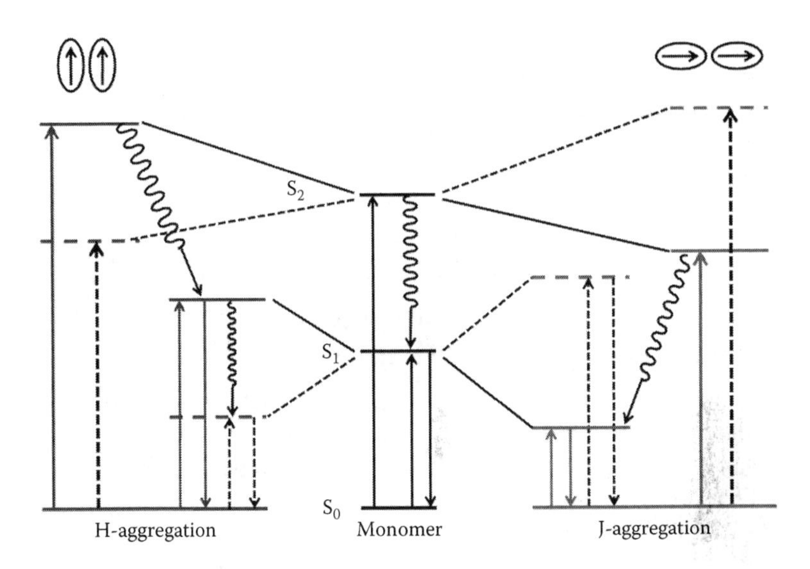

SCHEME 2.1 Modified energy diagram of monomer and aggregated states of porphyrin (not to scale). (Reprinted with permission from Mandal, S., Bhattacharyya, S., Borovkov, V., and Patra, A., Photophysical properties, self-assembly behavior, and energy transfer of porphyrin-based functional nanoparticles, *J. Phys. Chem. C*, 116, 11401–11407. Copyright 2012 American Chemical Society.)

already found that conjugated polymer nanoparticles have attracted considerable attention because of their multifunctional activities in bioimaging, biosensing, drug delivery, and photodynamic therapy, owing to their facile synthesis, less toxicity, higher photostability, and biocompatibility.[41–49] For example, McNeill and coworkers have demonstrated the use of metalloporphyrin (Pt(II)OEP)-doped polyfluorene nanoparticles for biological oxygen sensing.[50] Shen et al. have prepared tetraphenylporphyrin-doped conjugated polymer, poly [9,9-dibromohexylfluorene-2,7-yleneethylene-alt-1,4-(2,5-dimethoxy) phenylene] nanoparticles by using reprecipitation method for singlet oxygen generation under two-photon excitation.[51]

Another important aspect of the porphyrin-based system is porphyrin–quantum dot (QD) nanocomposites for photosensitization and photodynamic therapy because of efficient energy transfer (via FRET mechanism) from quantum dot to porphyrin, which facilitates the singlet oxygen generation.[52] The recovery of the fluorescence property of QD could be possible through the encapsulation of porphyrin molecules by macrocyclic cavity like cyclodextrin, crown ether, calix[n]arene, and cucurbit[n]uril, because of their unique abilities to encapsulate both cationic and hydrophobic moieties.[53–59] Keeping this in mind, porphyrin–QD–cucurbit[n]uril composite systems have been designed to find useful applications in sensor.[60,61]

In this chapter, we concentrate on the aggregation (molecular arrangement) behavior of porphyrin molecules and photophysical and photocatalytic properties of porphyrin-based functional nanomaterials. To begin with, we describe the general synthetic methodologies to prepare porphyrin-based functional nanomaterials.

Spectroscopic and microscopic methods are used to understand the structural changes due to formation of porphyrin-based functional nanomaterials, which eventually influence the physical properties. Finally, an outlook on the prospects of this research field is given. Such understanding will enable us to construct efficient devices for suitable applications.

2.2 EXPERIMENTAL PROCEDURES AND CHARACTERIZATION

2.2.1 Synthesis of Porphyrin Nanoparticles and Porphyrin-Doped Polymer Nanoparticles

Porphyrin nanoparticles of bis-zinc octaethylporphyrin [$(ZnOEP)_2$] were prepared by reprecipitation method.[62–64] Porphyrin was dissolved in dried tetrahydrofuran (THF) to maintain the required concentration of the porphyrin molecules. A 500 μL aliquot of this THF solution was rapidly injected into 10 mL double-distilled water under vigorous stirring for 10–15 min. Then, the obtained solution was ultrasonicated for 15 min. As a result, an aqueous solution of porphyrin nanoparticles was obtained. To avoid aging of the porphyrin nanoparticles, THF was removed from aqueous solution by partial vacuum evaporation for 1 h, followed by filtration through a 0.2 μm filter paper. Finally, a stable (3–4 days) aqueous solution of porphyrin nanoparticle was obtained.

The porphyrin-encapsulated polymer nanoparticle system was also prepared by reprecipitation method.[44,64] In this particular case, THF solutions of porphyrin and polymer were mixed thoroughly to maintain the required concentration of porphyrin and polymer, followed by ultrasonication for 10 min to obtain a clear mixed solution. Then, 500 μL aliquot of the THF solution was rapidly injected into double-distilled water under vigorous stirring for 10–15 min. Finally, porphyrin-encapsulated polymer nanoparticle was obtained by ultrasonication, vacuum evaporation, and filtration through 0.2 μm filter.

2.2.2 Synthesis of Various Morphologies of Porphyrin-Based Nano-/Microstructures

Different morphologies of meso-tetra (4-carboxyphenyl)porphyrin (TCPP) aggregates were synthesized by a typical acid–base neutralization strategy along with hydrophobic interaction method.[19] First, 0.004 g of TCPP was dissolved in 0.5 mL of 0.2 M NaOH solution. In another container, 0.0364 g of cetyltetra-amonium bromide (CTAB) was dissolved in 9.5 mL of 0.0105 M HCl solution. Then, the TCPP solution was added to the CTAB-containing solution and allowed for constant stirring (1200 rpm) at room temperature. The final concentration of TCPP was 0.5×10^{-3} M. The reaction was carried out in a 20 mL beaker covered with aluminum foil to avoid the photochemical reaction; 0.5 mL of this solution was taken out after 15 min, 1, 6, and 48 h to get different morphologies of surfactant-assisted TCPP aggregates. The samples are reproducible and stable for more than 15 days.

2.3 SYNTHESIS OF IL-ASSISTED ZnOEP NANOPARTICLES

The porphyrin nanoparticle was synthesized by simple miniemulsion method.[29] Zinc octaethylporphyrin (ZnOEP) was properly dissolved in dichloromethane (DCM) to maintain the 100 μM concentration of ZnOEP. A 500 μL aliquot of this DCM solution was rapidly injected into 2.75 mg of IL containing 10 mL of double-distilled water under vigorous stirring for 10–15 min, followed by ultrasonication for another 15 min to form stable miniemulsion containing small droplets of the ZnOEP solution. The DCM solvent was then evaporated by partial vacuum evaporation for 1 h at 65°C. Finally, we obtained a stable aqueous solution of ZnOEP nanoparticles, which was stable for a week.

2.4 SYNTHESIS OF VARIOUS COMPOSITIONS OF $Cd_{1-x}Zn_xS$ ALLOYS

$Cd_{1-x}Zn_xS$ QDs were synthesized following hot injection method.[61,65] For the preparation of $Cd_{0.31}Zn_{0.69}S$ QDs, 0.067 g (0.25 mMol) cadmium acetate, 0.109 g (0.5 mMol) zinc acetate, and 5 mL of oleylamine were firstly taken in a two-necked round-bottom flask. Then the mixture was heated to 150°C under continuous Ar gas flow for 20 min to make a clear solution. At this condition, an excess amount of sulfur solution containing 0.096 g (3.0 mMol) sulfur dissolved in 2.5 mL of oleylamine was quickly injected into the hot reaction mixture with mild stirring and the temperature was kept constant at 150°C for desired growth of QDs. After 3 h, the reaction was quenched by adding the excess amount of toluene at once. Then, the QDs were washed with ethanol, followed by centrifugation at 10,000 rpm and it was repeated twice. Finally, QDs were dissolved in toluene. The other two QDs, namely, $Cd_{0.52}Zn_{0.48}S$ and $Cd_{0.62}Zn_{0.38}S$, were synthesized following the same protocol just changing the amount of zinc acetate (0.25 mMol and 0.125 mMol, respectively). The QDs were transferred into water from toluene by using the ligand exchange phase transfer process.[49] Atomic absorption, x-ray diffraction (XRD), and energy dispersive x-ray analysis (EDAX) studies confirm the stoichiometric constituent of the different compositions of nanocrystals.[65]

2.5 INSTRUMENTATION

The morphological characterization of porphyrin nanoparticle, pure poly 9-venylcarbazole (PVK) nanoparticle, and porphyrin-encapsulated PVK nanoparticle was done by field emission scanning electron microscopy (FE-SEM, JEOL, JSM-6700F) and atomic force microscopy (AFM, VEECO, dicp-II). The morphological characterization and sizes of $Cd_{1-x}Zn_xS$ QD were measured by transmission electron microscopy (TEM) (JEOL-TEM-2010) operating voltage at 200 kV. Room temperature optical absorption spectra were taken by a UV-vis spectrophotometer (SHIMADZU). Room temperature photoluminescence spectra were recorded by a Fluoromax-P (HORIBA JOBIN YVON) photoluminescence spectrophotometer. For the time-correlated single photon counting (TCSPC) measurements, the samples were excited using picosecond diode laser (IBH Nanoled-07) in an IBH fluorocube apparatus. The typical full width at half maximum (FWHM) of the system response using a liquid scatter was about 90 ps. The repetition rate was 1 MHz. The fluorescence decays were collected at a Hamamatsu MCP photomultiplier (C487802). The fluorescence decays were analyzed using IBH DAS6 software. The following equation was used to analyze the experimental time-resolved fluorescence decays, $P(t)$[66]:

$$P(t) = b + \sum_{i}^{n} \alpha_i \exp\left(-\frac{t}{\tau_i}\right) \tag{2.1}$$

where

n is the number of discrete emissive species

b is a baseline correction ("dc" offset)

α_i and τ_i are the pre-exponential factors and excited-state fluorescence lifetimes associated with the ith component, respectively

For multi-exponential decays, the average lifetime, $\langle \tau \rangle$, was calculated from the following equation[18,66]:

$$\langle \tau \rangle = \sum_{i=1}^{n} \beta_i \tau_i \tag{2.2}$$

where $\beta_i = \alpha_i / \sum \alpha_i$ and β_i is the contribution of the decay component. For anisotropy measurements, a polarizer was placed before the sample. The analyzer was rotated by 90° at regular intervals and the parallel (I_{II}) and the perpendicular ($I_{\perp}$) components for the fluorescence decay were collected for equal times, alternatively. Then, $r(t)$ was calculated using the formula[66]:

$$r(t) = \frac{I_{II}(t) - GI_{II}(t)}{I_{\perp}(t) + 2GI_{\perp}(t)} \tag{2.3}$$

G factor of the setup was 0.41. The analysis of the time-resolved data was done using IBH DAS, version 6 decay analysis software. The same software was used to analyze the anisotropy data. Xenon lamp (wavelength > 400 nm) was used as visible light source to investigate the photocatalytic activity of the aggregated structures.

2.6 RESULTS AND DISCUSSION

2.6.1 Porphyrin Nanoparticles and Porphyrin-Encapsulated Polymer Nanoparticles

The morphologies of $(ZnOEP)_2$ porphyrin nanoparticles and $(ZnOEP)_2$ porphyrin-doped PVK nanoparticles are confirmed[52] by the SEM study (Figure 2.1). The average sizes are found to be 50 and 70 nm for porphyrin nanoparticle and porphyrin-doped PVK nanoparticles, respectively. A plausible mechanism of the formation of the nanoparticle is as follows: during the injection of THF solution of porphyrin into

FIGURE 2.1 FE-SEM images of porphyrin nanoparticle (A). Porphyrin-encapsulated PVK nanoparticle (B). (Reprinted with permission from Mandal, S., Bhattacharyya, S., Borovkov, V., and Patra, A., Porphyrin-based functional nanoparticles: Conformational and photophysical properties of bis-porphyrin and bis-porphyrin encapsulated polymer nanoparticles, *J. Phys. Chem. C*, 115, 24029–24036. Copyright 2011 American Chemical Society; Reprinted with permission from Mandal, S., Bhattacharyya, S., Borovkov, V., and Patra, A., Photophysical properties, self-assembly behavior, and energy transfer of porphyrin-based functional nanoparticles, *J. Phys. Chem. C*, 116, 11401–11407. Copyright 2012 American Chemical Society.)

water under the stirring condition, THF molecules quickly diffuse into bulk water and porphyrin molecules are exposed to the aqueous environment. Since water is a poor solvent for hydrophobic porphyrin, the molecules tend to aggregate to avoid contact with water, thus yielding nanoparticles.

In the case of the porphyrin-encapsulated PVK nanoparticle system, the concentration of porphyrin in the stock THF solution is much lower than that of PVK, thus resulting in PVK to serve as a size-determinant component of the whole system. It has been investigated that $(ZnOEP)_2$ exists as *syn* conformation in porphyrin nanoparticle form itself due to strong intermolecular π–π interactions between the two zinc porphyrin macrocycles.[64] However, the $(ZnOEP)_2$ molecules inside the PVK nanoparticle are in the *anti* conformation. The conformation (*syn/anti*) of $(ZnOEP)_2$ molecule encapsulated in PVK nanoparticle is essentially governed by the rigidity of the polymer matrix, although some weak binding interactions (an attraction interaction between a tertiary nitrogen atom of the carbazole moiety of the PVK and Zn ion of the porphyrin) may also play an additional role in stabilizing the *anti* conformation in the solid-state environment.[64] There is a considerable difference in the corresponding spectroscopic properties of *syn* and *anti* forms of $(ZnOEP)_2$ (shown in Figure 2.2).[64] The absorption Soret band of $(ZnOEP)_2$ in nanoparticle form is almost similar (maxima at 400 nm) with the $(ZnOEP)_2$ in DCM. It indicates the *syn* conformation of porphyrin in nanoparticle form. The Soret band arises at 420 nm with a shoulder at 437 nm, indicating the *anti* conformation of porphyrin when this porphyrin is encapsulated inside PVK nanoparticle. Furthermore, porphyrin nanoparticles [syn $(ZnOEP)_2$] exhibit two emission bands, Q^*_{X00} and Q^*_{X01}, with maxima at 620 and 660 nm (as a shoulder), respectively. On the other hand, the destruction of the intramolecular π–π interactions between two porphyrin moieties yields the corresponding *anti* conformation of $(ZnOEP)_2$ in the case of

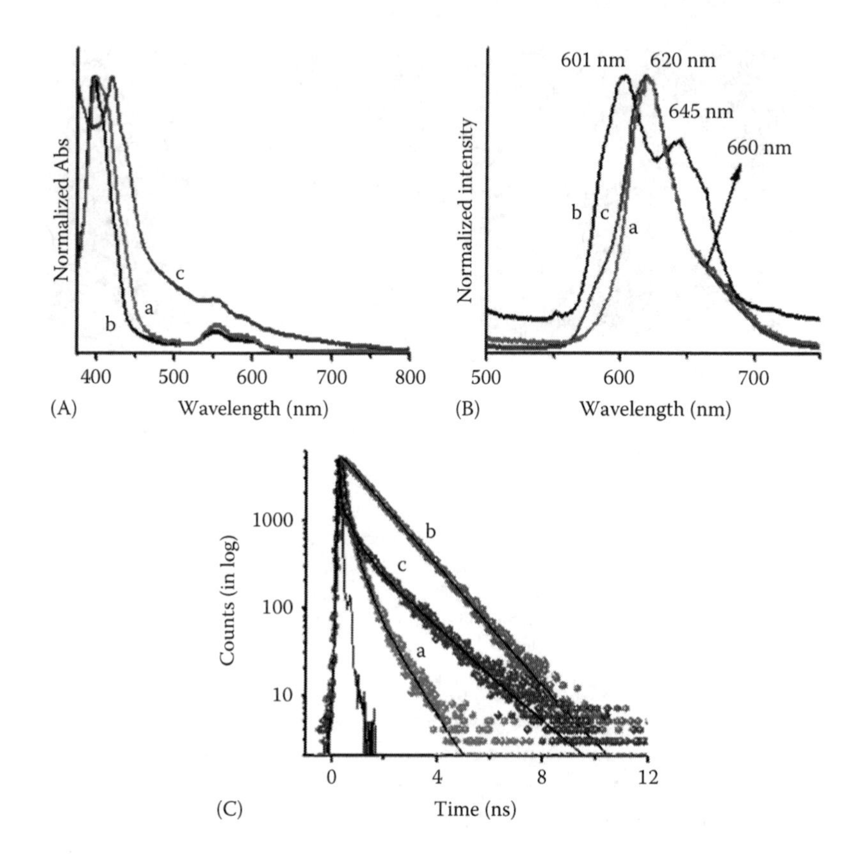

FIGURE 2.2 (A) Absorption, (B) photoluminescence, and (C) decay spectra of $(ZnOEP)_2$. a: in nanoparticle form, b: in DCM solution, and c: encapsulated in PVK nanoparticle. (Reprinted with permission from Mandal, S., Bhattacharyya, S., Borovkov, V., and Patra, A., Porphyrin-based functional nanoparticles: Conformational and photophysical properties of bis-porphyrin and bis-porphyrin encapsulated polymer nanoparticles, *J. Phys. Chem. C*, 115, 24029–24036. Copyright 2011 American Chemical Society.)

$(ZnOEP)_2$-doped PVK nanoparticle. A significant change in the fluorescence spectrum with two well-defined emission bands with the maxima at 601 and 645 nm is observed for the *anti* conformation of $(ZnOEP)_2$ inside PVK nanoparticle. Further insight in the excited-state relaxation pathways of the *syn* and *anti* conformational $(ZnOEP)_2$ is investigated by fluorescence decay dynamics. The decay curve of $(ZnOEP)_2$ in DCM is reasonably well fitted by a mono-exponential decay with a lifetime of 1.24 ns. The decay curves of $(ZnOEP)_2$ nanoparticle in water and $(ZnOEP)_2$ inside PVK nanoparticle are fitted by tri-exponential decay curves. The fitting components of the $(ZnOEP)_2$ nanoparticle systems are 0.06 ns (82%), 0.28 ns (15%), and 0.85 ns (3%) with the average decay time of 0.11 ns. In the case of $(ZnOEP)_2$ inside PVK nanoparticle, the corresponding components are 0.06 ns (39%), 0.65 ns (26%), and 1.90 ns (35%) and the average decay time is 0.68 ns. The radiative and nonradiative decay rates of different systems are calculated to evaluate their influence on decay time.

The observed emission lifetime (τ_{obs}) was combined with the fluorescence quantum yield (ϕ^0) to determine the radiative and nonradiative rate separately for all the systems. The following equations[66] were used to determine these rates:

$$k_r = \frac{\phi^0}{\tau} \tag{2.4}$$

$$k_{nr} = \frac{\left(1-\phi^0\right)}{\tau} \tag{2.5}$$

where

k_r, k_{nr} are the radiative and nonradiative rates constant, respectively
ϕ^0 is the quantum yield of the molecule
τ is the average decay time

It is seen from Table 2.1 that the nonradiative decay rates are 7.9×10^8, 9×10^9, and 14.5×10^8 s^{-1} for $(ZnOEP)_2$ in DCM, $(ZnOEP)_2$ nanoparticle, and in PVK nanoparticle, respectively. The nonradiative decay rate is highest in $(ZnOEP)_2$ nanoparticle.

As discussed, in water (poor solvent), hydrophobic $(ZnOEP)_2$ molecules tend to aggregate to form nanoparticles and hence the intermolecular interaction as well as π–π stacking is increased. The enhanced nonradiative rate is due to close proximity of the neighboring porphyrin molecules. It is interesting to note that the radiative decay rate increases from 0.11×10^7 to 2.00×10^7 s^{-1} for $(ZnOEP)_2$ nanoparticle and $(ZnOEP)_2$-encapsulated PVK nanoparticle, respectively, due to enhancement of hydrophobicity and refractive index in the polymer matrix. Analysis suggests that the conformations of bis-porphyrin nanoparticles and bis-porphyrin inside PVK nanoparticle are *syn* and *anti*, respectively. The molecular arrangement of $(ZnOEP)_2$ during the formation of nanoparticles is found be to modified.

2.6.2 Aggregated Porphyrin in Polymer Nanoparticles

The aggregation behavior of zinc octaethylporphyrin [ZnOEP] inside the conjugated polymer matrix in aqueous media is demonstrated here.[18] It is evident that the hypsochromic and bathochromic shifts of absorption spectra occur due to H- and J-aggregates, respectively.

ZnOEP in DCM solvent shows a very sharp Soret band at 400 nm and Q bands at 530 and 567 nm (Figure 2.3A), thus indicating a pure monomeric form. However, the bathochromic shifting of both Soret (by 12 nm) and Q bands (by 6–8 nm) is observed when ZnOEP is encapsulated inside PVK nanoparticle. Upon excitation at 405 nm, ZnOEP exhibits two emission bands at 570 nm (Q^*_{X00}) and 623 nm (Q^*_{X01}) in DCM solvent (Figure 2.3B), which are the characteristic peaks of ZnOEP chromophore in DCM solvent. In the case of ZnOEP-encapsulated PVK nanoparticle, small red-shifted emission peaks are observed at 575 nm (Q^*_{X00}) and 630 nm (Q^*_{X01}), which suggest that the ZnOEP molecules form the J-aggregated state due to intermolecular porphyrin–porphyrin (such as π-π/hydrophobic) interactions because of increment of local concentration of ZnOEP inside the PVK matrix. Time-resolved fluorescence

TABLE 2.1
Fluorescence Decay Parameters, Radiative and Nonradiative Decay Rates of $(ZnOEP)_2$ in Various Systems (Excitation Wavelength at 405 nm)

Systems	λ_{em} (nm)	τ_1 (ns)(a_1)	τ_2 (ns)(a_2)	τ_3 (ns)(a_3)	$<\tau>$ (ns)	QY $(ZnOEP)_2$	$\kappa_r \times 10^{-7}$ (s^{-1})	$\kappa_{nr} \times 10^{-8}$ (s^{-1})
$(ZnOEP)_2$ nanoparticle in water	620	0.28(0.15)	0.06(0.82)	0.85(0.03)	0.11	0.010	0.11	90.0
$(ZnOEP)_2$ inside PVK nanoparticle	601	0.65(0.26)	1.90(0.35)	0.06(0.39)	0.68	0.014	2.00	14.5
$(ZnOEP)_2$ in DCM	620	1.24(1)	—	—	1.24	0.024	1.9	7.9

Source: Reprinted with permission from Mandal, S., Bhattacharyya, S., Borovkov, V., and Patra, A., Porphyrin-based functional nanoparticles: Conformational and photophysical properties of bis-porphyrin and bis-porphyrin encapsulated polymer nanoparticles, *J. Phys. Chem. C*, 115, 24029–24036. Copyright 2011 American Chemical Society.

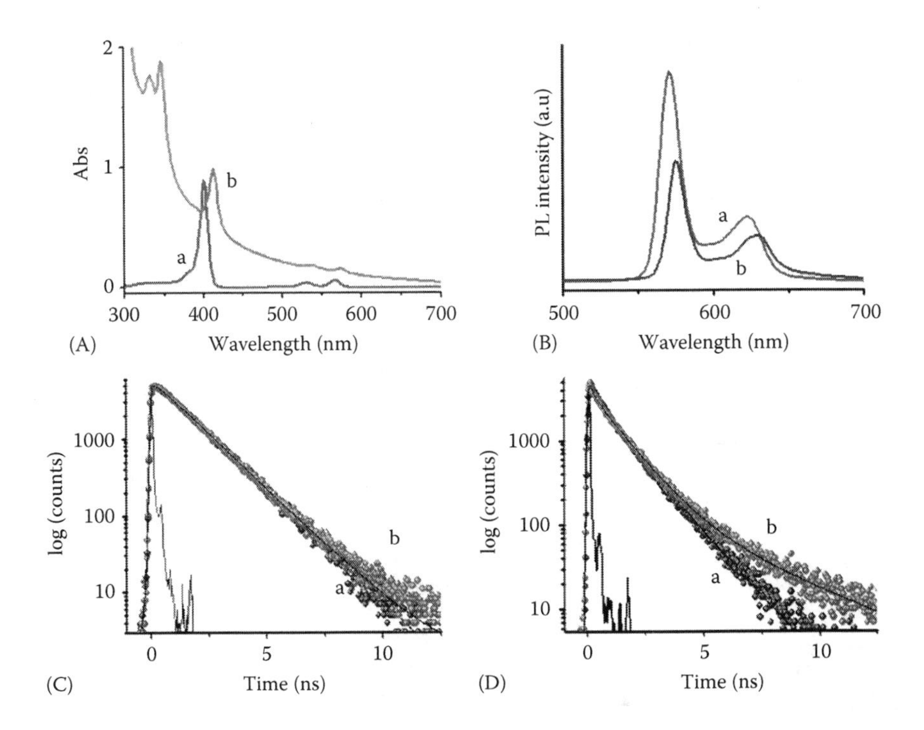

FIGURE 2.3 (A) Absorption spectra, and (B) photoluminescence spectra (a) in DCM solution, (b) encapsulated in PVK nanoparticle. Time-resolved fluorescence spectra (C) ZnOEP in DCM [(a) emission: 570 nm; (b) emission: 630 nm], (D) ZnOEP encapsulated in PVK nanoparticle [(a) emission: 575 nm; (b) emission: 630 nm] at an excitation wavelength of 405 nm. (Reprinted with permission from Mandal, S., Bhattacharyya, S., Borovkov, V., and Patra, A., Photophysical properties, self-assembly behavior, and energy transfer of porphyrin-based functional nanoparticles, *J. Phys. Chem. C*, 116, 11401–11407. Copyright 2012 American Chemical Society.)

decay time study reveals that the average decay of ZnOEP inside PVK nanoparticles is decreased. Indeed, the decay time of aggregated molecules is much shorter than that of its monomeric form (Figure 2.3C and D). For example, Verma et al. have reported the lifetime of 9 and 0.64 ns for the corresponding monomeric and J-aggregated forms of 5,10,15-trisphenyl-20-(3,4-dihydroxybenzene)porphyrin.[67] Analysis suggests that ZnOEP molecules tend to form J-type aggregates inside the conjugated PVK matrix in aqueous media.

2.6.3 Photoinduced Energy Transfer from Polymer Nanoparticle to Porphyrin

Porphyrin-doped PVK nanoparticle system is a photoactive energy-donor system, where the photoinduced energy transfer process is highly expected because there is a perfect spectral overlap between the absorption spectrum of porphyrin and the emission spectrum of PVK nanoparticles.[18,64] In general, fluorescence resonance energy transfer (FRET) is a process involving the radiation-less (nonradiative) transfer of energy from a "donor" fluorophore to an appropriate "acceptor" counterpart. This process arises from the dipole–dipole interactions and strongly depends upon the center-to-center distance between energy donor and energy acceptor. According to the Förster theory,[68] the rate of the energy transfer for an isolated single donor–acceptor pair separated by a distance r is given by the following equation:

$$k_T(r) = \frac{1}{\tau_D}\left(\frac{R_0}{r}\right)^6 \tag{2.6}$$

where

τ_D is the lifetime of the donor in the absence of the acceptor

R_0 is known as the Förster distance, the distance at which the transfer rate $k_T(r)$ is equal to the decay rate of the donor in absence of the acceptor

The Förster distance (R_0) is defined as

$$R_0^6 = \frac{9000(In10)\kappa^2\phi_D}{128\pi^5 Nn^4} J(\lambda) \tag{2.7}$$

where

ϕ_D is the quantum yield of donor in the absence of acceptor

N is Avogadro's number

n is the refractive index of medium

$J(\lambda)$ is the spectral overlap integral, which is defined as

$$J(\lambda) = \int_0^\infty F_D(\lambda)\varepsilon_A(\lambda)\lambda^4 d\lambda \tag{2.8}$$

where

$F_D(\lambda)$ is the normalized emission spectrum of donor

$\varepsilon_A(\lambda)$ is the absorption coefficient of acceptor at the wavelength λ (in nm)

κ^2 is the orientation factor of two dipoles interacting

The value of κ^2 depends on the relative orientation of the donor and acceptor dipoles. For randomly oriented dipoles, $\kappa^2 = 2/3$ and it varies between 0 and 4 for the cases of orthogonal and parallel dipoles, respectively. The calculated overlap integrals are found to be 1.23×10^{16} M^{-1} cm^{-1} nm^4 and 1.42×10^{14} M^{-1} cm^{-1} nm^4 for ZnOEP and $(ZnOEP)_2$-doped PVK nanoparticle, and the corresponding donor–acceptor Förster distances are 49.0 Å and 23.3 Å, respectively. Figure 2.4A shows a strong photoluminescence quenching (82%) of PVK nanoparticle for 2.5 wt% doped ZnOEP inside the PVK matrix.

Furthermore, time-resolved spectroscopic investigation suggests that the energy transfer efficiency from the PVK energy donor to energy acceptor porphyrin molecules [$(ZnOEP)_2$ and ZnOEP] is more than 90% (depicted in Figure 2.4B).[18,64] The high efficiency of energy transfer compared to photoluminescence quenching is apparently not only due to the dipole–dipole interaction between the donor and

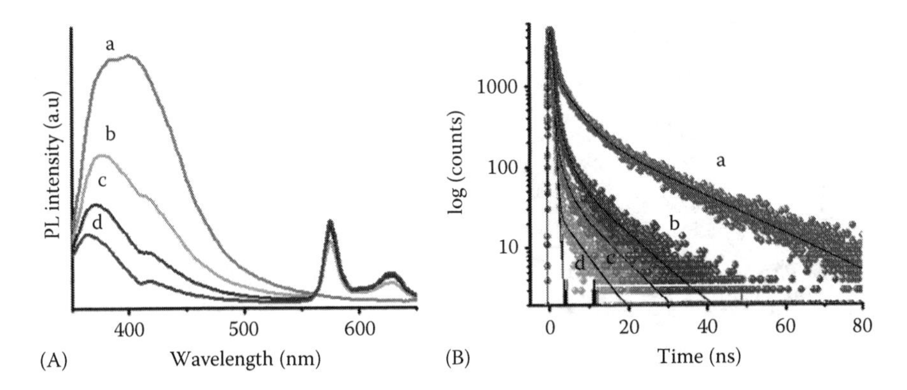

FIGURE 2.4 (A) Fluorescence spectra and (B) time-resolved fluorescence spectra [(a) pure PVK nanoparticle; (b–d) ZnOEP-doped PVK nanoparticle; (b) 0.31 wt%; (c) 1.15 wt%; (d) 2.5 wt%)] at an excitation wavelength of 340 nm. (Reprinted with permission from Mandal, S., Bhattacharyya, S., Borovkov, V., and Patra, A., Photophysical properties, self-assembly behavior, and energy transfer of porphyrin-based functional nanoparticles, *J. Phys. Chem. C*, 116, 11401–11407. Copyright 2012 American Chemical Society.)

acceptor but also due to excitonic energy diffusion throughout the polymer chain, which may sufficiently increase the energy transfer effectiveness.[69,70] This significant high efficiency of energy transfer in the porphyrin-doped polymer nanoparticles opens up new prediction for potential applications as light-harvesting systems and other photo-driven devices.

2.6.4 Porphyrin Nanoaggregates and Singlet Oxygen Generation

Here, we demonstrate the fabrication of ionic liquid–assisted ZnOEP nanoaggregates and generation of efficient amount of singlet oxygen by the photosensitizer, ZnOEP nanoaggregates.[29] The ZnOEP nanoaggregates have been synthesized by typical miniemulsion technique and well characterized by SEM, AFM, and steady-state and time-resolved spectroscopy. SEM image shows that the size of spherical particles of ZnOEP molecules is 65 nm (Figure 2.5A). These particles make the microaggregates by assembling themselves. In UV-vis absorption, ZnOEP in DCM solvent shows a very sharp Soret band at 400 nm and Q bands at 530 and 567 nm, indicating a pure monomeric form. However, the hypsochromic shifting of Soret band (by 11 nm) and bathochromic shifting of Q bands (by 17 and 25 nm) are observed for ZnOEP nanoaggregate (Figure 2.5B). A large difference in the spectral shape of ZnOEP nanoaggregate and ZnOEP monomer is observed. In particular, the calculated FWHM values are 35 and 12 nm for ZnOEP nanoaggregate and ZnOEP monomer, respectively. The observed steady-state spectral changes (blue shift and spectral broadening) suggest that the ZnOEP molecules form the H-aggregated state due to the increment of local concentration of ZnOEP inside the IL micelle.

Again, large red-shifted emission (by 25 and 24 nm) and lower fluorescence intensity of ZnOEP nanoaggregate in comparison to ZnOEP in DCM are due to the

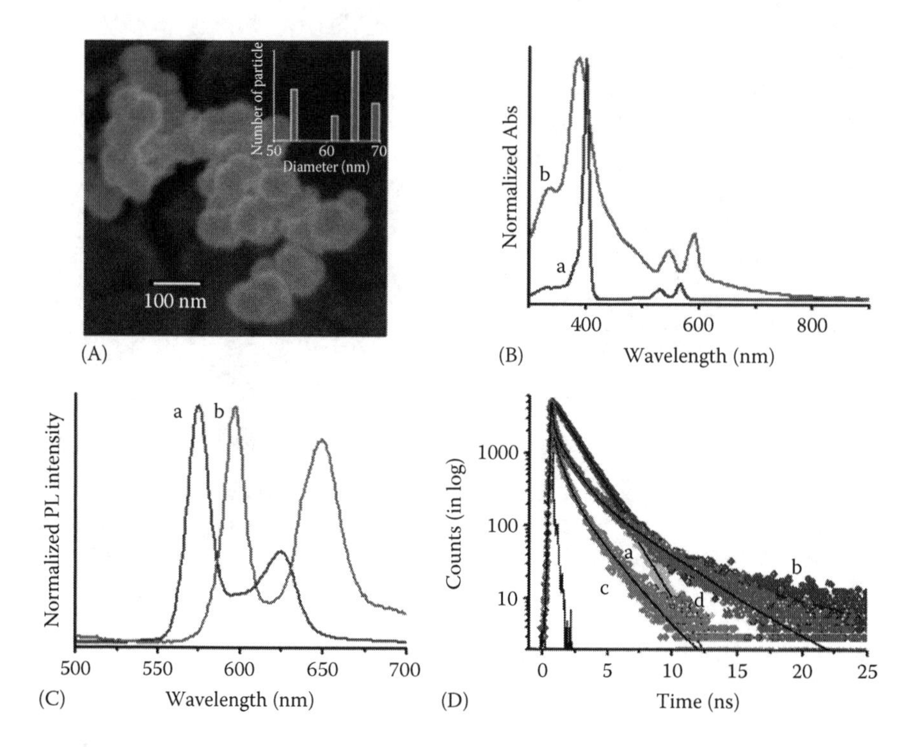

FIGURE 2.5 (A) SEM images of ZnOEP nanoaggregates, (B) UV–vis spectra [(a) ZnOEP in DCM and (b) ZnOEP nanoaggregates], (C) fluorescence spectra [(a) ZnOEP in DCM and (b) ZnOEP nanoaggregates], and (D) photoluminescence decay curves of ZnOEP at an excitation wavelength of 375 nm [(a) ZnOEP in DCM (λ_{em} = 572 nm), (b) ZnOEP in DCM (λ_{em} = 623 nm), (c) ZnOEP nanoparticle in water (λ_{em} = 597 nm), and (d) ZnOEP nanoparticle in water (λ_{em} = 623 nm)]. (Mandal, S., Kundu, S., Bhattacharyya, S., and Patra, A., Photophysical properties of ionic liquid-assisted porphyrin nanoaggregate-nickel phthalocyanine conjugates and singlet oxygen generation, *J. Mater. Chem. C*, 2, 8691–8699. Reproduced by permission of The Royal Society of Chemistry.)

H-aggregation, which is induced by intermolecular porphyrin–porphyrin (such as π–π/hydrophobic) interactions and the hydrophobic interaction between porphyrin and long hydrophobic chains of ILs (Figure 2.5C). In this case, the emission spectrum of H-aggregated ZnOEP nanoaggregate is red shifted with respect to the ZnOEP monomer because the emission occurs from the low-energy excited state of H-aggregated species, which are transitionally forbidden. Furthermore, time-resolved spectroscopic data support the H-aggregation of the porphyrin nanoaggregates. In the case of ZnOEP in DCM, the decay time of ZnOEP is 1.48 and 1.56 ns at the emission wavelength of 572 and 623 nm, respectively. On the other hand, the average decay time of ZnOEP nanoaggregate is 0.15 and 0.46 ns at the emission wavelength of 597 and 647 nm, respectively. After the formation of nanoaggregate, the average decay time of the ZnOEP decreases in both emission bands (Figure 2.5D). The decrease of decay time of ZnOEP in nanoaggregate confirms the H-aggregation of porphyrin molecules.[29,67]

It is interesting to note that these porphyrin nanoaggregates are quite efficient at generating photoinduced singlet oxygen. The nanoaggregates exhibit emission at the near infrared (NIR) region, 1270 nm under the excitation wavelength of 375 nm, which is the characteristic emission of singlet oxygen (Figure 2.6A). The ZnOEP nanoaggregate generates singlet oxygen by following the proposed pathways depicted in Figure 2.6B. During the excitation of ZnOEP nanoaggregate at 375 nm, the nanoaggregates are promoted to first excited singlet (S_1) state for Q-bands and second excited singlet (S_2) state for B-bands. This de-excitation from S_2 to S_1 state occurs via internal conversion. Then, the nanoaggregate undergoes intersystem crossing from S_1 state to the triplet (T_1) state by a nonradiative relaxation process. The triplet state of the nanoparticle eventually transfers energy nonradiatively toward the triplet oxygen and converts to the singlet oxygen.[71] Therefore, these IL-assisted ZnOEP nanoaggregates may open up further prospects for application in photodynamic therapy.

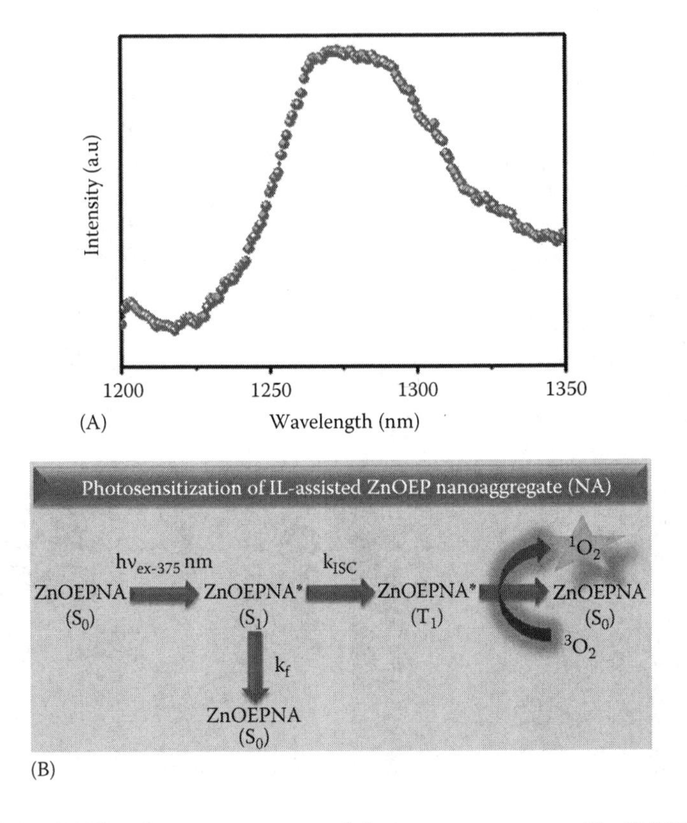

FIGURE 2.6 (A) Phosphorescence spectra of singlet oxygen generated by ZnOEP nanoaggregate at an excitation wavelength of 375 nm. (B) Schematic representation of pathways of singlet oxygen generation. (Mandal, S., Kundu, S., Bhattacharyya, S., and Patra, A., Photophysical properties of ionic liquid-assisted porphyrin nanoaggregate-nickel phthalocyanine conjugates and singlet oxygen generation, *J. Mater. Chem. C*, 2, 8691–8699. Reproduced by permission of The Royal Society of Chemistry.)

2.6.5 Porphyrin Nano-/Microaggregates for Photocatalytic Activity

Again, we have demonstrated the synthesis of surfactant-assisted different morphologies of TCPP assembled structures (spherical particle, rod, flakes, and flower).[16] The assembled structures are well characterized by scanning electron microscopy (SEM) and X-ray diffraction (XRD). The SEM images of the different morphologies of TCPP are shown in Figure 2.7. The porphyrin molecules formed nearly spherical nanoaggregates with the average diameter of 100 nm when the stirring time was 15 min. The spherical nanoaggregates are transformed into rod-shaped aggregates with the length of 230 and 73 nm width when the stirring time was 1 h. As the stirring time extended to 6 h, the formation of flakes-shaped microaggregates is observed. The average size of the flakes is 3.5 μm with 0.4–0.5 μm width. Finally, when the stirring time is extended to 48 h, the flakes are assembled to form the flower-shaped aggregates. The average size of the flowers is about 4.5 μm. A plausible mechanism has been proposed for the formation of the different morphologies of nano-/microstructures in Figure 2.8. In general, TCPP molecules are insoluble in water. The carboxylic groups of TCPP are de-protonated in alkaline solution to form tetra-carboxylate anion $[TCPP]^{-4}$, which is soluble in water and forms a homogeneous solution.

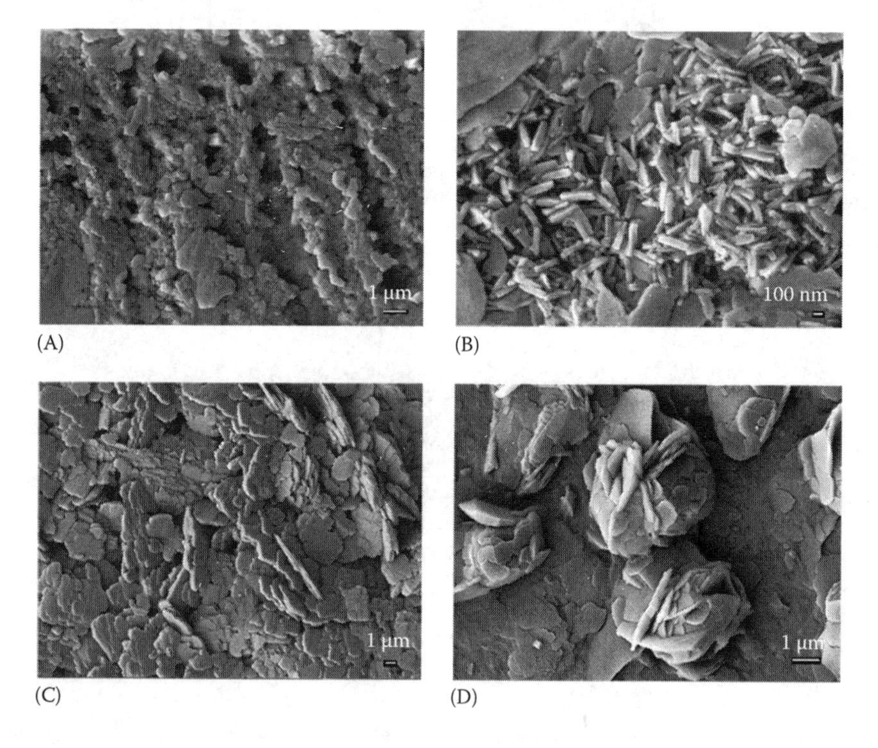

FIGURE 2.7 SEM images of surfactant-assisted different morphology of TCPP aggregates: (A) particles, (B) rods, (C) flakes, and (D) flowers. (Reprinted with permission from Mandal, S., Nayak, S.K., Mallampalli, S., and Patra, A., Surfactant-assisted porphyrin based hierarchical nano/micro assemblies and their efficient photocatalytic behavior, *ACS Appl. Mater. Interfaces*, 6, 130–136. Copyright 2014 American Chemical Society.)

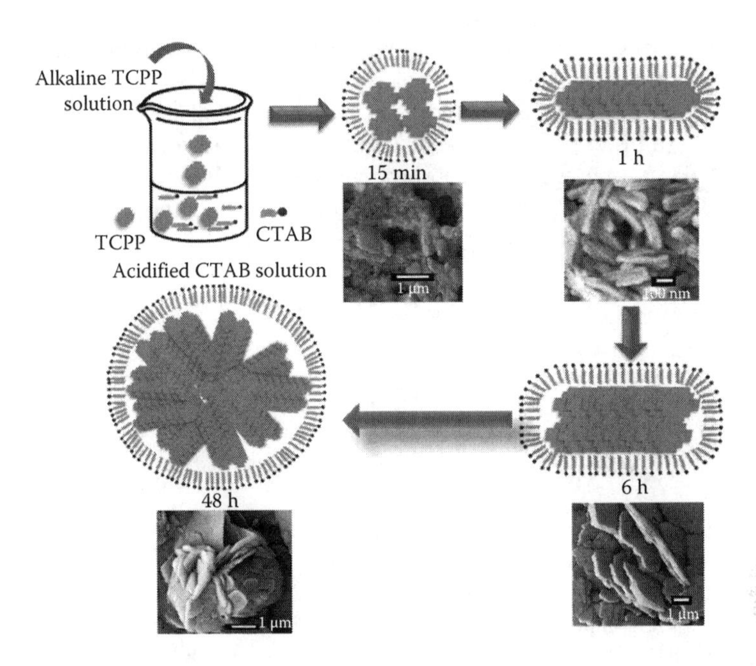

FIGURE 2.8 Schematic representation of plausible mechanism of different morphologies. (Reprinted with permission from Mandal, S., Nayak, S.K., Mallampalli, S., and Patra, A., Surfactant-assisted porphyrin based hierarchical nano/micro assemblies and their efficient photocatalytic behavior, *ACS Appl. Mater. Interfaces*, 6, 130–136. Copyright 2014 American Chemical Society.)

The encapsulation process into the micelle occurs by adding this $[TCPP]^{-4}$ basic aqueous solution to acidic aqueous solution of CTAB under vigorous stirring. Here, the cationic surfactant might act as a stabilizer via the electrostatic interaction between the carboxylate anion in the porphyrin molecules and the cationic surfactant. Subsequently, the surfactant might affect the formation mechanism as well as the micelle formation for dispersion of the porphyrin aggregation. During acid–base neutralization, the $[TCPP]^{-4}$ anion becomes hydrophobic TCPPs after protonation. These hydrophobic TCPP molecules are encapsulated into the hydrophobic micelle because of hydrophobic–hydrophobic interaction between porphyrin and CTAB molecules. The number of available porphyrin molecules in the micelle interior is increased with increasing stirring time, which might help to form different morphologies of TCPP aggregated structures, namely, sphere, rod, flakes, and flowers. The formation of assembled structures is driven by noncovalent interactions such as hydrophobic–hydrophobic and aromatic π–π stacking between the molecules.[36,72] The XRD data did not show refraction peak in case of TCPP molecules, indicating the noncrystalline behavior of TCPP molecules. However, strong refraction peaks are observed in different morphologies of porphyrin aggregates, indicating the crystalline nature may be due to aromatic π–π stacking between the porphyrin molecules.[19] The photophysical properties of the nano-/microstructures suggest that the structures are the results of J- and H-aggregation of the porphyrin molecules in interior of the micelle. A small spectral shifting with broadening of the absorption and emission band is due to the aggregation of the porphyrin.

It has been observed that the different morphologies of nano-/microaggregated structures have different photocatalytic activities. The photocatalytic performance is illustrated quantitatively by the absorption values at 557 nm of the Rhodamine B (RhB) at different times under visible light irradiation. The degradation rate can be calculated by using the following expression:

$$D\% = \frac{C_0 - C}{C_0} \times 100\% \tag{2.9}$$

where

C_0 is the initial concentration of RhB

C is the concentration at time t

The calculated degradation rates of RhB are 56%, 81%, 79%, and 71% for the spherical, rod, flakes, and flower-shaped TCPP aggregated structures, respectively. The C/C_0 versus time plot of RhB in the presence of different morphologies of TCPP aggregates clearly indicates that the photocatalytic activity depends upon the morphology of the TCPP-aggregated structures (Figure 2.9). In case of J-aggregation of π-conjugated organic molecules, a strong intermolecular π electronic coupling occurs between the coherently aligned chromophores which enhance the coherent electronic delocalization over the aggregated molecules due to the strong intermolecular π–π interactions. In the nonaggregated monomers, the π-conjugated electrons are locally delocalized over their macrocyclic plane, and hence, a very weak

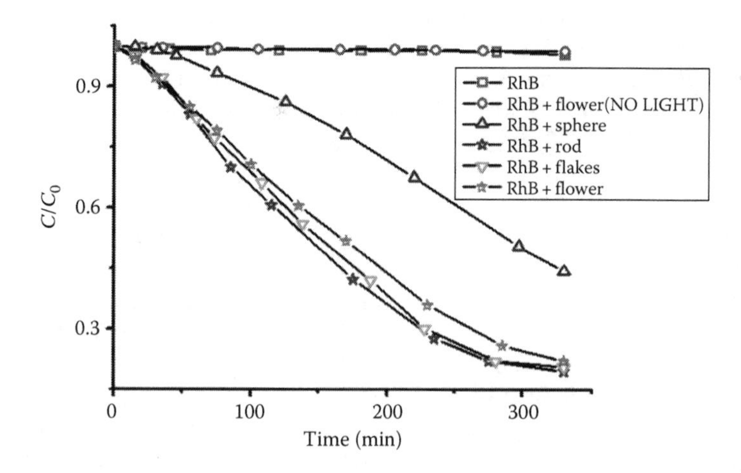

FIGURE 2.9 Photocatalytic activity plot (C/C_0 vs. Time) of surfactant-assisted different morphologies of TCPP aggregates. (Reprinted with permission from Mandal, S., Nayak, S.K., Mallampalli, S., and Patra, A., Surfactant-assisted porphyrin based hierarchical nano/micro assemblies and their efficient photocatalytic behavior, *ACS Appl. Mater. Interfaces*, 6, 130–136. Copyright 2014 American Chemical Society.)

intermolecular π–π interaction exists. To compare with the J-aggregated porphyrin units, porphyrin monomers are not good photo-semiconductors. A mechanism has been proposed to explain the aggregation-induced photocatalytic of the different morphologies of TCPP assemblies. Most of the porphyrin molecules exist as monomeric species in case of spherical shape. They are in contact with each other by means of forming complexes with CTAB molecules. In case of rod-shaped TCPP aggregates, the porphyrin molecules are cooperatively aligned as J-aggregates with a slipped co-facial arrangement. As a result, rod-shaped aggregates have the stronger intermolecular π–π interactions than the spherical aggregates. The J-aggregated structures enhance the coherent electronic delocalization and become an efficient photo-semiconductor. Recently, Guo et al. also reported that nanofibers have more photocatalytic efficiency than spherical nanoaggregates of ZnTPyP, as the ZnTPyP nanofibers are good photo-semiconductors.[73]

Similarly, the photocatalytic activities of the TCPP-based flakes and flowers have been demonstrated. In these two cases, the photocatalytic efficiency is lower than the rod but larger than spherical-shaped TCPP aggregates. It is evident from spectroscopic data that the porphyrin molecules are in both J- and H-aggregations in flakes and flower-shaped aggregates. It is proposed that the J-aggregates exhibit better photocatalytic efficiency due to the coherent electronic delocalization, which could facilitate the electron transfer process.[73,74] However, the H-aggregated molecules may diminish the intermolecular π–π interactions in the flake and flower-shaped aggregates. As a result, the photocatalytic efficiency of the flakes and flower-shaped aggregates is lower than that of the rod-shaped TCPP aggregates.

2.6.6 Porphyrin–QD Nanocomposites: Recovery of Luminescence of QD

Three different compositions of $Cd_{1-x}Zn_xS$ QD with the same size of 5 ± 0.5 nm have been synthesized and well characterized by TEM, absorption, and emission spectra (Figure 2.10A and B).[61] The $Cd_{1-x}Zn_xS$ QD-APTPP system was prepared by the electrostatic interaction between negatively charged QD and positively charged APTPP and was confirmed by zeta potential measurements. The photoluminescence quenching of the QD has shown to occur in the presence of APTPP. The Stern–Volmer plot implies the presence of both dynamic and static quenching of the QD.[66] The dynamic quenching is due to the Fröster resonance energy transfer from QD to APTPP, which is confirmed by the fluorescence decay time study. The energy transfer efficiencies are found to be 64%, 85%, and 90% for $Cd_{0.62}Zn_{0.38}S$, $Cd_{0.52}Zn_{0.48}S$, and $Cd_{0.31}Zn_{0.69}S$, respectively. The energy transfer efficiency increases with the increase of zinc/cadmium ratio (Figure 2.10C and D). This is because of the fact that the emission spectra of QD are blue shifted with the increase in the $Cd_{1-x}Zn_xS$ QD, and as a result a better overlap with the absorption spectra of the APTPP molecules is observed. A higher value of overlap integral as well as strong interaction between $Cd_{1-x}Zn_xS$ QD and APTPP enhances the energy transfer efficiency. Therefore, the energy transfer efficiency can be controlled by changing the compositions of the $Cd_{1-x}Zn_xS$ QD.

Indeed, particular macrocyclic receptors are known to bind cationic ligands by their negatively charged surface and hydrophobic moieties by their hydrophobic

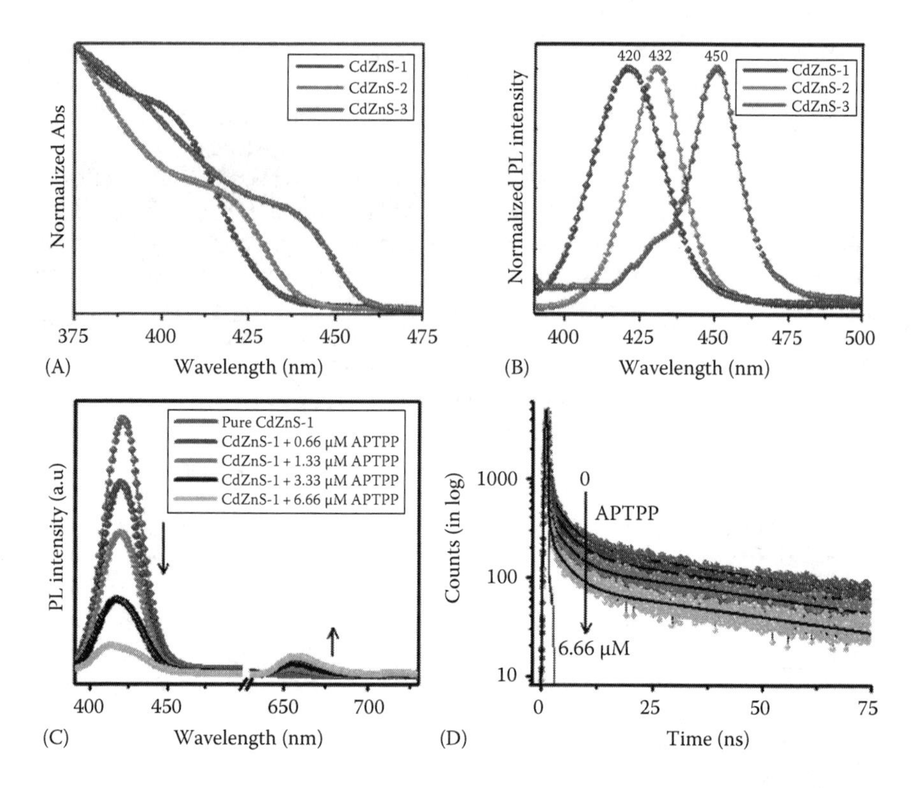

FIGURE 2.10 (A) Absorption spectra and (B) emission spectra of three CdZnS QDs; (C) fluorescence spectra and (D) fluorescence decay profiles of CdZnS-1 in the presence of different concentrations of APTPP. (Reprinted with permission from Mandal, S., Rahaman, M., Sadhu, S., Nayak, S.K., and Patra, A., Fluorescence wwitching of quantum dot in quantum Dot-Porphyrin-Cucurbit[7]Uril assemblies, *J. Phys. Chem. C*, 117, 3069–3077. Copyright 2013 American Chemical Society.)

cavity in aqueous solutions. The porphyrins of QD–porphyrin composites are encapsulated within macrocyclic receptors (cucurbit[7]uril) to restore the fluorescence of QD. The change imposed on photoluminescence of the $Cd_1{-}_xZn_xS$ QD by APTPP has been reversed with the addition of an excess amount of CB[7]. It is interesting to note that the photoluminescence quenching of QD in presence of APTPP is further restored to its original intensity after addition of CB[7]uril, which is shown in Figure 2.11A. A plot of F/F_0 of QD versus concentration of CB[7]uril reveals that the maximum fluorescence intensity (>85%) of QD is restored at 70 μM concentration of CB[7]uril and above (Figure 2.11B).

It suggests that QD–porphyrin assemblies are destroyed in the presence of CB[7]uril and QDs are isolated due to encapsulation of porphyrin molecules within CB[7]uril. To further confirm the encapsulation of porphyrin within CB[7]uril, the photoluminescence spectra of porphyrin are taken in water, and after encapsulation within CB[7]uril, at the excitation wavelength of 405 nm, which corresponds to the direct excitation of APTPP molecules (Figure 2.11C). The blue shift (~6 nm) of the emission

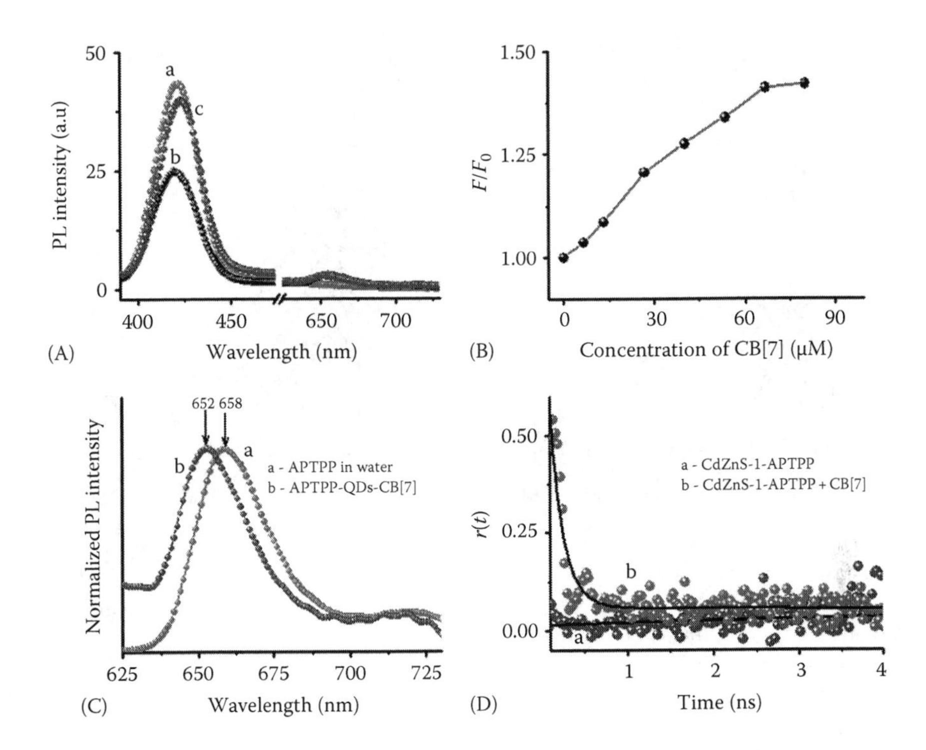

FIGURE 2.11 (A) Fluorescence spectra of CdZnS-1 in the absence and in the presence of APTPP and CB[7]uril [(a) pure QDs, (b) QDs+APTPP, (c) QDs+APTPP+CB[7]uril], (B) plot of F/F_0 of QD versus concentration of CB[7]uril, (C) fluorescence spectra of APTPP in water and CB[7]uril at an excitation wavelength of 405 nm, and (D) time-resolved anisotropy spectra of CdZnS-1-APTPP and CdZnS1-APTPP+CB[7]uril. (Reprinted with permission from Mandal, S., Rahaman, M., Sadhu, S., Nayak, S.K., and Patra, A., Fluorescence switching of quantum dot in quantum Dot-Porphyrin-Cucurbit[7]Uril assemblies, *J. Phys. Chem. C*, 117, 3069–3077. Copyright 2013 American Chemical Society.)

spectrum of APTPP-QDs-CB[7]uril system with respect to APTPP in water confirms the encapsulation of amino-phenyl and phenyl moieties of APTPP molecules.

In addition, the fluorescence intensity of APTPP increases with the increase of CB[7]uril concentration at the excitation wavelength of 405 nm. The nonlinear behavior of the double reciprocal plot implies that the stoichiometry of APTPP-CB[7] uril system is not 1:1. Furthermore, the time-resolved anisotropy measurement of APTPP-QD and APTPP-QD-CB[7]uril systems confirms the encapsulation of the amino-phenyl and phenyl moieties of APTPP molecules. In the presence of CB[7] uril, the rotation of the porphyrin molecules will be restricted due to the encapsulation. Time-resolved anisotropy decays analysis gives better insight into the dynamics of the APTPP molecules. The anisotropy decay spectra (depicted in Figure 2.11D) clearly show that the anisotropy decay of APTPP molecules increases in the presence of CB[7]uril. The increment of correlation time constant strongly implies the

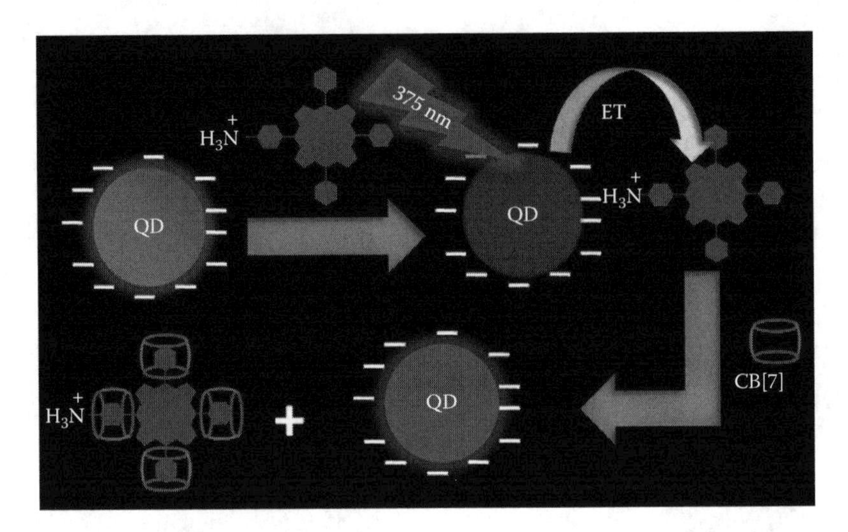

FIGURE 2.12 Schematic representation of fluorescence switching of QD in the presence of APTPP and CB[7]uril. (Reprinted with permission from Mandal, S., Rahaman, M., Sadhu, S., Nayak, S.K., and Patra, A., Fluorescence switching of quantum dot in quantum Dot-Porphyrin-Cucurbit[7]Uril assemblies, *J. Phys. Chem. C*, 117, 3069–3077. Copyright 2013 American Chemical Society.)

encapsulation of the amino-phenyl and phenyl moieties of APTPP molecules in the hydrophobic nanocavity of CB[7]uril. Therefore, CB[7]uril encapsulates the APTPP and prevents the energy transfer process that restores photoluminescence properties of $Cd_1{-}_xZn_xS$ QD. The turn off-on process of $Cd_1{-}_xZn_xS$ QD is presented in the Figure 2.12. This phenomenon may be applicable as luminescent probe for imaging and sensing with exceptional performances.

2.7 CONCLUSION

In summary, a few aspects of porphyrin-based functional nanomaterials have been demonstrated in this chapter. $(ZnOEP)_2$ exhibits *syn* conformation in nanoparticle form due to the π–π stacking between the two moieties of the bis-porphyrin, whereas *anti* conformation is stable inside PVK nanoparticle because of the rigid environment inside PVK nanoparticle. ZnOEP molecules are J-aggregated inside PVK nanoparticles and an efficient energy transfer occurs from PVK nanoparticle to porphyrin molecules. IL-assisted porphyrin nanoaggregate generates singlet oxygen, which may be useful in photodynamic therapy. Furthermore, the fabrication of different morphologies of TCPP nanomaterials has been explored. The nanomaterials are formed due to the J- and H-aggregation of the porphyrin molecules. These porphyrin nanomaterials, especially rod-shaped nanostructure, have high efficiency of photo degradation with RhB under visible light irradiation. In addition, porphyrin–QD composite system has also been demonstrated. In this system, energy transfer from alloy QDs to porphyrin molecules depends upon the compositions of the alloy QD. The luminescence property of the QD has been recovered using a very fascinating microcavity, CB[7]uril, due to the encapsulation of porphyrin molecules inside CB[7]uril. This system may have potential application in biological sensing.

Based on the interesting findings of the porphyrin-based functional nanomaterials, there are several issues that need to be explored in future to reveal the practical usability of these results in the field of photonics, optoelectronics, and sensor chemistry. Porphyrin-encapsulated semiconducting nanomaterials can be used in the artificial light-harvesting system by proper modification that will enhance the unidirectional energy transfer from polymer to porphyrin molecules. A new composite system, porphyrin nanostructure–graphene can be designed for application in photocatalysis and solar cell devices. Porphyrin–quantum dot composite systems may be used in photodynamic therapy and biological sensing. Therefore, ultrafast spectroscopic investigation is required to study the fundamental spectroscopic properties of these composite systems in details.

ACKNOWLEDGMENT

SERB-DST and "DAE-SRC Outstanding Investigator Award" are gratefully acknowledged for financial support. SM thanks CSIR for awarding fellowship.

REFERENCES

1. Lehn, J.-M. 2002. Toward self-organization and complex matter. *Science* 295: 2400–2403.
2. Jang, J. and Oh, J. H. 2003. Facile fabrication of photochromic dye-conducting polymer core-shell nanomaterials and their photoluminescence. *Adv. Mater.* 15: 977–980.
3. Qi, L. 2010. Colloidal chemical approaches to inorganic micro- and nanostructures with controlled morphologies and patterns. *Coord. Chem. Rev.* 254: 1054–1071.
4. Lin, C., Zhu, W., Yang, H., An, Q., Tao, C.-a., Li, W., Cui, J., Li, Z., and Li, G. 2011. Facile fabrication of stimuli-responsive polymer capsules with gated pores and tunable shell thickness and composite. *Angew. Chem. Int. Ed.* 50: 4947–4951.
5. Xiao, J. and Qi, L. 2011. Surfactant-assisted, shape-controlled synthesis of gold nanocrystals. *Nanoscale* 3: 1383–1396.
6. Ding, Q., Miao, Y.-E., and Liu, T. 2013. Morphology and photocatalytic property of hierarchical polyimide/Zno fibers prepared via a direct ion-exchange process. *ACS Appl. Mater. Interfaces* 5: 5617–5622.
7. Leishman, C. W. and McHale, J. L. 2015. Light-harvesting properties and morphology of porphyrin nanostructures depend on ionic species inducing aggregation. *J. Phys. Chem. C* 119: 28167–28181.
8. Yamamoto, S., Mori, S., Wagner, P., Mozer, A. J., and Kimura, M. 2016. A novel covalently linked Zn phthalocyanine-Zn porphyrin dyad for dye-sensitized solar cells. *Isr. J. Chem.* 56: 175–180.
9. Lemon, C. M., Karnas, E., Han, X., Bruns, O. T., Kempa, T. J., Fukumura, D., Bawendi, M. G., Jain, R. K., Duda, D. G., and Nocera, D. G. 2015. Micelle-encapsulated quantum dot-porphyrin assemblies as in vivo two-photon oxygen sensors. *J. Am. Chem. Soc.* 137: 9832–9842.
10. Gong, X., Milic, T., Xu, C., Batteas, J. D., and Drain, C. M. 2002. Preparation and characterization of porphyrin nanoparticles. *J. Am. Chem. Soc.* 124: 14290–14291.
11. Hasobe, T., Imahori, H., Fukuzumi, S., and Kamat, P. V. 2003. Nanostructured assembly of porphyrin clusters for light energy conversion. *J. Mater. Chem.* 13: 2515–2520.

12. Jiang, L., Fu, Y., Li, H., and Hu, W. 2008. Single-crystalline, size, and orientation controllable nanowires and ultralong microwires of organic semiconductor with strong photoswitching property. *J. Am. Chem. Soc.* 130: 3937–3941.
13. Drain, C. M., Varotto, A., and Radivojevic, I. 2009. Self-organized porphyrinic materials. *Chem. Rev.* 109: 1630–1658.
14. Yan, Y., Lin, Y., Qiao, Y., and Huang, J. 2011. Construction and application of tunable one-dimensional soft supramolecular assemblies. *Soft Matter* 7: 6385–6398.
15. Kadish, K. M. and Smith, K. M. 2000. In *The Porphyrin Handbook,* Guilard, R. (ed.). Academic Press: San Diego, CA. 6: 44–130.
16. Schwab, A. D., Smith, D. E., Bond-Watts, B., Johnston, D. E., Hone, J., Johnson, A. T., De Paula, J. C., and Smith, W. F. 2004. Photoconductivity of self-assembled porphyrin nanorods. *Nano Lett.* 4: 1261–1265.
17. Arai, T., Tanaka, M., and Kawakami, H. 2012. Porphyrin-containing electrospun nanofibers: Positional control of porphyrin molecules in nanofibers and their catalytic application. *ACS Appl. Mater. Interfaces* 4: 5453–5457.
18. Mandal, S., Bhattacharyya, S., Borovkov, V., and Patra, A. 2012. Photophysical properties, self-assembly behavior, and energy transfer of porphyrin-based functional nanoparticles. *J. Phys. Chem. C* 116: 11401–11407.
19. Mandal, S., Nayak, S. K., Mallampalli, S., and Patra, A. 2014. Surfactant-assisted porphyrin based hierarchical nano/micro assemblies and their efficient photocatalytic behavior. *ACS Appl. Mater. Interfaces* 6: 130–136.
20. Maggini, L. and Bonifazi, D. 2012. Hierarchised luminescent organic architectures: Design, synthesis, self-assembly, self-organisation and functions. *Chem. Soc. Rev.* 41: 211–241.
21. Li, S., Chang, K., Sun, K., Tang, Y., Cui, N., Wang, Y., Qin, W., Xu, H., and Wu, C. 2016. Amplified singlet oxygen generation in semiconductor polymer dots for photodynamic cancer therapy. *ACS Appl. Mater. Interfaces* 8: 3624–3634.
22. Medforth, C. J., Wang, Z., Martin, K. E., Song, Y., Jacobsen, J. L., and Shelnutt, J. A. 2009. Self-assembled porphyrin nanostructures. *Chem. Commun.* 46(47): 7261–7277.
23. Bera, R., Mandal, S., Mondal, B., Jana, B., Nayak, S. K., and Patra, A. 2016. Graphene–porphyrin nanorod composites for solar light harvesting. *ACS Sustain. Chem. Eng.* 4: 1562–1568.
24. Sheng, N., Zong, S., Cao, W., Jiang, J., Wang, Z., and Cui, Y. 2015. Water dispersible and biocompatible porphyrin-based nanospheres for biophotonics applications: A novel surfactant and polyelectrolyte-based fabrication strategy for modifying hydrophobic porphyrins. *ACS Appl. Mater. Interfaces* 7: 19718–19725.
25. Schwab, A. D., Smith, D. E., Rich, C. S., Young, E. R., Smith, W. F., and De Paula, J. C. 2003. Porphyrin nanorods. *J. Phys. Chem. B* 107: 11339–11345.
26. Wang, Z., Medforth, C. J., and Shelnutt, J. A. 2004. Self-metallization of photocatalytic porphyrin nanotubes. *J. Am. Chem. Soc.* 126: 16720–16721.
27. Wang, H., Song, Y., Medforth, C. J., and Shelnutt, J. A. 2006. Interfacial synthesis of dendritic platinum nanoshells templated on benzene nanodroplets stabilized in water by a photocatalytic lipoporphyrin. *J. Am. Chem. Soc.* 128: 9284–9285.
28. Yoon, S. M., Hwang, I.-C., Kim, K. S., and Choi, H. C. 2009. Synthesis of single-crystal tetra(4-pyridyl)porphyrin rectangular nanotubes in the vapor phase. *Angew. Chem. Int. Ed.* 48: 2506–2509.
29. Mandal, S., Kundu, S., Bhattacharyya, S., and Patra, A. 2014. Photophysical properties of ionic liquid-assisted porphyrin nanoaggregate-nickel phthalocyanine conjugates and singlet oxygen generation. *J. Mater. Chem. C* 2: 8691–8699.
30. Ikeda, M., Takeuchi, M., and Shinkai, S. 2003. Unusual emission properties of a triphenylene-based organogel system. *Chem. Commun.* 1354–1355.
31. Estroff, L. A. and Hamilton, A. D. 2004. Water gelation by small organic molecules. *Chem. Rev.* 104: 1201–1217.

32. Ryu, J.-H. and Lee, M. 2005. Transformation of isotropic fluid to nematic gel triggered by dynamic bridging of supramolecular nanocylinders. *J. Am. Chem. Soc.* 127: 14170–14171.
33. Lee, E., Kim, J.-K., and Lee, M. 2008. Lateral association of cylindrical nanofibers into flat ribbons triggered by "Molecular Glue". *Angew. Chem. Int. Ed.* 47: 6375–6378.
34. Chen, L., Revel, S., Morris, K., and Adams, D. J. 2010. Energy transfer in self-assembled dipeptide hydrogels. *Chem. Commun.* 46: 4267–4269.
35. Lu, G., Zhang, X., Cai, X., and Jiang, J. 2009. Tuning the morphology of self-assembled nanostructures of amphiphilic tetra(P-hydroxyphenyl)porphyrins with hydrogen bonding and metal-ligand coordination bonding. *J. Mater. Chem.* 19: 2417–2424.
36. Ma, J., Zhang, W., Li, Z., Lin, Q., Xu, J., and Han, Y. 2016. Competition of major forces dominating the structures of porphyrin assembly. *Cryst. Growth Des.* DOI:10.1021/acs.cggd.5b01502. 16(4): 1942–1947.
37. Kasha, M. 1963. Energy transfer mechanisms and the molecular exciton model for molecular aggregates. *Radiat. Res.* 20: 55–70.
38. Smith, K. M. 1975. *Porphyrins and Metalloporphyrins*. Elsevier: Amsterdam, the Netherlands.
39. Arnold, S., Whitten, W. B., and Damask, A. C. 1971. Davydov splitting and band structure for triplet excitons in pyrene. *Phys. Rev. B* 3: 3452–3457.
40. Kasha, M., Rawls, H. R., and El-Bayoumi, M. A. 1965. Exciton model in molecular spectroscopy. *Pure Appl. Chem.* 11: 371–392.
41. Wu, C., Szymanski, C., Cain, Z., and McNeill, J. 2007. Conjugated polymer dots for multiphoton fluorescence imaging. *J. Am. Chem. Soc.*, 129: 12904–12905.
42. Wu, C., Bull, B., Szymanski, C., Christensen, K., and McNeill, J. 2008. Multicolor conjugated polymer dots for biological fluorescence imaging. *ACS Nano* 2: 2415–2423.
43. Abdul Rahim, N. A., McDaniel, W., Bardon, K., Srinivasan, S., Vickerman, V., So, P. T. C., and Moon, J. H. 2009. Conjugated polymer nanoparticles for two-photon imaging of endothelial cells in a tissue model. *Adv. Mater.* 21: 3492–3496.
44. Tuncel, D. and Demir, H. V. 2010. Conjugated polymer nanoparticles. *Nanoscale* 2: 484–494.
45. Bhattacharyya, S., Barman, M. K., Baidya, A., and Patra, A. 2014. Singlet oxygen generation from polymer nanoparticles–photosensitizer conjugates using fret cascade. *J. Phys. Chem. C* 118: 9733–9740.
46. Bhattacharyya, S., Paramanik, B., and Patra, A. 2011. Energy transfer and confined motion of dyes trapped in semiconducting conjugated polymer nanoparticles. *J. Phys. Chem. C* 115: 20832–20839.
47. Bhattacharyya, S., Prashanthi, S., Bangal, P. R., and Patra, A. 2013. Photophysics and dynamics of dye-doped conjugated polymer nanoparticles by time-resolved and fluorescence correlation spectroscopy. *J. Phys. Chem. C* 117: 26750–26759.
48. Elsabahy, M., Heo, G. S., Lim, S.-M., Sun, G., and Wooley, K. L. 2015. Polymeric nanostructures for imaging and therapy. *Chem. Rev.* 115: 10967–11011.
49. Dmitriev, R. I., Borisov, S. M., Düssmann, H., Sun, S., Müller, B. J., Prehn, J., Baklaushev, V. P., Klimant, I., and Papkovsky, D. B. 2015. Versatile conjugated polymer nanoparticles for high-resolution O_2 imaging in cells and 3d tissue models. *ACS Nano* 9: 5275–5288.
50. Wu, C., Bull, B., Christensen, K., and McNeill, J. 2009. Ratiometric single-nanoparticle oxygen sensors for biological imaging. *Angew. Chem. Int. Ed.* 48: 2741–2745.
51. Shen, X., He, F., Wu, J., Xu, G. Q., Yao, S. Q., and Xu, Q.-H. 2011. Enhanced two-photon singlet oxygen generation by photosensitizer-doped conjugated polymer nanoparticles. *Langmuir* 27: 1739–1744.
52. Lemon, C. M., Karnas, E., Bawendi, M. G., and Nocera, D. G. 2013. Two-photon oxygen sensing with quantum dot-porphyrin conjugates. *Inorg. Chem.* 52: 10394–10406.

53. Pluth, M. D. and Raymond, K. N. 2007. Reversible guest exchange mechanisms in supramolecular host-guest assemblies. *Chem. Soc. Rev.* 36: 161–171.
54. Praetorius, A., Bailey, D. M., Schwarzlose, T., and Nau, W. M. 2008. Design of a fluorescent dye for indicator displacement from cucurbiturils: A macrocycle-responsive fluorescent switch operating through a pKa shift. *Org. Lett.* 10: 4089–4092.
55. Ghale, G., Ramalingam, V., Urbach, A. R., and Nau, W. M. 2011. Determining protease substrate selectivity and inhibition by label-free supramolecular tandem enzyme assays. *J. Am. Chem. Soc.*133: 7528–7535.
56. Montes-Navajas, P. and García, H. 2010. Long-lived charge separation in gold nanoparticles encapsulated inside Cucurbit[7]Uril and its relevance for photocatalysis. *J. Phys. Chem. C* 114: 18847–18852.
57. Gupta, M., Maity, D. K., Singh, M. K., Nayak, S. K., and Ray, A. K. 2012. Supramolecular interaction of coumarin 1 dye with Cucurbit[7]Uril as host: Combined experimental and theoretical study. *J. Phys. Chem. B* 116: 5551–5558.
58. Lan, Y., Loh, X. J., Geng, J., Walsh, Z., and Scherman, O. A. 2012. A supramolecular route towards core-shell polymeric microspheres in water via Cucurbit[8]Uril complexation. *Chem. Commun.* 48: 8757–8759.
59. Barrow, S. J., Kasera, S., Rowland, M. J., del Barrio, J., and Scherman, O. A. 2015. Cucurbituril-based molecular recognition. *Chem. Rev.* 115: 12320–12406.
60. Yildiz, I., Tomasulo, M., and Raymo, F. M. 2006. A mechanism to signal receptor-substrate interactions with luminescent quantum dots. *Proc. Natl. Acad. Sci.* 103: 11457–11460.
61. Mandal, S., Rahaman, M., Sadhu, S., Nayak, S. K., and Patra, A. 2013. Fluorescence switching of quantum dot in quantum Dot-Porphyrin-Cucurbit[7]Uril assemblies. *J. Phys. Chem. C* 117: 3069–3077.
62. Kashani-Motlagh, M. M., Rahimi, R., and Kachousangi, M. J. 2010. Ultrasonic method for the preparation of organic porphyrin nanoparticles. *Molecules* 15: 280–287.
63. Bhattacharyya, S., Sen, T., and Patra, A. 2010. Host-guest energy transfer: Semiconducting polymer nanoparticles and Au nanoparticles. *J. Phys. Chem. C* 114: 11787–11795.
64. Mandal, S., Bhattacharyya, S., Borovkov, V., and Patra, A. 2011. Porphyrin-based functional nanoparticles: Conformational and photophysical properties of bis-porphyrin and bis-porphyrin encapsulated polymer nanoparticles. *J. Phys. Chem. C* 115: 24029–24036.
65. Sadhu, S. and Patra, A. 2008. Synthesis and spectroscopic study of high quality alloy Cdxzn1-Xs nanocrystals. *J. Chem. Sci.* 120: 557–564.
66. Lakowicz, J. R. 1999. *Principles of Fluorescence Spectroscopy*. Kluwer Academic/Plenum Publishers: New York.
67. Verma, S., Ghosh, A., Das, A., and Ghosh, H. N. 2010. Ultrafast exciton dynamics of J- and H-aggregates of the porphyrin-catechol in aqueous solution. *J. Phys. Chem. B* 114: 8327–8334.
68. Forster, T. 1959. Transfer mechanisms of electronic excitation. *Discuss. Faraday Soc.* 27: 7–17.
69. Wu, C., Zheng, Y., Szymanski, C., and McNeill, J. 2008. Energy transfer in a nanoscale multichromophoric system: Fluorescent dye-doped conjugated polymer nanoparticles. *J. Phys. Chem. C* 112: 1772–1781.
70. Hwang, I. and Scholes, G. D. 2010. Electronic energy transfer and quantum-coherence in ∏-conjugated polymers. *Chem. Mater.* 23: 610–620.
71. Prasad, P. N. 2004. *Introduction to Biophotonics*. John Wiley & Sons, New York.
72. Bai, F., Sun, Z., Wu, H., Haddad, R. E., Coker, E. N., Huang, J. Y., Rodriguez, M. A., and Fan, H. 2011. Porous one-dimensional nanostructures through confined cooperative self-assembly. *Nano Lett.* 11: 5196–5200.

73. Guo, P., Chen, P., Ma, W., and Liu, M. 2012. Morphology-dependent supramolecular photocatalytic performance of porphyrin nanoassemblies: From molecule to artificial supramolecular nanoantenna. *J. Mater. Chem.* 22: 20243–20249.
74. Kano, H. and Kobayashi, T. 2002. Time-resolved fluorescence and absorption spectroscopies of porphyrin J-aggregates. *J. Chem. Phys.* 116: 184–195.

3 Solubility of Conjugated Polymers
Current Status and Future Prospects and Strategies

Crystal A. Young, Andrew C. Jamison, and T. Randall Lee

CONTENTS

3.1 INTRODUCTION AND HISTORICAL PERSPECTIVE

Conjugated polymers have been the focus of numerous studies for their potential application in organic electronic devices, such as organic light-emitting diodes (OLEDs), organic photovoltaics (OPVs), and organic thin-film transistors (OTFTs).[1–3] The use of conjugated polymer films in these devices has been appealing because of the potential for low-cost facile film processing. Alternative routes to forming an active layer in organic electronics include the use of a variety of conjugated oligomers and small molecules. However, such small molecule devices require the use of chemical vapor deposition under a high vacuum, which makes the production of large devices incredibly expensive, while conjugated polymer systems have been designed for manipulation via much more cost-effective methods, such as inkjet printing.[4]

Trans-polyacetylene (see Figure 3.1), the simplest of the conjugated polymers, is insoluble and intractable. Hideki Shirakawa, Alan J. Heeger, and Alan G. MacDiarmid made modifications to its structure and shared the Nobel Prize in chemistry in 2000 for exploring its electronic properties.[5] With these developments, the interest in polyacetylene (PA) has grown. However, to be a useful material, derivatives of this polymer format

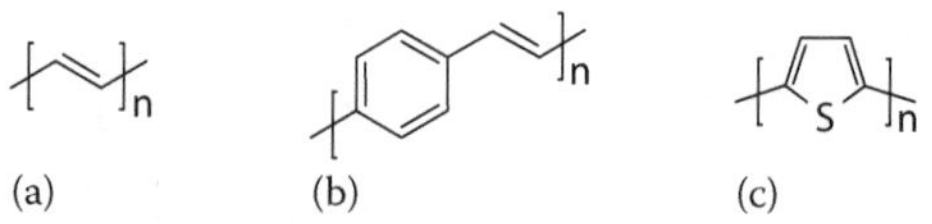

FIGURE 3.1 Chemical structures of common conjugated polymers: (a) polyacetylene (PA), (b) poly(phenylene vinylene) (PPV), and (c) polythiophene (PT).

must be easily and efficiently processed during device preparation; consequently, many studies have focused on modifications that will increase the solubility of its derivatives.[6] Since the first report of poly(phenylene vinylene) (PPV) in an OLED device by Tang and Van Slyke in 1987, there have been thousands of publications addressing conjugated polymer issues such as bandgap modification, the role of film morphology, and charge-transfer processes, with each of these concerns ultimately effecting device efficiency.[7–10] Polymer solubility is a concern that is intertwined with all of these parameters; an adjustment to modify one characteristic of the polymer might create another problem in using the polymer. And to characterize and evaluate the performance of these materials, many solution-based measurements are needed as well as thin-film measurements that require the polymers to be dissolved in a compatible solvent.

In reports focused on conjugated polymer research, it is common to find the phrase "readily soluble in common organic solvents" for investigations describing new polymers for use in organic electronics, but it is difficult to find studies that quantify "readily soluble," a lack of definition that adds to the application problems for these polymers.[11] But the absence of such clarity has not deterred researchers from tackling the concern. The challenges associated with polymer processability have been approached from several directions, some of which address the issue of polymer solubility through structural changes, such as the introduction of solubilizing substituents or the use of dopants. Alternatively, soluble precursors that lack the desirable properties but are more easily processed can be used and then treated after film formation to produce the desired conjugated polymer. Each approach not only affects the processability of the conjugated polymer but also necessarily alters the desired characteristics that influence device performance. To minimize structural intervention in the desired morphological properties, and any deleterious effects on device performance, it is necessary to find techniques that will maximize both the solvency and successful incorporation of the polymer for commercial applications. Therefore, in this review, the modifications that affect conjugated polymer solubility will be discussed in the context of their impact on device performance.

The majority of the applications for conjugated polymers that will be reviewed herein will be for OLEDs and OPVs. While solubility is important for both applications, the effectiveness of a polymer for each device is measured in widely different ways. Successful conjugated polymers for use in organic light-emitting diodes must have high photoluminescent (PL) efficiency and low turn-on voltage. For OPVs, power conversion efficiency (PCE) is an important parameter that combines several factors used to define the overall performance of the device. While numerous studies of OLEDs have focused on increasing the bandgap to enable the realization of blue-emitting devices, OPV studies are often more concerned with finding low bandgap materials to maximize the low-wavelength light harvested from the solar spectrum.[12] Therefore, for both types of device applications, any changes to structure for enhancing the solubility should produce no significant change in the underlying electronic characteristics of the conjugated chain. Stability under operating conditions is also a major concern for all conjugated polymers for use in devices, and techniques to improve solubility cannot be successful if they produce less stable polymers. Altering the synthetic method, the polymer structure, and device fabrication procedures to enhance the solubility of the conjugated polymer might also have unexpected and unintended effects on device efficiency, making it necessary to have several options to overcome solubility problems to allow for the successful tailoring of each conjugated polymer to its unique application.

3.2 SOLUBLE PRECURSORS

Due to the insolubility of unsubstituted conjugated polymers, several synthetic methods were developed to produce soluble precursor polymers, which upon treatment provide the desired properties in the final product. The first of these methods is the Wessling–Zimmerman route, which involves a sulfonium precursor now named for the authors.[13] The general synthetic pathway begins with the functionalization of a *para*-xylylene derivative bearing an appropriate leaving group, followed by a polymerization step (Scheme 3.1 shows the Wessling–Zimmerman route to an unfunctionalized PPV). The saturated precursor polymer can then be cast into a film, which, following treatment (chemical, thermal, or irradiative), will afford the desired conjugated film. While the first polymer to be created by the sulfonium precursor route was PPV, many bishalomethyl aromatics are compatible with these reactions. The first step involves the addition of a dialkyl sulfide in methanol affording the sulfonium salt. Upon addition of a base such as potassium *tert*-butoxide, the precursor polymer is formed. This polymer can be purified by dialysis to remove any monomeric or oligomeric impurities. It can then be spin- or dip-coated onto a substrate as part of an organic electronic device. To afford the fully conjugated product, elimination is most often achieved by thermal treatment ranging from 180°C to 300°C, depending on the dialkyl sulfide used and the counterion.

Modification of the leaving group has a significant impact on the solubility, stability, and subsequent treatment required to form the desired conjugated product. The use of sulfonium salts in the Wessling–Zimmerman route affords water-soluble precursors. While the molecular weight cannot be characterized directly, it is possible to use size-exclusion chromatography as a characterization method following the neutralization of the polymer by refluxing in methanol. One study reports number average molecular weights slightly below 100,000 with polydispersities greater than 5.[15]

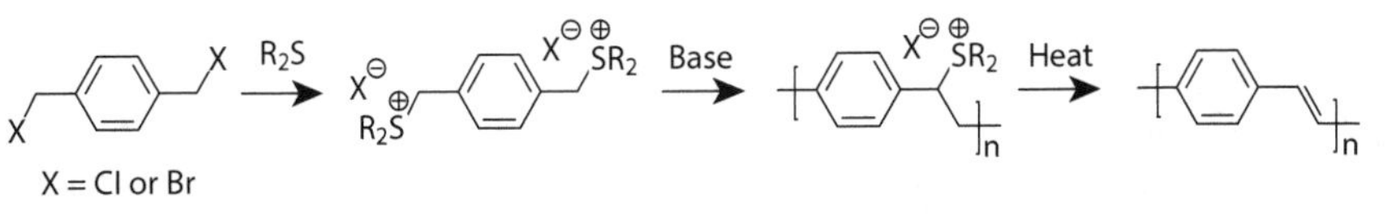

SCHEME 3.1 The Wessling–Zimmerman route to poly(phenylene vinylene). (From Burn, P.L. et al., *J. Chem. Soc., Perkin Trans.*, 1, 3225, 1992.)

This indirect measure of molecular weight is useful because fully conjugated PPV cannot be analyzed by any solution-based measurement protocol due to its insolubility. While many different sulfonium salts have been investigated, those incorporating tetrahydrothiophene are the most popular due to considerations regarding material costs and the minimization of side reactions.[16] Furthermore, changing the structure of the sulfonium salt to affect polymerization can also include changing the counterion. In cases where bromide is substituted for chloride, elimination can occur at temperatures as low as 100°C.[14] Other leaving groups such as sulfinyls, xanthates, halogens, and phosphonium salts have all been investigated.[17,18] Another concern that plays a role in the use of this method in the preparation of conjugated polymers is that PPV, while relatively stable, is vulnerable to photooxidation. Solutions of the soluble sulfonium precursor in methanol, however, have been shown to be stable for months.[14]

The methods of precursor treatment vary based on the precursor polymer chosen. Almost all procedures require heating (100°C–300°C). While PPV is thermally stable, the caustic acids that are released can be detrimental to other materials in the device, such as the indium tin oxide (ITO) anode used in OLEDs.[19,20] Morgado et al. have shown that doping the PPV with indium chloride can increase the electroluminescent (EL) efficiency of a PPV precursor polymer in the presence of ITO despite a reduction in PL efficiency (see Figure 3.2), which the authors attribute to

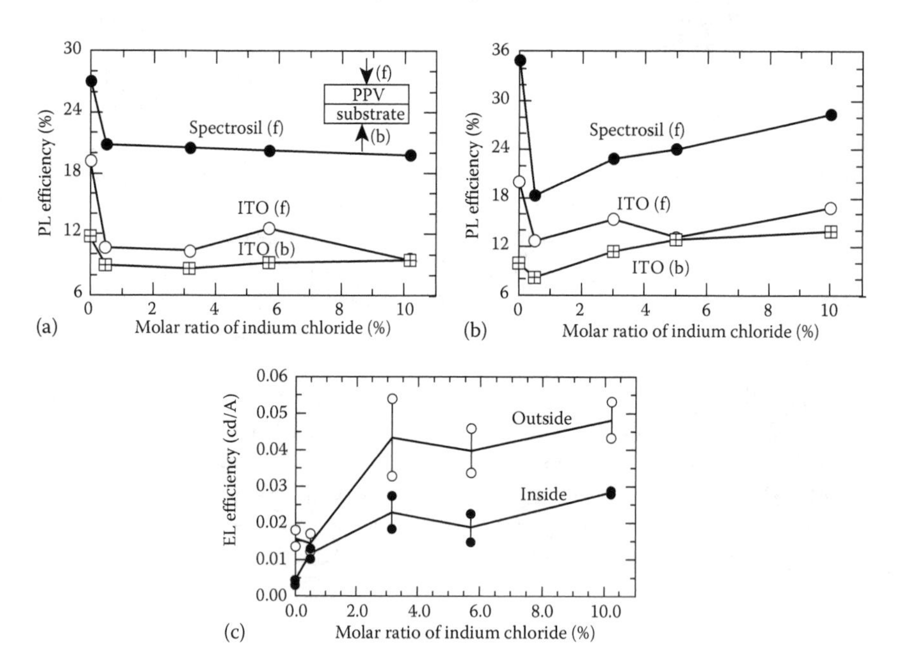

FIGURE 3.2 Photoluminescent (PL) efficiency of PPV with different molar ratios of $InCl_3$ in the precursor solution: (a) conversion in atmosphere—the inset shows the "front" (f) and "back" (b) excitation geometry; (b) conversion in inert atmosphere. (c) Electroluminescent (EL) efficiency of OLED device with different molar ratios of $InCl_3$ in the precursor solution, showing conversion inside and outside the glove box. (From Morgado, J. et al., *Synthetic Met.*, 122, 119, 2001.)

the salt-induced changes in disorder and conjugation length.[21] While methods have been developed to eliminate the need for thermal treatment, those which use acids are unlikely to be suitable for electronic devices because few devices are made with electrodes that are stable against degradation by acids.[22,23] Alternative methods that require no thermal or chemical treatment use irradiation with UV light, laser light, or microwaves.[24–30] While UV light can induce cross-linking that might be deleterious to device performance, it has been successfully used for device patterning. Other methods of patterning have been accomplished with scanning tunneling microscopy, lithography, and screen inkjet printing.[31,32] An interesting complication for the conversion of the precursor to the conjugated product is that of incomplete conversion. While it is difficult to analyze the film once posttreatment has been performed, evidence pointing to saturated defects can be observed. The presence of these defects has been shown to improve device performance in some cases by reducing the effective conjugation length. In some studies, the soluble precursors have been intentionally partially reduced to limit the length of conjugation for the conjugated film.[33] Other parameters that can be controlled and modified using the soluble precursor method include molecular weight and *cis/trans* geometry in the final film.[34]

While the soluble precursor method allowed for the first solution processable films for conjugated polymer devices, this method is severely limited in scope. Many types of conjugated vinylenes are readily accessed by this method, such as polynaphthalene vinylene, polybiphenyl vinylene, and polyphenyl butadiene. All of these polymers share the vinyl moiety. In the case of polymers without vinyl linkages, such as polythiophene, the soluble precursor method has been ineffective. One proposed mechanism to extend this methodology to other conjugated polymer systems involves a quinodimethane intermediate as shown in Scheme 3.2. If the quinodimethane intermediate is unstable, the polymerization might occur with lower yields or not occur at all.[35,36] In the case of poly(naphthalene vinylenes) (PNVs), aromaticity is retained in the unsubstituted ring of the poly(1,5-naphthalene vinylene) (1,5-PNV) intermediate, while it is lost entirely in the poly(2,6-naphthalene vinylene) (2,6-PNV) intermediate. These factors likely give rise to the much lower yield of the 2,6-PNV polymerization via this method.

Device performance is also sensitive to treatment conditions. And while the precursor polymers are usually simple to polymerize and easy to store, the monomer synthesis can be costly and difficult. These limitations have necessitated alternate approaches to improving the solubility of conjugated polymers.

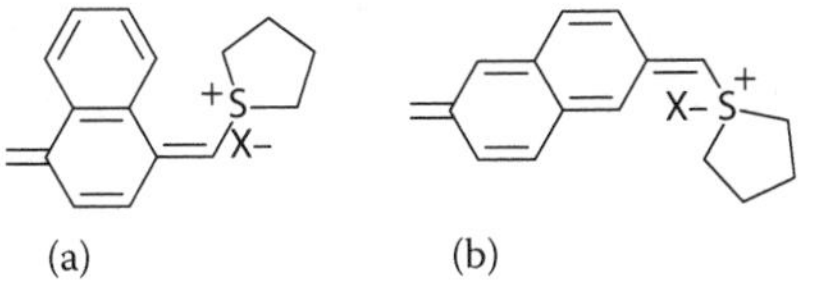

+S X− X− S+

(a) (b)

SCHEME 3.2 Quinodimethane intermediates of poly(1,5-naphthalene vinylene) (a) and poly(2,6-naphthalene vinylene) (b).

3.3 ALKYL AND ALKOXY DERIVATIVES

Unsubstituted conjugated polymers are insoluble, but derivatives modified with long alkyl or alkoxy chains can be solution-processed for incorporation in devices. The choice of solubilizing moiety not only alters processability but also can be used to modify film morphology, optical and electronic properties, and overall device performance. Indeed, all of these aspects are entangled with one another. The addition of bulky side chains can disrupt π–π stacking, which is desirous in some device applications, while unwanted in others. Furthermore, the inclusion of electron-donating groups integrated into the polymer structure can alter the HOMO–LUMO gap, which can produce either favorable or unfavorable results in device performance.

Common conjugated polymer derivatives with favorable solubility characteristics involve alkoxy chains attached to the aromatic rings of the main chain of the polymer backbone. The number of chains attached per repeat unit has a marked effect on solubility. With an increase in the number of attachments, solubility increases, but the polymerization yields are adversely affected, as well as the device performance for such polymers in OLEDs when tetra-substituted.[37]

The optimal chain length for alkoxy substituents needed for improving device performance has also been studied. Methoxy substituents added to PPV fail to produce an improvement in solubility.[38] Ethoxy derivatives have been reported as well but have only been synthesized by the soluble precursor route.[37] In general, it appears that short chains fail to improve the solubility of PPV; however, once the chain length is increased past that of decyloxy, new problems become apparent.[39] Doi et al. noted that dodecyloxy- and tetradecyloxy-substituted conjugated polymers formed substandard films with device performance suffering from the formation of pinhole defects, a problem that likely was caused by poor polymer solubility. Although these authors discuss PL and EL intensity, they make no comment on properties that might affect solubility, such as conjugation length or molecular weight. In a separate study that focused on topics other than chain length comparisons, Tan et al. attributed improved solubility to dodecyloxy chains on alternating rings.[40]

Several studies suggest that the incorporation of side chains with branches increases solubility more than the inclusion of straight chains of an equivalent number of carbons.[41–43] Additionally, branched chains have been shown to disrupt π–π stacking and interchain charge transfer, which can improve some device efficiencies, while adversely effecting performance in others. Perhaps the most extensively studied conjugated polymer is poly[2-methoxy-5-(2-ethylhexyloxy)-1,4-phenylene vinylene] (MEH-PPV), an asymmetric structure that features a methoxy substitution as well as an ethylhexyloxy branch. This polymer has been synthesized by many different routes (e.g., soluble precursor, Gilch, and Horner–Emmons).[44–46] While entire reviews have been published on the subject of MEH-PPV alone, this polymer provides no clear-cut solution to the solubility problem. Some synthetic methods can lead to the creation of insoluble gels that are no more tractable than the unsubstituted PPV.[44] Yet the MEH derivatives of other types of conjugated polymers have proven quite successful at generating soluble derivatives.[47] Alternative branched side chains of different lengths have also been investigated; a common example being poly[2-methoxy-5-(3′,7′-dimethyloctyloxy)-1,4-phenylene vinylene]

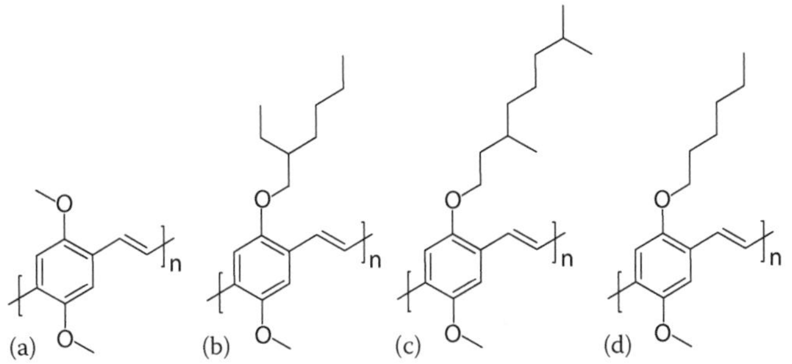

FIGURE 3.3 Symmetrical, unsymmetrical branched, and straight-chain alkoxy-substituted PPVs: (a) DM-PPV, (b) MEH-PPV, (c) MDMO-PPV, and (d) MH-PPV.

or MDMO-PPV, which is shown in Figure 3.3. While comparative studies between straight and branched chains have found that branched derivatives are more soluble, the incorporation of straight-chain substituents has been more widely pursued for polythiophene derivatives used in OPV devices due to the decreased order in the branched derivative films, which reduces interchain interactions, leading to improved device efficiency.

The use of alkoxy side chains impacts more than just the polymer solubility: the presence of a strong electron-donating group directly attached to the conjugated system can change both device performance and the emissive character of the polymer. The most critical impact is on the electronic configuration, leading to changes in the bandgap by lowering the highest occupied molecular orbital (HOMO). Therefore, the placement of these groups along the chain can be used to tune the bandgap. However, when a large bandgap is required, such as with blue light-emitting devices, it is prudent to eliminate any strong electron-donating groups along the conjugated chain. Another drawback in the use of alkoxy substituents has been the enhanced photodegradation of the conjugated polymer by singlet oxidation. Cumpston and Jensen have proposed that alkoxy substituents promote the 1,2-cycloaddition of singlet oxygen to the vinyl bonds of the polymer backbone, leading to the oxidative cleavage of the polymer chain as shown, for example, in Figure 3.4.[48–50]

To address solubility concerns, without the incorporation of such alkoxy substituents, alkyl chain substituents have also been investigated; examples of this approach can be found for polythiophene, polyfluorene, and poly(phenylene vinylene) derivatives. For alkyl PPV derivatives synthesized by precursor routes, the resulting yields and solubilities typically have been low.[17,51–53] For these systems, as higher degrees of conjugation are achieved through elimination, device performance suffers.[54] The increased conversion to the fully conjugated product results in a reduced PL quantum yield as well as an increased PL maxima as illustrated in Figure 3.5. Studies involving poly-3-alkyl thiophene have shown that chains shorter than hexyl substituents fail to produce a solubility enhancement sufficient for many applications, while those longer than hexyl substituents introduce steric bulk that can be detrimental to polymerization yields or π–π interactions.[55,56] As with alkoxy chains, the use of branched or

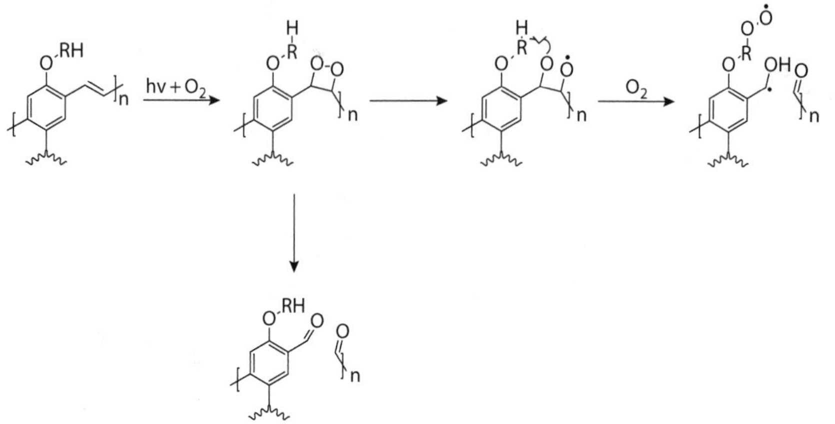

FIGURE 3.4 Proposed mechanism for the oxidation of alkoxy-substituted PPVs. (From Cumpston, B.H. and Jensen, K.F., *Synthetic Met.*, 73, 195, 1995; Cumpston, B.H. et al., *J. Appl. Phys.*, 81, 3716, 1997.)

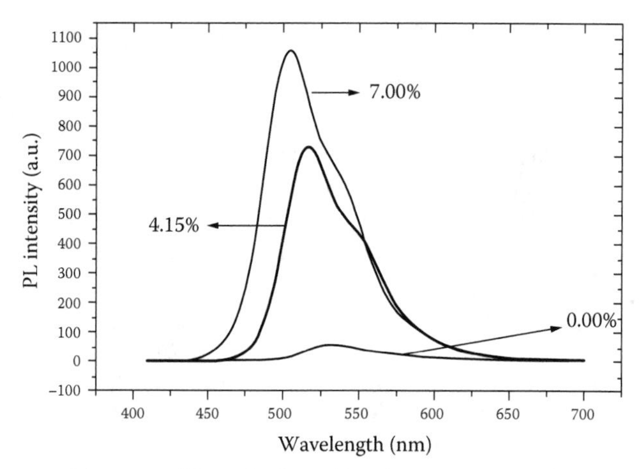

PL spectra of films of polymer(c) with 0.00%, 4.15%, and 7.00% Cl

FIGURE 3.5 Photoluminescent (PL) spectra of prepared thin films of poly(2-dodecyl-5-methyl-1,4-phenylene vinylene) with different levels of chlorine content. (From Zheng, J. et al., *J. Appl. Polym. Sci.*, 80, 1299, 2001.)

straight alkyl chains on mono-, di-, or multi-substituted rings or vinyl moieties can lend improved solubility to the polymer.[57–60]

Other substituents similar to alkyl and alkoxy chains have also been investigated. Amino-substituted chains and silyl side chains produce solubility improvements that are analogous to the alkoxy and alkyl derivatives.[61,62] Similar to the alkyl derivatives, silyl derivatives with long alkyl chains are more soluble than those with short bulky chains. Furthermore, aryl side chains can cause the precursor route to polymerization to fail from insolubility; however, aryl side chains can impart solubility if additional solubilizing groups are attached to the pendant rings.[63] And the use of other bulky groups like the trimethylsilyl (TMS) group on polyacetylene can solubilize the polymer chains, but when too many substitutions are introduced, the conjugation length can be disrupted.[64,65]

3.4 IONIZABLE FUNCTIONALITIES AND SOLUBILITY IN WATER

The development of the first water-soluble conjugated polymers by Patil et al. in 1987 followed shortly after their work on organic-soluble poly(3-alkyl thiophene) analogues.[66] While a large number of water-soluble conjugated polymers have found use in biosensing, some have been incorporated into organic electronic devices as hole transporting, electron transporting, or emissive layers.[67] Water-soluble conjugated polymers can also be used in devices in conjunction with organic soluble polymers since the solution processing will not cause intermixing between layers or cause the first layer to be washed away with the deposition of the next. Poly(3,4-ethylenedioxythiophene) (PEDOT) as seen in Figure 3.6, while insoluble by itself, when mixed with the polyelectrolyte polystyrene sulfonate (PSS) forms a water-soluble mixture ubiquitous in organic electronics as a conducting buffer between the anode and emissive or hole-transporting layer.[68]

To impart solubility in aqueous media, charged side chains can be attached to the rigid hydrophobic backbone. Sulfonic pendant polymers are commonly prepared by electropolymerization of the water-soluble monomer.[66,67,71] Water-soluble sulfonates can also be synthesized using Suzuki coupling and other palladium-catalyzed reactions.[72,73] The Wessling precursor route is also compatible with sulfonic acids, yielding the salt after polymerization.[74] While polymers formed with sulfonic groups make up the majority of reports on anionic conjugated polymers, similar systems utilizing carboxylic groups have also been reported from the hydrolysis of organic-soluble ester pendants.[75–77] Some carboxylic acid pendants provide water solubility to the polymer, while only the salts of others produce this characteristic, dependent in part

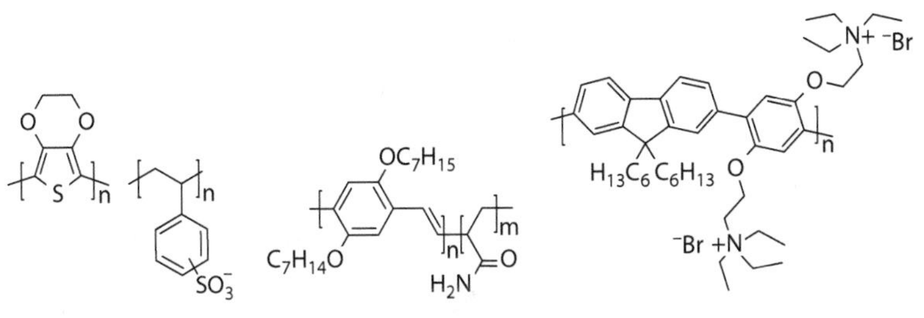

FIGURE 3.6 Anionic (sulfonate), neutral (acrylamide), and cationic (quaternary ammonium salts) water-soluble conjugated polymers used in organic electronic devices: (a) PEDOT:PSS, (b) PPV-PAm, and (c) PF-alt-PPV+. (From Groenendaal, L.B. et al., *Adv. Mater.*, 12, 481, 2000; Zhang, R. et al., *Chem. Commun.*, 823, 2000; Liu, B. et al., *Macromolecules*, 35, 4975, 2002.)

upon the fundamental polymer framework itself.[67] These anionic conjugated polymers include carbonyl moieties, which have been promoted as fluorescent quenchers that reduce PL yield in OLED devices.[78] Wagaman et al. have suggested that only carbonyl defects along the polymer backbone are responsible for such quenching, with pendant carbonyls still exhibiting high quantum yields.[75] To avoid these functional groups but still retain water solubility, cationic conjugated polymers have also been developed. Quaternary ammonium salts are formed after polymerization of the neutral precursor. Partial quaternization can be employed to tune the solubility of the polymer, as well as its optical properties, as can be seen with the spectra in Figure 3.7.[70] Moreover, incorporating cationic and anionic conjugated polymers in the same device allows for the creation of interesting layer-by-layer assemblies.[79,80]

A variety of neutral water-soluble polymers have also been reported. Acrylamide blocks within the polymer decrease the polymer's solubility in chloroform, dimethyl formamide, and tetrahydrofuran, relative to an alkoxy PPV homopolymer; however, polymer compositions approaching 82 wt% acrylamide content impart water solubility.[69] More recently, poly(ethylene oxide) (PEO) has been incorporated in polythiophene block copolymers to modify its solubility.[81] While a previous study using oligo(ethylene oxide) blocks in poly(phenylene vinylene) only mentions solubility in chloroform, larger PEO units have been found to be soluble in aqueous solvents.[82] Yet gains in solubility through structural changes might prove detrimental if the introduction of nonconjugated portions in the backbone also reduces the effective conjugation length, resulting in altered optical properties. While it has been proposed that the ionic repulsion of charged side chains helps reduce molecular aggregation, self-aggregation is still evident in ionic polymers—a phenomenon that has generated

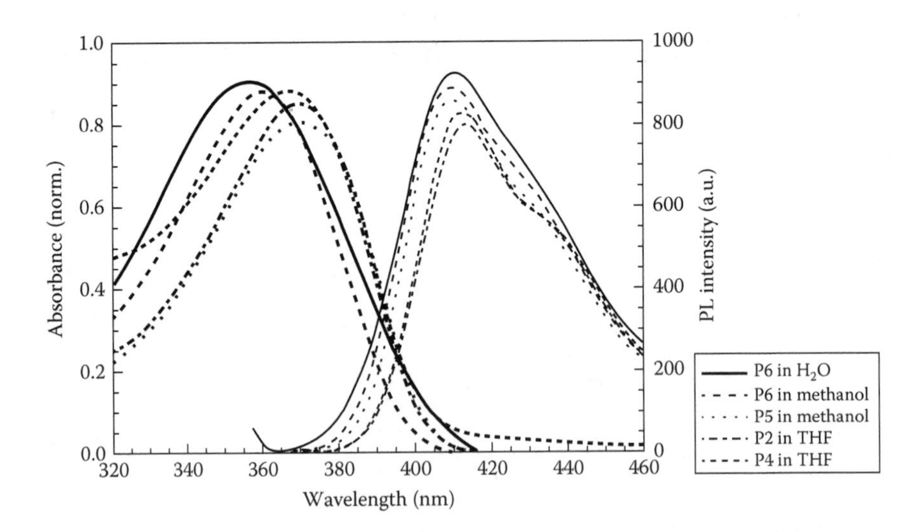

FIGURE 3.7 UV-vis absorption and PL emission spectra of polyfluorene derivatives with different degrees of ammonium salt quaternization (P2 about 0%, P4 about 30%, P5 about 60%, and P6 about 80%) in the solvents identified in the legend. (From Liu, B. et al., *Macromolecules*, 35, 4975, 2002.)

investigations into the use of PEO in combination with MEH-PPV and a separate ionic transport species.[74,83,84] The incorporation of neutral side chains for imparting water solubility has generally been limited to applications in biological sensors and will not be examined further in this review.[85]

Much like the charged soluble precursors mentioned previously, it is difficult to measure the molecular weight of water-soluble conjugated particles.[76,80] If the neutral polymer can be characterized before quaternization in the case of ammonium salts, or before hydrolysis in the case of carboxylates, indirect measurements of molecular weight might be determined. Gel permeation chromatography (GPC) measurements in aqueous media with sulfonated polystyrene standards can be obtained for some compounds, and viscosity measurements can be used to estimate molecular weight in samples with limited solubility.[69]

3.5 STRUCTURAL EFFECTS

The geometry of a polymer chain is another parameter that can affect solubility. However, modifying the geometry can have significant consequences on the optical properties of the conjugated polymer as well. The simplest conjugated polymer, polyacetylene, is a prime example. As the ratio of *cis* to *trans* double bonds increases, the resistivity and the bandgap increase as well; however, neither isomer of polyacetylene is soluble.[86] Yet these factors must be considered when altering the geometry of other polymers in an effort to increase polymer solubility. In a study by Anders et al., the *cis/trans* ratios in the cyclopolymerization of 1,6-heptadiynes were controlled by altering the concentration of the Schrock initiator.[87] An all-*trans* polymer was found to be less soluble than a polymer with a 50/50 ratio of *cis* to *trans* in the conjugated chain. Another study involving the ring-opening metathesis polymerization (ROMP) of substituted cyclooctatetraene (COT) produced a mixture of *cis* and *trans* along the chain, forming a soluble polyacetylene derivative.[88] After isomerizing the chains to an all-*trans* conformation, the products became insoluble. The authors attribute the decrease in solubility to the less conformationally flexible *trans* isomer, as compared to the more amorphous *cis* isomer.[89–91] A study of PPV derivatives in OLEDs comments on the solubility of poly(*para*-phenylene vinylene-alt-3-*tert*-butyl-*meta*-phenylene vinylene) synthesized by the Wittig condensation before and after *cis/trans* isomerization.[92] Based on analysis by ^{1}H NMR spectroscopy, the authors estimated that 62% of the alkene bonds had *trans* configurations before treatment with iodine and full conversion to the *trans* isomer after treatment. While both isomers are soluble, the authors note the solubility was "noticeably higher" before isomerization. However, in device studies, the *trans* isomer was found to give higher luminance and efficiency, offering a prime example of the failure of single-parameter optimization for maximizing overall performance.

In PPV derivatives, altering the substitution pattern of the ring is another way to influence solubility. Similar to *cis/trans* modifications, *meta*-linked PPVs (P*m*PV; see Figure 3.8) exhibit different optical properties from *para*-linked PPVs (P*p*PV). Notably, P*m*PV derivatives have increased bandgaps due to disruption of the conjugation in the chain.[93,94] While this change might be desirable in some applications (i.e., blue OLEDs), it renders the polymers less effective as the active layer in OPVs

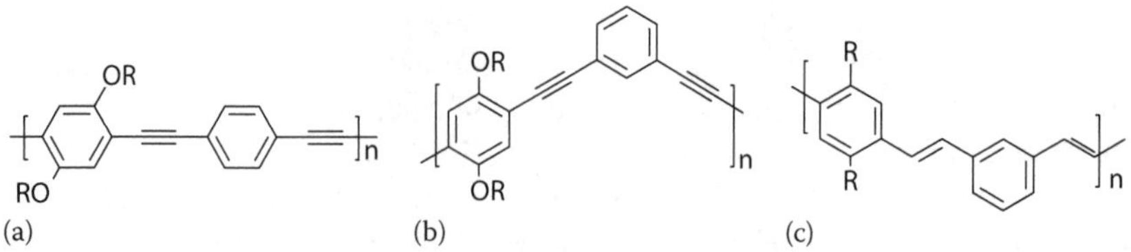

OR
RO
(a)
OR
OR
(b)
R
R
(c)

FIGURE 3.8 Conjugated polymers showing *para-* and *meta*-linked phenylene moieties: (a) P*p*PV, (b) *m*-PPE, and (c) P*m*PV. (From Pang, Y. et al., *J. Mater. Chem.*, 8, 1687, 1998.)

where low bandgap materials are the most effective at harvesting sunlight. *Meta*-linked poly(phenylene ethynylene) (PPE) exhibits improved solubility, possibly because the kinks introduced in the structure break the rigid linear rodlike conformation of the *para*-linked PPE derivatives, as shown in Figure 3.8.[95] A study by Yang et al. incorporating a 2,5-thienylene unit also introduced a kink in the structure of PPE, an adjustment that significantly changed the geometry and increased the solubility.[96] While the authors attributed the increased solubility to the kinks introduced by the thienylene, the change in geometry might be only one factor. Another study incorporating 2,5-thienylene derivatives yielded only insoluble products until an alternating copolymer with iptycene was prepared.[97] In a more recent report by Yang et al., instead of using thienyl groups to alter the geometry, *meta*-phenylene units were analyzed. Even though the *para* derivative had only 1/3 the molecular weight of the *meta*, the *meta* derivative had a higher solubility of 0.1 g/mL in chloroform ($CHCl_3$). The authors failed to quantify the results for the *para* derivative, instead indicated it had "low solubility," which provides convincing evidence that the *meta* geometry is indeed contributing to improved solubility, even if 2,5-thienylene is incapable of rendering solubility to a polymer.

In a later study by the same group, P*m*PV derivatives were analyzed.[98] The authors attributed increased solubility to the *meta*-linked rings because both the butoxy- and hexyloxy-substituted derivatives exhibited similar solubilities. The Wittig condensation procedure used to make the polymers yielded a mixture of *cis* and *trans* P*m*PVs, and after conversion to an all-*trans* system, the authors commented on an increased chain stiffness based on the Mark–Houwink coefficient as determined by an online viscometer in a size exclusion chromatography (SEC) system, but they failed to mention any change in solubility. While the influence of the *cis/trans* isomerization might be significant in other systems, the *meta* linkage itself appears to provide an ample improvement in solubility in this case. Recent work in the Lee group, which has not yet been published, has shown P*m*PV derivatives to be more soluble than P*p*PV derivatives, but due to varying molecular weights generated for the two types of structures during synthesis, it is difficult to say which factor is responsible for the increase in solubility. Studies of alternating copolymers of P*m*PV and P*p*PV with polyfluorene derivatives have also been reported, but no comments are given about their relative solubilities, or the solubilities reported for these copolymers are described as being the same (2% in $CHCl_3$).[99,100] In reports concerning hyperbranched P*m*PV derivatives, Lin et al. attributed only the absorption properties of the polymer to the *meta* linkages in the hyperbranched poly(3,5-bisvinylic) benzene.[101] But another property

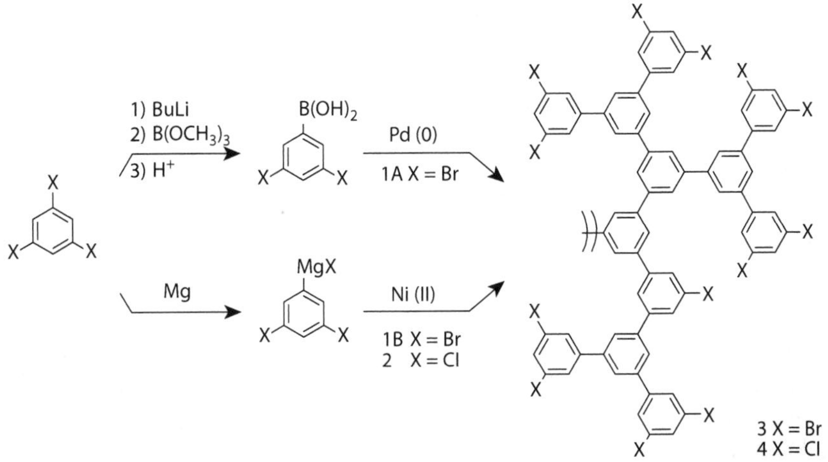

SCHEME 3.3 Stepwise preparation of hyperbranched polyphenylene from 1,3,5-tribromobenzene. (From Kim, Y.H. and Webster, O.W., *Macromolecules*, 25, 5561, 1992.)

that might be affected by these substituents is the solubility, which showed improvement with increased disorder. An earlier report studying similar structures does attribute solubility to the disordered nature of the hyperbranched polymer. An example of this polymer system is shown in Scheme 3.3.[102]

Altering the structure of conjugated polymers to improve solubility can also be achieved by using bulky substituents. As noted earlier, unsubstituted polyacetylene is insoluble, and copolymers of acetylene with methyl-substituted acetylene exhibit reduced conjugation length and conductivity due to steric repulsion.[103] Gorman et al. showed with ROMP of a set of substituted cyclooctatetraenes that substituted polyacetylene derivatives can be synthesized that attain "partial solubility."[88] As found in previous studies, increasing the chain length of the substituents can produce an increase in solubility, and the *cis* isomers exhibit higher solubility than the *trans* isomers. A simple methyl group provided no improvement in solubility, but as the steric bulk of the side groups increased, the polymer solubility increased, while simultaneously reducing the effective conjugation length. In a related report, the proximity of the bulky group to the backbone was investigated.[104] Secondary and tertiary substituents α to the polyacetylene chain caused disorder in chain alignments by twisting the main chain out of planarity; therefore, these systems maintained solubility even after *trans* isomerization. When a bulky group was moved to a β position (i.e., *trans*-poly-O(*tert*-butyldimethylsilylcyclooctatetraene [*trans*-PRCOT], as shown in Figure 3.9), only extremely low molecular weight products (<1000 kDa) were soluble.[105] The importance of the α/β position shows that this method of improving solubility is not simply due to incorporating chains to provide sufficient entropic drive to facilitate solvation, but that the steric bulk twisting of the main chain is primarily responsible for the improved solubility. However, the incorporation of substituents with steric bulk in the structure of the monomer to impart solubility in the resulting polymer will necessarily reduce the conjugation length, implying a maximum soluble conjugation length.

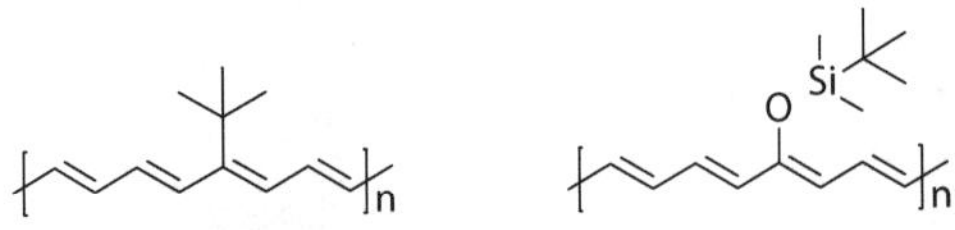

(a) (b)

FIGURE 3.9 *Trans*-PRCOT derivatives showing branching at the α position in the *more soluble* poly(*tert*-butylcyclooctatetraene) (P-*t*-Bu-COT) (a) and branching at the β position in the *less soluble* poly-O(*tert*-butyldimethylsilylcyclooctatetraene) (P-O-*t*-BuDMS-COT) (b). (From Gorman, C.B. et al., *J. Am. Chem. Soc.*, 115, 1397, 1993.)

Other types of conjugated polymers have also been produced using derivations of the aromatic structures along the polymer backbone to improve solubility, but the associated reports are less systematic in the determination of the root causes of the improved solubility. Tetrafluoro-substituted PPV exhibited solubility as high as 20 mg/mL in $CHCl_3$, but the authors fail to comment on the cause.[106] While it is possible that the presence of the fluorine atoms on the aromatic rings twists the backbone out of planarity, there are several other factors that might be influencing the solubility such as residual saturated precursor, the influence of the highly electronegative atoms on solvent interactions with the polymer chain, and the presence of a mixture of *cis* and *trans* isomers. Furthermore, a report regarding a bulky monomer formed from iptycene claimed incorporating 2,5-thienylenes in an alternating iptycene copolymer impart solubility, while 2,5-thienylene homopolymer derivatives were totally insoluble, thus the steric bulk of the alternating iptycene unit was responsible for the improved solubility.[96]

3.6 MOLECULAR WEIGHT CONSIDERATIONS

The molecular weight of conjugated polymers might be the most readily understood factor effecting solubility. As molecular weight increases, solubility decreases. This relationship, while simple, is intricately tied to film morphology and device performance. Importantly, it can be difficult to draw conclusions based on the comparison of other parameters if there are even minor differences in the polydispersity index (PDI) for the polymers under evaluation. The corollary to the observation that increased molecular weight leads to decreased solubility is that the solubility of the targeted polymer chain also impacts the resulting molecular weight of the final polymer obtained.[107] Wet synthesis methods require polymers to have good solubility during polymerization to form high molecular weight polymers. If the solubility is too low, oligomers will precipitate out of the solution before reaching high molecular weights. While the simplest way to increase product solubility might be to decrease the average molecular weight, low molecular weight polymers often form poor films, which can adversely affect device performance. However, if the molecular weight is too great for the individual polymer chains to be properly solvated, the resulting films that are produced might be too thin, if they can be formed at all. Polymers with large PDI have been shown to exhibit reduced charge mobility in devices—a shortcoming that might be caused by a larger portion of the lower molecular weight fractions being solvated, while the higher molecular weight fractions remain aggregated.[108]

In a report by Osaka et al., organic photovoltaic devices using thiazolothiazole derivatives exhibited a maximum power conversion efficiency (PCE) for a number average molecular weight (M_n) of 33 kDa. The authors found that increased molecular ordering in high molecular weight polymers caused aggregation, which accounted for artificial peaks at shorter retention times in GPC measurements.[109] In prior studies, changing the concentrations of the sample caused a significant change in the GPC trace, which was attributed to an increased aggregate content at higher concentrations.[108,110] Bronstein et al. attributed improved PCE to increased molecular weight but failed to postulate why the highest molecular weight sample (73 kDa) exhibited decreased crystallinity and lower PCE. The 73 kDa polymer was soluble only in hot chlorobenzene, and the experimental section makes no comment on the need to heat the polymer solutions prior to spin coating. It is possible that the improved PCE of the lower molecular weight polymer might be attributable to the influence of the enhanced solubility during device preparation. These findings, however, are in stark contrast to others that show crystallinity decreases as molecular weight increases.[111–115]

Variation in molecular weight can also have a significant impact on the optical properties of conjugated polymer systems. The effective conjugation length has been widely studied for conjugated polymers in both theoretical and experimental work.[116,117] If the molecular weight of the polymer is similar to the maximum exciton delocalization limit, the molecular weight will alter the electronic properties of the polymer. In a study by Zen et al., several different solvents were used to extract fractions from a sample of poly(3-hexylthiophene-2,5-diyl) (P3HT).[111] As the lower molecular weight fractions were systematically removed from the polymer sample (those with highest solubility for each solvent), the absorption maxima exhibited a distinct red shift for the residual sample, as seen in Figure 3.10.

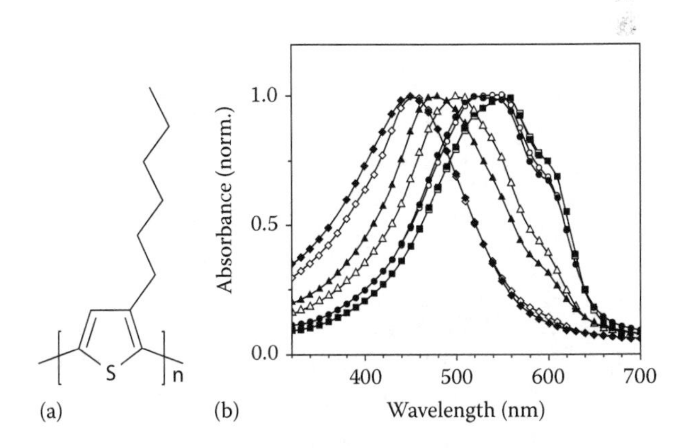

FIGURE 3.10 (a) Chemical structure of poly(3-hexylthiophene-2,5-diyl). (b) Solid-state spectra of the extracted polymer using ethyl acetate (◇, ◆), hexane (△, ▲), dichloromethane (○, ●), and chloroform (□, ■). Open and solid symbols indicate, respectively, UV-visible spectroscopic data collected for polymer layers before and after annealing for 5 min at 150°C. (From Zen, A. et al., *Adv. Funct. Mater.*, 14, 757, 2004.)

Ordering is generally considered desirable in OPVs and organic field-effect transistors (OFETs), but true crystalline domains have been proposed to be charge-carrier traps. Changing solubility by altering molecular weight also affects the delicate balance between low aggregation (low molecular weight) and high ordering (high molecular weight) of the chains. Of the studies that have analyzed the impact of molecular weight on device performance, some mention that the solubility of the higher molecular weight samples is reduced, but the analyses associated with these reports make no connection between solubility and performance.[118] Ballantyne et al. found that low molecular weight samples exhibit higher electron and hole mobility, but they comment that the molecular weight impacts many other parameters that influence device performance (i.e., film thickness).[119] The plot displayed in Figure 3.11 demonstrates the correlation between the molecular weight of the P3HT polymer samples and the hole mobility associated with each sample.

It is difficult to make any conclusive statement about an ideal molecular weight for a polymer system since so many factors can be affected when molecular weight is varied. Most research reports for conjugated polymers lack solubility analyses that provide quantification of this parameter and rely on the use of nonspecific terms such as "high" or "low"; however, it is also misleading to claim an optimal molecular weight for every application and every polymer, even though some authors have made such assertions.[107,120]

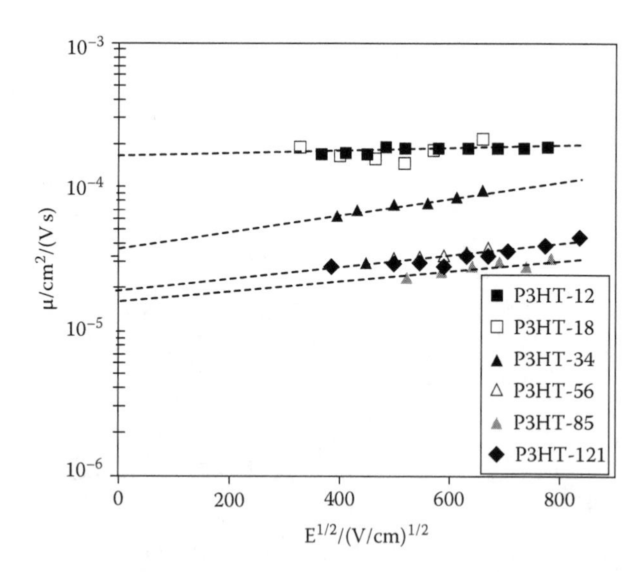

FIGURE 3.11 Poole–Frenkel plots of time-of-flight mobility as a function of the applied electric field for holes in films produced from P3HT composed of different molecular weights. The numbers associated with each P3HT symbol on the legend are the molecular weight (M_n) for each sample, rounded to the nearest whole number in kDa. (From Ballantyne, A.M. et al., *Adv. Funct. Mater.*, 18, 2373–2380, 2008.)

3.7 SOLVENT EFFECTS

Both the choice of polymerization solvent and the device fabrication solvent can influence device performance. If the polymer has a high solubility in the polymerization solvent, higher molecular weight polymers can be isolated. In a report by Tangonan et al., the synthesis of MEH-PPV was examined in both tetrahydrofuran (THF) and dimethylformamide (DMF).[44] MEH-PPV has exhibited a higher solubility in THF (1 w/v %)[121] than DMF,[122] as shown by the higher molecular weight polymers produced in THF solutions. Insoluble gels were produced by polymerizations carried out in THF when the reactants were introduced all at once, while soluble products could be isolated from similarly structured polymerizations in DMF. Even changing the volume of the polymerization solvent had significant effects on the solubility of the final product. Another study by Trouillet et al. compared DMF, toluene, and pyridine in the polymerization of an alternating alkylthiophene oligomer and Ru(II) complex copolymer.[123] The authors expected to achieve high molecular weights in toluene and pyridine due to their lower polarity and relative ability to solvate the thiophene oligomeric units but found instead that the yields were reduced and higher amounts of byproducts were formed in these less polar solvents.

While some researchers fail to fully characterize the polymer solubility by describing the use of "common organic solvents" (with little or no elaboration on which solvents or to what extent), others have noted useful details such as a simple change from monochlorinated benzene (CB) to *ortho*-dichlorobenzene (ODCB) can alter device performance.[124] Rispens et al. have reported that poly[2-methoxy-5-(3′,7′-dimethyloctyloxy)-1,4-phenylene vinylene] (MDMO-PPV) is more soluble in ODCB than CB, yet in OPV devices using a bulk heterojunction architecture, CB was shown to be the best solvent.[125] The authors attribute the improvement in power conversion efficiency (PCE) in this report to the fact that the resulting domain sizes were slightly larger in CB than in ODCB. They also fabricated devices cast in xylenes, which led to even larger domain sizes than CB, but even lower PCE than devices cast from ODCB. A similar study by Shaheen et al. comparing toluene to CB in MDMO-PPV OPV devices showed that CB had an incident photon to converted electron (IPCE) three times higher than devices cast from toluene, as illustrated in Figure 3.12.[126] The authors noted that MDMO-PPV is twice as soluble in CB as it is in toluene, leading to a surface roughness five times greater for films cast in toluene.

Solvent additives (i.e., low levels of a cosolvent) have also been used to improve device performance by altering the solubility of conjugated polymers in casting solutions. A study introducing a small amount (1.2%) of toluene, xylene, or CB to $CHCl_3$ solutions of MDMO-PPV showed only moderate improvement in PCE for added CB and a reduction in PCE for toluene and xylenes as compared to neat $CHCl_3$.[127] Other studies have shown that neat CB solutions produce better results than $CHCl_3$, which is ascribed to the increased solubility of MDMO-PPV in CB.[128] Additives other than common casting solvents have also been studied, such as 1,8-diiodooctane (DIO). In a report by Chu et al. using the alternating copolymer of dithienosilole and thienopyrrole-4,6-dione, OPV devices fabricated from CB solutions with 3% DIO achieved PCE of 7.3%, where a neat CB solution had a PCE below 1%, as shown in Figure 3.13.[129] In this same study, ODCB solutions were also tested, but the effect

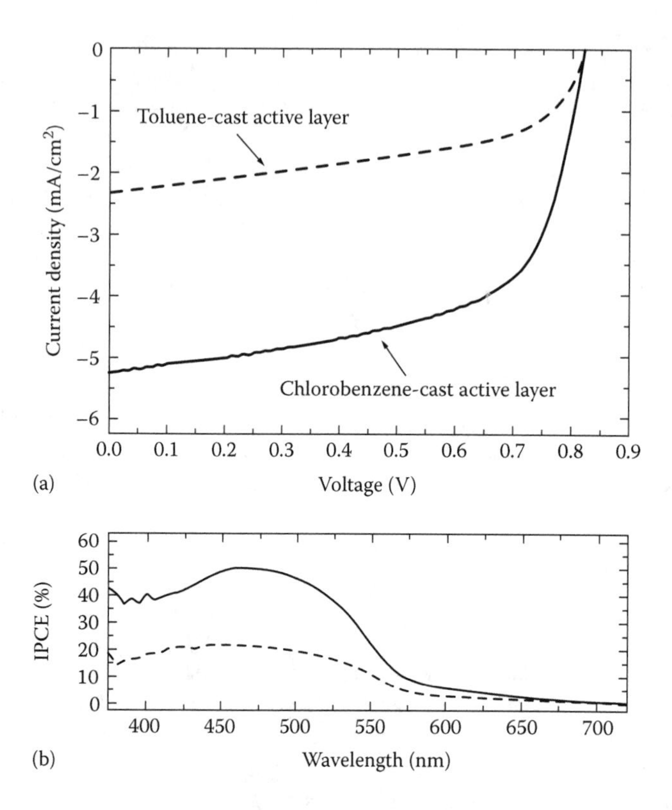

FIGURE 3.12 (a) *J–V* characteristics and (b) incident photon to converted electron (IPCE) spectra for OPV devices with an active layer that was spin-coated from a toluene solution (dashed line) and from a chlorobenzene solution (solid line). (From Shaheen, S.E. et al., *Appl. Phys. Lett.*, 78, 841, 2001.)

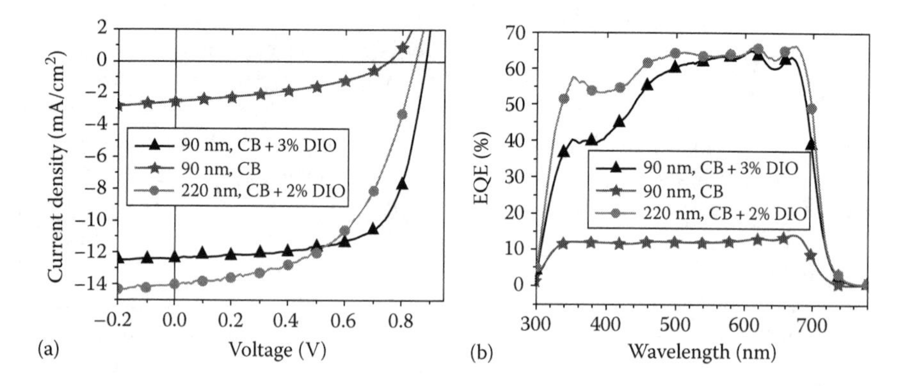

FIGURE 3.13 (a) Current density–voltage characteristics and (b) external quantum efficiency curves of OPVs prepared with CB with no additive and CB with 2 or 3 vol% DIO. (From Chu, T.-Y. et al., *J. Am. Chem. Soc.*, 133, 4250, 2011.)

of the DIO was not mentioned, only that CB-cast films with DIO performed better than ODCB films with DIO. A comparison by Lee et al. of 1,8-di(R)octane additives (where R is a functional group) to CB used to solvate a complex blend of conjugated structures showed that dichloro, dicyano, and diacetate groups reduced device efficiency, while dithiol, diiodo, and dibromo derivatives provided an improvement.[130] The authors also found that one criterion for choosing a successful processing additive was that it needed to possess a higher boiling point than the host solvent and provide selective solvency for a key structural feature for the blended system. And based on transmission electron microscopy (TEM) and atomic force microscopy (AFM) images, the authors concluded that the additives that improved device performance decreased the solubility of the conjugated polymer only slightly relative to preparations using neat CB, while those that reduced efficiency decreased the solubility to a greater extent.

The boiling point of effective casting solvents can determine if a polymer can be processed or not. In a study comparing molecular weight, larger, less soluble compounds were still able to be solvated by ODCB at high temperatures. Lower boiling point solvents like THF and $CHCl_3$ were found to be suitable for conjugated polymers with smaller molecular weights. However, with higher boiling points, removal of the solvent from the film can have detrimental dewetting effects.[131] Films that could be formed with relatively low boiling solvents (e.g., chloroform, toluene, chlorobenzene, and *p*-xylene) produced superior films over those cast with ODCB or 1,2,4-trichlorobenzene.

The treatment of polymer solutions prior to film deposition is a critical experimental parameter that is often neglected in published works. An example of how such issues are often downplayed in the literature can be found in a report by McCulloch et al., where a polythiophene derivative having "high solubility" was spin-cast from a "hot chlorobenzene" solution.[107] The authors failed to mention any reason other than poor solubility for using a hot casting solution.

Solvent polarity is another factor that can influence conjugated polymer solubility; however, the systematic application of this parameter has proven to be somewhat elusive. In a study analyzing aggregate formation of MEH-PPV films cast from THF, CB, and a mixture of the two solvents, the authors call solvents with higher polarity, such as the *more polar* THF, "poorer" solvents than those with lower polarity, such as the *less polar* CB.[132] Their conclusions are drawn from UV-visible (UV-Vis) and PL spectra in dilute solutions, showing a slight red shift for the CB/THF mixture and a further red shift for the pure CB solution. The authors attribute the increase in wavelength to an increased conjugation length, meaning longer conjugated units are present in the CB solution than in the THF solutions. Light-scattering experiments showed that the hydrodynamic volume of the polymer decreased in THF solutions relative to those with a THF/CB mix or CB, agreeing with the assertion by the authors (based on UV-Vis and PL measurements) that the polymer adopts a more extended conformation in CB. This conclusion is surprising because the authors found that the films cast from the less polar CB formed more aggregates than those cast from THF. The aggregation of polymers in films cast from CB is thought to be higher because in solution there are more opportunities for interchain interactions. It has been proposed that aggregates improve charge transfer in devices, thereby improving external quantum efficiency; consequently, while aggregates might act as exciton traps, reducing luminescent efficiency, a delicate balance must be made to ensure high luminescence and low turn-on voltages in OLED devices.[133]

The oversimplification that the *more polar* solvents are worse than the *less polar* solvents is directly contradicted by the findings that MDMO-PPV, which has a similar structure to MEH-PPV, was shown to be more soluble in the *more polar* CB than the *less polar* toluene, as determined from the results obtained by AFM measurements of surface roughness as seen in Figure 3.14.[126] MDMO-PPV films cast from CB were smoother than those cast from toluene and had a higher PCE in OPV devices. In a separate study, where it was determined that conjugated polymers in xylene exhibited a larger hydrodynamic volume as compared to the *more polar* CB, the difference in size was most likely due to molecular weight differences.[134] Additionally, P3HT is more soluble in polar $CHCl_3$, thiophene, and 1,2,4-trichlorobenzene, and less soluble in the *less polar* solvents of xylenes and cyclohexylbenzene, as indicated by solution aggregation and gelation after several hours in the latter solvents.[112] P3HT transistors spin-coated from the more polar solvents showed reduced charge-transfer characteristics based on the amount of aggregation as seen in Figure 3.15.

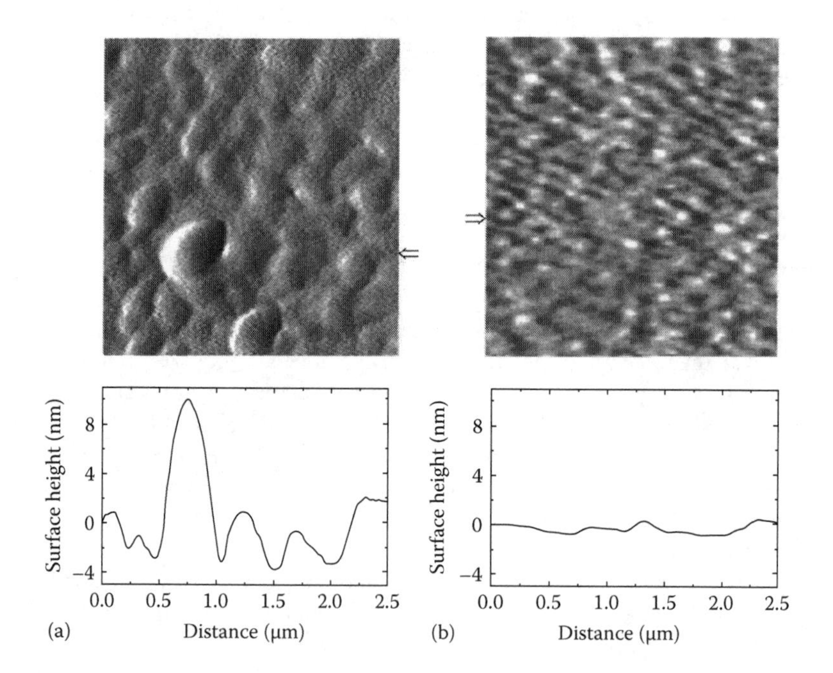

FIGURE 3.14 AFM images showing the surface morphology of MDMO-PPV:phenyl-C_{61}-butyric acid methyl ester (PCBM) blended films (1:4 by wt.) with a thickness of approximately 100 nm and the corresponding cross sections. (a) A film spin-coated from a toluene solution. (b) A film spin-coated from a chlorobenzene solution. The images show the first derivative of the actual surface heights. The cross sections of the true surface heights for the films were taken horizontally from the points indicated by the arrows. (From Shaheen, S.E. et al., *Appl. Phys. Lett.*, 78, 841, 2001.)

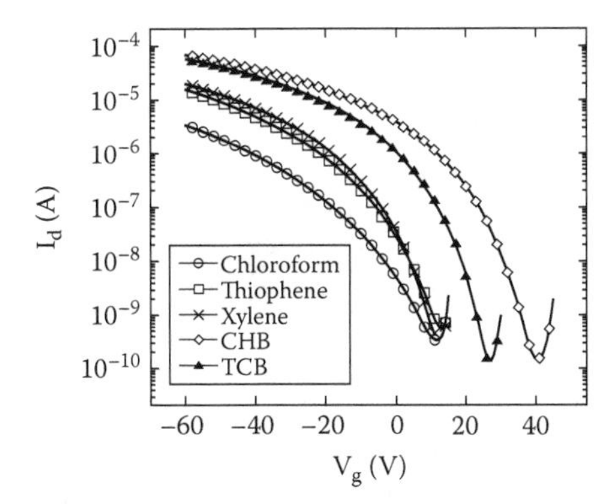

FIGURE 3.15 Transfer characteristics of drain current (I_d) versus gate voltage (V_g) of P3HT transistors spin-coated from solvents identified in the legend. (From Chang, J.-F. et al., *Chem. Mater.*, 16, 4772, 2004.)

3.8 CONCLUSIONS

While many soluble polymers have been synthesized for use in organic electronic devices, it is difficult to find any agreement on a quantifiable measure of what makes for "good" solubility. This has created a situation where the literature provides only limited assistance in determining how to address this issue when designing, producing, and utilizing conjugated polymers in research and industry. The soluble precursor route to preparing fundamentally insoluble conjugated polymers allows for the formation of their films, but this process requires further treatment of the product. This method is advantageous because the precursor polymers can be characterized; however, following treatment to afford the conjugated products, characterization becomes difficult. The range of polymers that can be synthesized via this route is also limited by the structure and stability of the precursors. The use of alkyl and alkoxy derivatives has been the most common method to address the solubility problem for conjugated systems. A wide variety of solubilizing chains have been investigated to improve these polymers for active layer applications, but each variation in length, branching, and atomic content might introduce complications to device preparation and performance. While water-soluble conjugated polymers have been achieved using ionizable functionalities, in organic electronic devices, they are most commonly used in conjunction with offsetting hydrophobic polymers to prevent mixing between layers in the fabrication process.

Unfortunately, the introduction of the ionizable side chains has been linked to film formation issues associated with polymer aggregation. Structural modifications to the polymer chain in the form of *cis* or *trans* isomers and *para*- or *meta*-linked phenyl rings not only introduce geometric irregularities that can improve solubility, but also greatly affect the electronic properties of the conjugated system. Similarly, the use of bulky substituents to distort the geometry of the conjugated chains has been shown to impart solubility and can also disrupt the polymerization process and limit the conjugation length of the polymer. Modification of the molecular weight of the polymer has a direct effect on the solubility without impacting the optical properties as long as a minimum conjugation length is met, but changes in film morphology might enhance or diminish device performance. The choice of polymerization solvent can also be used to affect changes in molecular weight, therefore the properties of the resulting polymer, just as the choice of casting solvent, can be used to modify film morphology. Additionally, minor changes in solvent polarity, vapor pressure, boiling point, or additives can also have significant impacts on device performance.

This review sheds light on the factors affecting the solubility of conjugated polymers, showing that they are numerous and frequently interdependent. The methods currently used to solvate these unique assemblies often become entangled in structural changes to the polymer chain along with the incorporation of increasingly sophisticated side chains, leading to complications during polymer synthesis and film organization, factors that greatly impact device performance. Consequently, these concerns must be considered in the design of any new solvent-compatible conjugated polymer for use in organic electronic devices.

ACKNOWLEDGMENTS

We thank the Robert A. Welch Foundation (Grant No. E-1320) and the Texas Center for Superconductivity at the University of Houston for generously supporting this research.

REFERENCES

1. Xu, X.; Han, B.; Chen, J.; Peng, J.; Wu, H.; Cao, Y. 2,7-Carbazole-1,4-phenylene copolymers with polar side chains for cathode modifications in polymer light-emitting diodes. *Macromolecules* 2011, *44*, 4204–4212.
2. Ding, T.; Zhao, B.; Shen, P.; Lu, J.; Li, H.; Tan, S. Synthesis, characterization, and photophysical properties of novel poly(*p*-phenylene vinylene) derivatives with conjugated thiophene as side chains. *J. Appl. Polym. Sci.* 2011, *120*, 3387–3394.
3. Qu, H.; Luo, J.; Zhang, X.; Chi, C. Dicarboxylic imide-substituted poly(*p*-phenylene vinylenes) with high electron affinity. *J. Polym. Sci. Pol. A, Polym. Chem.* 2010, *48*, 186–194.
4. Khim, D.; Lee, W.-H.; Baeg, K.-J.; Kim, D.-Y.; Kang, I.-N.; Noh, Y.-Y. Highly stable printed polymer field-effect transistors and inverters via polyselenophene conjugated polymers. *J. Mater. Chem.* 2012, *22*, 12774–12783.
5. Shirakawa, H.; Louis, E. J.; MacDiarmid, A. G.; Chiang, C. K.; Heeger, A. J. Synthesis of electrically conducting organic polymers: Halogen derivatives of polyacetylene, $(CH)_x$. *J. Chem. Soc., Chem. Commun.* 1977, 578–580.
6. Luo, K.; Kim, S. J.; Cartwright, A. N.; Rzayev, J. Soluble polyacetylene derivatives by chain-growth polymerization of dienes. *Macromolecules* 2011, *44*, 4665–4671.
7. Tang, C. W.; VanSlyke, S. A. Organic electroluminescent diodes. *Appl. Phys. Lett.* 1987, *51*, 913–915.
8. Zhou, H.; Yang, L.; Liu, S.; You, W. A tale of current and voltage: Interplay of band gap and energy levels of conjugated polymers in bulk heterojunction solar cells. *Macromolecules* 2010, *43*, 10390–10396.

9. Nguyen, T.-Q.; Martini, I. B.; Liu, J.; Schwartz, B. J. Controlling interchain interactions in conjugated polymers: The effects of chain morphology on exiton-exiton annihilation an aggregation in MEH-PPV films. *J. Phys. Chem. B* 2000, *104*, 237–255.
10. Bolinger, J. C.; Traub, M. C.; Brazard, J.; Adachi, T.; Barbara, P. F.; Bout, D. A. V. Conformation and energy transfer in single conjugated polymers. *Acc. Chem. Res.* 2012, *45*, 1992–2001.
11. Ginsburg, E. J.; Gorman, C. B.; Marder, S. R.; Grubbs, R. H. Poly(trimethylsilylcyclooctatetraene): A soluble conjugated polyacetylene via olefin metathesis. *J. Am. Chem. Soc.* 1989, *111*, 7622–7624.
12. Steckler, T. T.; Zhang, X.; Hwang, J.; Honeyager, R.; Ohira, S.; Zhang, X.-H.; Grant, A. et al. A spray-processable, low band gap, and ambipolar donor-acceptor conjugated polymer. *J. Am. Chem. Soc.* 2009, *131*, 2824–2826.
13. Wessling, R. A. The polymerization of xylylene bisdialkyl sulfonium salts. *J. Polym. Sci., Part C Polym. Symp.* 1985, *72*, 55–66.
14. Burn, P. L.; Bradley, D. D. C.; Friend, R. H.; Halliday, D. A.; Holmes, A. B.; Jackson, R. W.; Kraft, A. Precursor route chemistry and electronic properties of poly(*p*-phenylene-vinylene), poly[(2,5-dimethyl-*p*-phenylene)vinylene] and poly[(2,5-dimethoxy-*p*-phenylene) vinylene]. *J. Chem. Soc. Perkin Trans. 1* 1992, 3225–3231.
15. Halliday, D. A.; Burn, P. L.; Friend, R. H.; Bradley, D. D. C.; Holmes, A. B. Determination of the average molecular weight of poly(*p*-phenylenevinylene). *Synth. Met.* 1993, *55*, 902–907.
16. Lenz, R. W.; Han, C. C.; Stenger-Smith, J.; Karasz, F. E. Preparation of poly(phenylene vinylene) from cycloalkylene sulfonium salt monomers and polymers. *J. Polym. Sci. A Polym. Chem.* 1988, *26*, 3241–3249.
17. Bijnens, W.; Van Der Borght, M.; Manca, J.; De Ceuninck, W.; De Schepper, L.; Vanderzande, D.; Gelan, J.; Stals, L. A new precursor to electroconducting conjugated polymers: Synthesis and opto-electrical properties of luminescent devices based on these PPV derivatives. *Opt. Mater.* 1998, *9*, 150–153.
18. Lutsen, L.; Duyssens, I.; Penxten, H.; Vanderzande, D. Synthesis of blue light-emitting poly(*p*-arylene vinylene) derivative starting from new soluble polymeric precursors. *Synth. Met.* 2003, *139*, 589–592.
19. Lo, S.-C.; Sheridan, A. K.; Samuel, I. D. W.; Burn, P. L. Comparison of the electronic properties of poly[2-(2′-ethylhexyloxy)-1,4-phenylenevinylene] prepared by different precursor polymer routes. *J. Mater. Chem.* 1999, *9*, 2165–2170.
20. Lo, S.-C.; Pålsson, L.-O.; Kilitziraki, M.; Burn, P. L.; Samuel, I. D. W. Control of polymer–electrode interactions: The effect of leaving group on the optical properties and device characteristics of EHPPV. *J. Mater. Chem.* 2001, *11*, 2228–2231.
21. Morgado, J.; Thomas, D. S.; Friend, R. H.; Cacialli, F. Alteration of the photo and electroluminescent properties of poly(*p*-phenylene vinylene) upon addition of indium chloride. *Synth. Met.* 2001, *122*, 119–121.
22. Massardier, V.; Guyot, A.; Tran, V. H. Direct conversion of sulfonium precursors into poly(*p*-phenylene vinylene) by acids. *Polymer* 1994, *35*, 1561–1563.
23. Xia, Y.; MacDiarmid, A. G.; Epstein, A. J. Room temperature synthesis of poly(2,5-dimethoxy-l,4-phenylenevinylene) by the chloride sulfonium salt route. *J. Adv. Mater.* 1994, *6*, 293–295.
24. Taguchi, S.; Tanaka, T. Deep UV photolithography for forming fine patterns with conjugated polymers. EP Patent 261991, Sumimoto Chemical Co., Ltd., Tokyo, Japan, March 30, 1998.
25. Schmid, W.; Dankesreiter, R.; Gmeiner, J.; Vogtmann, T.; Schwoerer, M. Photolithography with poly-(*p*-phenylene vinylene) (PPV) prepared by the precursor route. *Acta Polym.* 1993, *44*, 208–210.

26. Torres-Filho, A.; Lenz, R. W. Electrical, thermal, and photo properties of poly(phenylene vinylene) precursors: I. Laser-induced elimination reactions in precursor polymer films. *J. Polym. Sci. B Polym. Phys.* 1993, *31*, 959–970.
27. Paik, S. Y.; Kwon, S. H.; Kwon, O. J.; Yoo, J. S.; Han, M. K. Conversion of precursor polymer into poly(*p*-phenylene vinylene) by XeCl excimer laser. *Synth. Met.* 2002, *129*, 101–105.
28. Bullot, J.; Dulieu, B.; Lefrant, S. Photochemical conversion of polyphenylenevinylene. *Synth. Met.* 1993, *61*, 211–215.
29. Torres-Filho, A.; Lenz, R. W. Electrical, thermal, and photo properties of poly(phenylene vinylene) precursors II. Microwave-induced elimination reactions in precursor polymer films. *J. Appl. Polym. Sci.* 1994, *52*, 377–386.
30. Kang, W.-B.; Yu, N.; Tokida, A. Process for patterning poly(arylenevinylene) polymer films by irradiation with light. EP Patent 700235, Hoeschst A.-G., Frankfurt, Germany, March 6, 1996.
31. Holmes, A. B.; Bradley, D. D. C.; Kraft, A.; Burn, P. L.; Brown, A.; Friend, R. H. Semiconductive copolymers for use in luminescent devices. US Patent 5401827A, Cambridge Display Technology Ltd., Cambridge, U.K., March 28, 1995.
32. Gymer, R. W.; Friend, R. H.; Ahmed, H.; Burn, P. L.; Kraft, A. M.; Holmes, A. B. The fabrication and assessment of optical waveguides in poly(*p*-phenylenevinylene/poly2,5-dimethoxy-*p*-phenylenevinylene) copolymer. *Synth. Met.* 1993, *57*, 3683–3688.
33. Webster, G. R.; Whitelegg, S. A.; Bradley, D. D. C.; Burn, P. L. Control of conjugation in poly(arylenevinylene)s. *Synth. Met.* 2001, *119*, 269–270.
34. Padmanaban, G.; Ramakrishnan, S. An improved method for the control of conjugation length in MEHPPV via a xanthate precursor route. *Synth. Met.* 2001, *119*, 533–534.
35. Garay, R. O.; Sarimbalis, M. N.; Montani, R. S.; Hernandez, S. A. Synthesis of conjugated polymers. Polymerizability studies of bis-sulfonium salts. *Des. Monomers Polym.* 2000, *3*, 231–244.
36. Onoda, M.; Uchida, M.; Ohmori, Y.; Yoshino, K. Organic electroluminescence devices using poly(arylene vinylene) conducting polymers. *Jpn. J. Appl. Phys., Part 1* 1993, *32*, 3895–3899.
37. Delmotte, A.; Biesemans, M.; Van Mele, B.; Gielen, M.; Bouman, M. M.; Meijer, E. W. Selective elimination in dialkoxy-PPV precursors yielding polymers with isolated tetraalkoxy-stilbene units. *Synth. Met.* 1995, *68*, 269–273.
38. Greenham, N. C.; Brown, A. R.; Burroughes, J. H.; Bradley, D. D. C.; Friend, R. H.; Burn, P. L.; Kraft, A.; Holmes, A. B. Electroluminescence from multilayer conjugated polymer devices—Spatial control of exciton formation and emission. *Proc. SPIE*, 1993, *1910*, 111–119.
39. Doi, S.; Kuwabara, M.; Noguchi, T.; Ohnishi, T. Organic electroluminescent devices having poly(dialkoxy-*p*-phenylene vinylenes) as a light emitting material. *Synth. Met.* 1993, *57*, 4174–4179.
40. Tan, H.; Chan, L.; Wang, X.; Xie, H.; Gao, G.; Yao, J. Studies on spectral properties of poly[(2,5-bis(dodecyloxy)-phenylenevinylene)-co-(*p*-phenylene-vinylene)]. *Proc. SPIE*, 1996, *2892*, 198–203.
41. Braun, D.; Heeger, A. J.; Kroemer, H. Improved efficiency in semiconducting polymer light-emitting diodes. *J. Electron. Mater.* 1991, *20*, 945–948.
42. Egbe, D. A. M.; Neugebauer, H.; Sariciftci, N. S. Alkoxy-substituted poly(arylene-ethynylene)-alt-poly(arylene-vinylene)s: Synthesis, electroluminescence and photovoltaic applications. *J. Mater. Chem.* 2011, *21*, 1338–1349.
43. Mayukh, M.; Jung, I. H.; He, F.; Yu, L. Incremental optimization in donor polymers for bulk heterojunction organic solar cells exhibiting high performance. *J. Polym. Sci. B Polym. Phys.* 2012, *50*, 1057–1070.

44. Parekh, B. P.; Tangonan, A. A.; Newaz, S. S.; Sanduja, S. K.; Ashraf, A. Q.; Krishnamoorti, R.; Lee, T. R. Use of DMF as solvent allows for the facile synthesis of soluble MEH-PPV. *Macromolecules* 2004, *37*, 8883–8887.
45. Samuel, I. D. W.; Crystall, B.; Rumbles, G.; Bum, P. L.; Holmes, A. B.; Friend, R. H. The efficiency and time-dependence of luminescence from poly(*p*-phenylene vinylene) and derivatives. *Chem. Phys. Lett.* 1993, *213*, 472–478.
46. Kraft, A.; Grimsdale, A. C.; Holmes, A. B. Electroluminescent conjugated polymers—Seeing polymers in a new light. *Angew. Chem. Int. Ed.* 1998, *37*, 402–428.
47. Liang, Y.; Feng, D.; Wu, Y.; Tsai, S.-T.; Li, G.; Ray, C.; Yu, L. Highly efficient solar cell polymers developed via fine-tuning of structural and electronic properties. *J. Am. Chem. Soc.* 2009, *131*, 7792–7799.
48. Cumpston, B. H.; Jensen, K. F. Photo-oxidation of polymers used in electroluminescent devices. *Synth. Met.* 1995, *73*, 195–199.
49. Cumpston, B. H.; Parker, I. D.; Jensen, K. F. In situ characterization of the oxidative degradation of a polymeric light emitting device. *J. Appl. Phys.* 1997, *81*, 3716–3720.
50. Scurlock, R. D.; Wang, B.; Ogilby, P. R.; Sheats, J. R.; Clough, R. L. Singlet oxygen as a reactive intermediate in the photodegradation of an electroluminescent polymer. *J. Am. Chem. Soc.* 1995, *117*, 10194–10202.
51. Li, A.-K.; Janarthanan, N.; Hsu, C.-S. Synthesis of substituted poly(arylene vinylene) films by vapor deposition polymerization. *Polym. Bull.* 2000, *45*, 129–135.
52. Sonoda, Y.; Kaeriyama, K. Preparation of poly(2,5-diheptyl-1,4-phenylene-vinylene) by sulfonium salt pyrolysis. *Bull. Chem. Soc. Jpn.* 1992, *65*, 853–857.
53. Sonoda, Y.; Nakao, Y.; Kaeriyama, K. Preparation and properties of poly(1,4-phenylenevinylene) derivatives. *Synth. Met.* 1993, *55*, 918–923.
54. Zheng, J.; He, G.; Yang, C.; Huang, L.; Li, Y. Synthesis and properties of alkyl-substituted poly(1,4-phenylenevinylene) derivatives. *J. Appl. Polym. Sci.* 2001, *80*, 1299–1304.
55. Arbizzani, C.; Bongini, A.; Mastragostino, M.; Zanelli, A.; Barbarella, G.; Zambianchi, M. Lateral deposition of polypyrrole lines over insulating gaps. Towards the development of polymer-based electronic devices. *Adv. Mater.* 1995, *7*, 571–574.
56. Paloheimo, J.; Stubb, H.; Yli-Lahti, P.; Kuivalainen, P. Field-effect conduction in polyalkylthiophenes. *Synth. Met.* 1991, *41*, 563–566.
57. Wang, D.; Wu, Z. New polymerization catalyzed by palladium complexes: Synthesis of poly(*p*-phenylenevinylene) derivatives. *Chem. Commun.* 1999, 529–530.
58. Hsieh, B. R.; Yu, Y. Electroluminescent polymer compositions and processes thereof. US Patent 5945502, Xerox Corp., Norwalk, CT, August 31, 1999.
59. Hsieh, B. R.; Yu, Y.; Forsythe, E. W.; Schaaf, G. M.; Feld, W. A. A new family of highly emissive soluble poly(*p*-phenylene vinylene) derivatives. A step toward fully conjugated blue-emitting poly(*p*-phenylene vinylenes). *J. Am. Chem. Soc.* 1998, *120*, 231–232.
60. Li, A.-K.; Yang, S.-S.; Jean, W.-Y.; Hsu, C.-S.; Hsieh, B. R. Poly(2,3-diphenylphenylene vinylene) derivatives having liquid crystalline side groups. *Chem. Mater.* 2000, *12*, 2741–2744.
61. Stenger-Smith, J. D.; Zarras, P.; Merwin, L. H.; Shaheen, S. E.; Kippelen, B.; Peyghambarian, N. Synthesis and Characterization of Poly(2,5-bis(*N*-methyl-*N*-hexylamino)phenylene vinylene), a Conjugated Polymer for Light-Emitting Diodes. *Macromolecules* 1998, *31*, 7566–7569.
62. Jin, Y.; Kim, J.; Lee, S.; Kim, J. Y.; Park, S. H.; Lee, K.; Suh, H. Synthesis and characterization of poly(2,5-bis(N-methyl-N-hexylamino)phenylene vinylene), a conjugated polymer for light-emitting diodes. *Macromolecules* 2004, *37*, 6711–6715.
63. Spreitzer, H.; Becker, H.; Kluge, E.; Kreuder, W.; Schenk, H.; Demandt, R.; Schoo, H. Soluble phenyl-substituted PPVs—New materials for highly efficient polymer LEDs. *Adv. Mater.* 1998, *10*, 1340–1342.

64. Ginsburg, E. J.; Gorman, C. B.; Marder, S. R.; Grubbs, R. H. Poly(trimethylsilylcyclooc tatetraene): A soluble conjugated polyacetylene via olefin metathesis. *J. Am. Chem. Soc.* 1989, *111*, 7621–7622.
65. Petit, M. A.; Soum A. H.; Leclerc, M.; Prud'homme, R. E. Properties of iodine complexes of monosubstituted polyacetylenes. *J. Polym. Sci. B Polym. Phys.* 1987, *25*, 423–433.
66. Patil, A. O.; Ikenoue, Y.; Wudl, F.; Heeger, A. J. Water-soluble conducting polymers. *J. Am. Chem. Soc.* 1987, *109*, 1858–1859.
67. Peng, Z.; Xu, B.; Zhang, J.; Pan, Y. Synthesis and optical properties of water-soluble poly(*p*-phenylenevinylene)s. *Chem. Commun.* 1999, 1855–1856.
68. Groenendaal, L. B.; Jonas, F.; Freitag, D.; Pielartzik, H.; Reynolds, J. R. Poly(3,4-ethylenedioxythiophene) and its derivatives: Past, present, and future. *Adv. Mater.* 2000, *12*, 481–494.
69. Zhang, R.; Zhang, G.; Shen, J. A new approach for the synthesis of conjugated/non-conjugated poly(phenylene vinylene)–polyacrylamide copolymers. *Chem. Commun.* 2000, 823–824.
70. Liu, B.; Yu W.-L.; Lai, Y.-H.; Huang, W. Blue-light-emitting cationic water-soluble polyfluorene derivatives with tunable quaternization degree. *Macromolecules* 2002, *35*, 4975–4982.
71. Udum, Y. A.; Pekmez, K.; Yildiz, A. Electrochemical synthesis of soluble sulfonated poly(3-methyl thiophene). *Eur. Polym. J.* 2004, *40*, 1057–1062.
72. Kim, S.; Jackiw, J.; Robinson, E.; Schanze, K. S.; Reynolds, J. R.; Baur, J.; Rubner, M. F.; Boil, D. Water soluble photo- and electroluminescent alkoxy-sulfonated poly(*p*-phenylenes) synthesized via palladium catalysis. *Macromolecules* 1998, *31*, 964–974.
73. Cimrová, V.; Schmidt, W.; Rulkens, R.; Schulze, M.; Meyer, W.; Neher, D. Efficient blue light emitting devices based on rigid-rod polyelectrolytes. *Adv. Mater.* 1996, *8*, 585–588.
74. Gu, Z.; Shen, Q.-D.; Zhang, J.; Yang, C.-Z.; Bao, Y.-J. Dual electroluminescence from a single-component light-emitting electrochemical cell, based on water-soluble conjugated polymer. *J. Appl. Polym. Sci.* 2006, *100*, 2930–2936.
75. Wagaman, M. W.; Grubbs, R. H. Synthesis of organic and water soluble poly(1,4-phenylenevinylenes) containing carboxyl groups: Living ring-opening metathesis polymerization (ROMP) of 2,3-dicarboxybarrelenes. *Macromolecules* 1997, *30*, 3978–3985.
76. Liu, B.; Yu, W.-L.; Lai, Y.-H.; Huang, W. Synthesis of polyfluorene derivatives through polymer reaction. *Opt. Mater.* 2002, *21*, 125–133.
77. Fujii, A.; Sonoda, T.; Fujisawa, T.; Ootake, R.; Yoshino, K. Synthesis and luminescent properties of water-soluble poly(*p*-phenylene vinylene). *Synth. Met.* 2001, *119*, 189–190.
78. Papadimitrakopoulos, F.; Konstadinidis, K.; Miller, T. M.; Opila, R.; Chandross, E. A.; Galvin, M. E. The role of carbonyl groups in the photoluminescence of poly(*p*-phenylenevinylene). *Chem. Mater.* 1994, *6*, 1563–1568.
79. Wu, H.; Huang, F.; Mo, Y.; Yang, W.; Wang, D.; Peng, J.; Cao, Y. Efficient electron injection from a bilayer cathode consisting of aluminum and alcohol-/water-soluble conjugated polymers. *Adv. Mater.* 2004, *16*, 1826–1830.
80. Ramey, M. B.; Hiller, J.; Rubner, M. F.; Tan, C.; Schanze, K. S.; Reynolds, J. R. Amplified fluorescence quenching and electroluminescence of a cationic poly(*p*-phenylene-co-thiophene) polyelectrolyte. *Macromolecules* 2005, *38*, 234–243.
81. Erothu, H.; Sohdi, A. A.; Kumar, A. C.; Sutherland, A. J.; Dagron-Lartigau, C.; Allal, A.; Hiorns, R. C.; Topham, P. D. Facile synthesis of poly(3-hexylthiophene)-block-poly(ethylene oxide) copolymers via steglich esterification. *Polym. Chem.* 2013, *4*, 3652–3655.

82. Cacialli, F.; Friend, R. H.; Feast, W. J.; Lovenich, P. W. Poly(distyrylbenzene-*block*-sexi(ethylene oxide)), a highly luminescent processable derivative of PPV. *Chem. Commun.* 2001, 1778–1779.
83. Pei, Q.; Yu, G.; Zhang, C.; Yang, Y.; Heeger, A. J. Polymer light-emitting electrochemical cells. *Science* 1995, *269*, 1086–1088.
84. Sorger, K.; von Ragué Schleyer, P.; Stalke, D. Dimeric [3,3-dimethyl-2-(trimethylsilyl) cyclopropenyl]-lithium–Tetramethylethylenediamine: Distortion of the cyclopropenyl geometry due to strong rehybridization at the lithiated carbon. *J. Am. Chem. Soc.* 1996, *118*, 1086–1091.
85. Zhu, C.; Liu, L.; Yang, Q.; Lv, F.; Wang, S. Water-soluble conjugated polymers for imaging, diagnosis, and therapy. *Chem. Rev.* 2012, *112*, 4687–4735.
86. Shirakawa, H.; Ito, T.; Ikeda, S. Electrical-properties of polyacetylene with various cis-trans compositions. *Macromol. Chem. Phys.* 1978, *179*, 1565–1573.
87. Anders, U.; Nuyken, O.; Buchmeiser, M. R.; Wurst, K. Stereoselective cyclopolymerization of 1,6-heptadiynes: Access to alternating *cis*-trans-1,2-(Cyclopent-1-enylene) vinylenes by fine-tuning of molybdenum imidoalkylidenes. *Angew. Chem. Int. Ed.* 2002, *41*, 4044–4047.
88. Gorman, C. B.; Ginsburg, E. J.; Marder, S. R.; Grubbs, R. H. Highly conjugated substituted polyacetylenes via the ring-opening metathesis polymerization of substituted cyclooctatetraenes. *Angew. Chem. Int. Ed. Engl. Adv. Mater.* 1989, *28*, 1571–1574.
89. Elert, M. L.; White, C. T. Helical versus planar *cis*-polyacetylene. *Phys. Rev. B* 1983, *28*, 7387–7389.
90. Rao, B. K.; Darsey, J. A.; Kestner, N. R. Existence of helical forms of polyacetylene. *Phys. Rev. B* 1985, *31*, 1187–1190.
91. Bates, F. S.; Baker, G. L. Polyacetylene single crystals. *Macromolecules* 1983, *16*, 1013–1015.
92. Sarker, A. M.; Elif Gürel, E.; Ding, L.; Styche, E.; Lahti, P. M.; Karasz, F. E. Light emitting poly(para-phenylenevinylene-alt-3-tert-butyl-meta-phenylenevinylenes). *Synth. Met.* 2003, *132*, 227–234.
93. Liao, L.; Pang, Y.; Ding, L.; Karasz, F. E. Blue-emitting soluble poly(m-phenylenevinylene) derivatives. *Macromolecules* 2001, *34*, 7300–7305.
94. Lipson, S. M.; O'Brien, D. F.; Byrne, H. J.; Davey, A. P.; Blau, W. J. Investigation of efficiency and photostability in polymer films. *Synth. Met.* 2000, *111–112*, 553–557.
95. Pang, Y.; Li, J.; Hu, B.; Karasz, F. E. A processible poly(phenyleneethynylene) with strong photoluminescence: Synthesis and characterization of poly[(*m*-phenyleneethynylene)-alt-(*p*-phenyleneethynylene)]. *Macromolecules* 1998, *31*, 6730–6732.
96. Pang, Y.; Li, J.; Barton, T. J. Processible poly[(*p*-phenylene ethynylene)-alt-(2,5-thienyleneethynylene)]s of high luminescence: Their synthesis and physical properties. *J. Mater. Chem.* 1998, *8*, 1687–1690.
97. Williams, V. E.; Swager, T. M. Iptycene-containing poly(aryleneethynylene)s. *Macromolecules* 2000, *33*, 4069–4073.
98. Pang, Y.; Li, J.; Hu, B.; Karasz, F. E. A highly luminescent poly[(*m*-phenylenevinylene)-alt-(*p*-phenylenevinylene)] with defined conjugation length and improved solubility. *Macromolecules* 1999, *32*, 3946–3950.
99. Cho, H. N.; Kim, D. Y.; Kim, J. K.; Kim, C. Y. Control of band gaps of conjugated polymers by copolymerization. *Synth. Met.* 1997, *91*, 293–296.
100. Cho, H. N.; Kim, D. Y.; Kim, Y. C.; Lee, J. Y.; Kim, C. Y. Blue and green light emission from new soluble alternating copolymers. *Adv. Mater.* 1997, *9*, 326–328.
101. Lin, T.; He, Q.; Bai, F.; Dai, L. Design, synthesis and photophysical properties of a hyperbranched conjugated polymer. *Thin Solid Films* 2000, *363*, 122–125.
102. Kim, Y. H.; Webster, O. W. Hyperbranched polyphenylenes. *Macromolecules* 1992, *25*, 5561–5572.

103. Chien, J. C. W.; Wnek, G. E.; Karasz, F. E.; Hirsch, J. A. Electrically conducting acetylene-methylacetylene copolymers. Synthesis and properties. *Macromolecules* 1981, *14*, 479–485.
104. Gorman, C. B.; Ginsburg, E. J.; Grubbs R. H. Soluble, highly conjugated derivatives of polyacetylene from the ring-opening metathesis polymerization of monosubstituted cyclooctatetraenes: Synthesis and the relationship between polymer structure and physical properties. *J. Am. Chem. Soc.* 1993, *115*, 1397–1409.
105. Moore, J. S.; Gorman, C. B.; Grubbs, R. H. Soluble, chiral polyacetylenes: Syntheses an investigation of their solution conformation. *J. Am. Chem. Soc.* 1991, *113*, 1704–1712.
106. Gan, L. H.; Wang, Y. M.; Xu, Y.; Goh, N. K.; Gan, Y. Y. Synthesis and characterization of poly(2,3,5,6-tetrafluorophenylenevinylene): A revisit. *Macromolecules* 2001, *34*, 6117–6120.
107. McCulloch, I.; Ashraf, R. S.; Biniek, L.; Bronstein, H.; Combe, C.; Donaghey, J. E.; James, D. I.; Nielsen, C. B.; Schroeder, B. C.; Zhang, W. Design of semiconducting indacenodithiophene polymers for high performance transistors and solar cells. *Acc. Chem. Res.* 2012, *45*, 714–722.
108. Bijleveld, J. C.; Gevaerts, V. S.; Di Nuzzo, D.; Turbiez, M.; Mathijssen, S. G. J.; de Leeuw, D. M.; Wienk, M. M.; Janssen, R. A. J. Efficient solar cells based on an easily accessible diketopyrrolopyrrole polymer. *Adv. Mater.* 2010, *22*, E242–E246.
109. Osaka, I.; Saito, M.; Mori, H.; Koganezawa, T.; Takimiya, K. Drastic change of molecular orientation in a thiazolothiazole copolymer by molecular-weight control and blending with $PC_{61}BM$ leads to high efficiencies in solar cells. *Adv. Mater.* 2012, *24*, 425–430.
110. Bronstein, H.; Chen, Z.; Ashraf, R. S.; Zhang, W.; Du, J.; Durrant, J. R.; Tuladhar, P. S. et al. Thieno[2,3-b]thiophene–diketopyrrolopyrrole-containing polymers for high-performance organic field-effect transistors and organic photovoltaic devices. *J. Am. Chem. Soc.* 2011, *133*, 3272–3275.
111. Zen, A.; Pflaum, J.; Hirschmann, S.; Zhuang, W.; Jaiser, F.; Asawapirom, U.; Rabe, J. P.; Scherf, U.; Neher, D.; Effect of molecular weight and annealing of poly(3-hexylthiophene)s on the performance of organic field-effect transistors. *Adv. Funct. Mater.* 2004, *14*, 757–764.
112. Chang, J.-F.; Sun, B.; Breiby, D. W.; Nielsen, M. M.; Sölling, T. I.; Giles, M.; McCulloch, I.; Sirringhaus, H. Enhanced mobility of poly(3-hexylthiophene) transistors by spin-coating from high-boiling-point solvents. *Chem. Mater.* 2004, *16*, 4772–4776.
113. Kline, R. J.; McGehee, M. D.; Kadnikova, E. N.; Liu, J.; Fréchet, J. M. J. Controlling the field-effect mobility of regioregular polythiophene by changing the molecular weight. *Adv. Mater.* 2003, *15*, 1519–1522.
114. Zen, A.; Saphiannikova, M.; Neher, D.; Grenzer, J.; Grigorian, S.; Pietsch, U.; Asawapirom, U.; Janietz, S.; Scherf, U.; Lieberwirth, I.; Wegner, G. Effect of molecular weight on the structure and crystallinity of poly(3-hexylthiophene). *Macromolecules* 2006, *39*, 2162–2171.
115. Kline, R. J.; McGehee, M. D.; Kadnikova, E. N.; Liu, J.; Fréchet, J. M. J.; Toney, M. F. Dependence of regioregular poly(3-hexylthiophene) film morphology and field-effect mobility on molecular weight. *Macromolecules* 2005, *38*, 3312–3319.
116. Rissler, J. Effective conjugation length of π-conjugated systems. *Chem. Phys. Lett.* 2004, *395*, 92–96.
117. Smith, T. M.; Hazelton, N.; Peteanu, L. A.; Wildeman, J. Electrofluorescence of MEH-PPV and its oligomers: Evidence for field-induced fluorescence quenching of single chains. *J. Phys. Chem. B* 2006, *110*, 7732–7742.
118. Chang, J.-F.; Clark, J.; Zhao, N.; Sirringhaus, H.; Breiby, D. W.; Andreasen, J. W.; Nielsen, M. M.; Giles, M.; Heeney, M.; McCulloch, I. Molecular-weight dependence of interchain polaron delocalization and exciton bandwidth in high-mobility conjugated polymers. *Phys. Rev. B* 2006, *74*, 115318/1–115318/12.

119. Ballantyne, A. M.; Chen, L.; Dane, J.; Hammant, T.; Braun, F. M.; Heeney, M.; Duffy, W.; McCulloch, I.; Bradley, D. D. C.; Nelson, J. The effect of poly(3-hexylthiophene) molecular weight on charge transport and the performance of polymer: Fullerene solar cells. *Adv. Funct. Mater.* 2008, *18*, 2373–2380.
120. Jen, K.-Y.; Miller, G. G.; Eisenbaumer, R. L. Highly conducting, soluble, and environmentally-stable poly(3-alkylthiophenes). *J. Chem. Soc. Chem. Commun.* 1986, 1346–1347.
121. Ou-Yang, W.-C.; Wu, T.-Y. Lin, Y.-C. Supramolecular structure of poly[2-methoxy-5-(2′-ethylhexyloxy)-1,4-phenylenevinylene] (MEH-PPV) probed using wide-angle x-ray diffraction and photoluminescence. *Iran. Polym. J.* 2009, *18*, 453–464.
122. Lee, T.-W.; Park, O. O.; Kim, J.; Kim, Y. C. Application of a novel fullerene-containing copolymer to electroluminescent devices. *Chem. Mater.* 2002, *14*, 4281–4285.
123. Trouillet, L.; De Nicola, A.; Guillerez, S. Synthesis and characterization of a new soluble, structurally well-defined conjugated polymer alternating regioregularly alkylated thiophene oligomer and 2,2′-bipyridine units: Metal-free form and Ru(II) complex. *Chem. Mater.* 2000, *12*, 1611–1621.
124. Liang, Y.; Yu, L. A new class of semiconducting polymers for bulk heterojunction solar cells with exceptionally high performance. *Acc. Chem. Res.* 2010, *43*, 1227–1236.
125. Rispens, M. T.; Meetsma, A.; Rittberger, R.; Brabec, C. J.; Serdar Sariciftci, N.; Hummelen, J. C. Influence of the solvent on the crystal structure of PCBM and the efficiency of MDMO-PPV:PCBM 'Plastic' solar cells. *Chem. Commun.* 2003, 2116–2118.
126. Shaheen, S. E.; Brabec, C. J.; Serdar Sariciftci, N.; Padinger, F.; Fromherz, T.; Hummelen, J. C. 2.5% Efficient organic plastic solar cells. *Appl. Phys. Lett.* 2001, *78*, 841–843.
127. Zhang, F.; Jespersen, K. G.; Björström, C.; Svensson, M.; Andersson, M. R.; Sundström, V.; Magnusson, K.; Moons, E.; Yartsev, A.; Inganäs, O. Influence of solvent mixing on the morphology and performance of solar cells based on polyfluorene copolymer/fullerene blends. *Adv. Funct. Mater.* 2006, *16*, 667–674.
128. Ma, W.; Yang, C.; Gong, X.; Lee, K.; Heeger, A. J. Thermally stable, efficient polymer solar cells with nanoscale control of the interpenetrating network morphology. *Adv. Funct. Mater.* 2005, *15*, 1617–1622.
129. Chu, T.-Y.; Lu, J.; Beaupré, S.; Zhang, Y.; Pouliot, J.-R.; Wakim, S.; Zhou, J.; Leclerc, M.; Li, Z.; Ding, J.; Tao, Y. Bulk heterojunction solar cells using thieno[3,4-*c*]pyrrole-4,6-dione and dithieno[3,2-b:2′,3′-*d*]silole copolymer with a power conversion efficiency of 7.3%. *J. Am. Chem. Soc.* 2011, *133*, 4250–4253.
130. Lee, J. K.; Ma, W. L.; Brabec, C. J.; Yuen, J.; Moon, J. S.; Kim, J. Y.; Lee, K.; Bazan, G. C.; Heeger, A. J. Processing additives for improved efficiency from bulk. *J. Am. Chem. Soc.* 2008, *130*, 3619–3623.
131. Green, R.; Morfa, A.; Ferguson, A. J.; Kopidakis, N.; Rumbles, G.; Shaheen, S. E. Performance of bulk heterojunction photovoltaic devices prepared by airbrush spray deposition. *Appl. Phys. Lett.* 2008, *92*, 033301/1–033301/3.
132. Nguyen, T.-Q.; Doan, V.; Schwartz, B. J. Conjugated polymer aggregates in solution: Control of interchain interactions. *J. Chem. Phys.* 1999, *110*, 4068–4078.
133. Zheng, W.; Angelopoulos, M.; Epstein, A. J.; MacDiarmid, A. G. Concentration dependence of aggregation of polyaniline in NMP solution and properties of resulting cast films. *Macromolecules*, 1997, *30*, 7634–7637.
134. Gettinger, C. L.; Heeger, A. J.; Drake, J. M.; Pine, D. J. A photoluminescence study of poly(phenylene vinylene) derivatives: The effect of intrinsic persistence length. *J. Chem. Phys.* 1994, *101*, 1673–1678.

Section II

Self-Assembly of Nanocrystals/ Supramolecular Architectures

4 Geometry and Entropy *New Tools for Assembling Hierarchical Mesoscale Structures from Nanoscale Assemblies of Molecules*

Timothy S. Gehan and D. Venkataraman

CONTENTS

4.1 INTRODUCTION

The field of supramolecular chemistry has provided us the tools and rules to organize molecules and macromolecules into discrete assemblies, morphologies, or extended solids. Thus far, the focus of this field has been on understanding and utilizing interactions between molecules. Intermolecular interactions such as π interactions, electrostatic interactions, hydrogen bonds, metal–ligand coordination bonds, and reversible covalent bonds have been used to design a fantastic array of molecular and macromolecular assemblies.[1–5] The next frontier in supramolecular chemistry is to develop the tools and rules to organize molecular and macromolecular assemblies into desired structures or morphologies. Thus, we need to shift our focus from interactions between *molecules* to interactions between *molecular assemblies*. With the ability to control molecules to molecular assemblies and molecular assemblies to mesoscale structures, we seek the ability to control and direct the assembly of molecules at multiple length scales.

Our inspiration to control the assembly of molecules at multiple length scales stems from functional hierarchical structures that exist in nature. The archetypical example of hierarchical structure is the tobacco mosaic virus wherein a single-stranded helix assembles into discrete helical assemblies, which self-assembles into a helical rod.[6] In other viruses, the protein subunits organize into higher-order structures resulting in virus particles with defined shapes such as an icosahedron in picornaviruses.[7,8] In materials chemistry, our need to control the molecular assembly at various length scales arises from the fact that most functional materials consist of molecules assembled in a hierarchy of length scales. The obvious question is: how we can exploit the underlying principles that result in hierarchical assemblies in nature to create hierarchical assemblies in materials?

4.1.1 Concept of Geometric Packing

In recent years, there is a burgeoning interest in using entropy and principles of geometry to guide the assembly of molecular assemblies. The idea of using geometry in chemistry goes back to the understanding of mineral structures and the packing of anions and cations. The radius ratio (R_{cation}/R_{anion}) rules, stated by Pauling, use principles of sphere packing to predict the structure of ionic crystals and the coordination number of the ions from the sizes of the constituent anion and cation is example of exploiting geometry to understand structure.[9] Similar to ionic crystals, two types of nanoparticles or colloidal microparticles can self-assemble into ordered supraparticle assemblies, also termed as binary nanoparticle superlattices (BNSL), at a much larger scale. Like in ionic crystals, the packing of the particles depends on the radius ratio of the particles ($\gamma = R_{small}/R_{large}$).[10–13] Unlike in ionic crystals, the self-assembly of particles occurs even if they do not have electrostatic attraction. We now know that the primary mechanism that drives assembly of spherical particles into crystalline arrays (or superlattices) is entropy maximization by adopting the most efficient packing. The classic example of entropy-driven crystallization is the formation of face-centered cubic (FCC) arrays of hard spheres (having billiard-ball-like interactions). The FCC lattice has the largest close-packed particle volume fraction, ϕ, of any ordered lattice containing spheres of the same size: $\phi = 0.74$. Hence, this lattice provides more free volume to each particle than a randomly close-packed structure that has a maximum $\phi \sim 0.64$. Thus, entropy and geometry can be powerful tools to organize molecular assemblies into specific structures or morphologies. Herein, we explore some of the key concepts and studies in this area. In the ensuing sections, we will use the term *microparticle* for a spherical particle with diameter between 0.1 and 100 μm and the term *nanoparticle* for a spherical particle with diameter between 1 and 100 nm.

4.2 MICROPARTICLE SUPERLATTICES

Pusey and coworkers prepared binary microparticle superlattices from the entropy driven self-assembly of poly(methylmethacrylate) (PMMA) microparticles.[14,15] First Pusey and coworkers demonstrated they could tune the fluid–solid-phase transition of a binary mixture of spheres (A and B) with a radius ratio (γ) of 0.61, using spheres of 407 nm and 670 nm.[14] By tuning the number ratio of the spheres, various crystalline solids were prepared including pure A, pure B, and a binary mixture of A and B (AB_{13}). Pusey and coworkers went on to change the radius ratio between the two spheres to 0.58.[15] In this work, they report the formation of AB_2 and AB_{13} binary microparticle superlattices and construct a phase diagram identifying the crystalline lattice for a select radius ratio of the binary microparticle solution. This work demonstrates that binary microparticle superlattices can be formed and the microparticle self-assembly can be easily tuned using the radius ratio and the volume fractions of the microparticles. Although these experimental results suggest that AB_2 and AB_{13} are thermodynamically stable equilibrium structures, since they can take months to form, there is no evidence that they are not metastable states.

To further understand the stability of AB_2 and AB_{13} structures, Frenkel and coworkers performed Monte Carlo simulations of these systems.[16,17] By calculating the free energy of the particles in the solid phase and the fluid phase, the authors determined the stability of the AB_{13} superlattice as a function of the radius ratio between the small and the large particles.[16] From their calculated phase diagram, they determined that a radius ratio of 0.56 was the central point for a stable AB_{13} superlattice and the crystal with a maximum packing fraction, as seen in Figure 4.1a. Frenkel and coworkers calculated phase diagram correlates well with Pusey and coworkers results for 0.58 and 0.61 radius ratios indicating 0.58 can result in a thermodynamically stable AB_{13} superlattice and 0.61 forms a metastable structure.[14,15] Frenkel and coworkers followed this by computing the stability of the AB_2 superlattice as a function of the particle radius ratio, as seen in Figure 4.1b. The computational results agree with the experimental results from Pusey and coworkers, since the AB_2 lattice is formed at the 0.58 radius ratio and not at 0.61.[14,15] The authors determined that the formation of these hard sphere superlattices "... is thought to be entropically driven with an increase in the free volume available to each of the spheres upon crystallization."[17,18] Frenkel and coworkers also prepared a more complete phase diagram for microparticles with a radius ratio of 0.58 as seen in Figure 4.2.

Monson and coworkers applied cell theory to calculate the thermodynamic properties of these binary microparticle mixtures.[19] Their results agreed well with Frenkel's phase diagram for a binary microparticle mixture with a radius ratio of 0.58, shown in Figure 4.2. Monson and coworkers used this method to further understand the stability of the AB, AB_2, and AB_{13} superlattices, developing a general chart for regions of radius ratio that each superlattice is stable seen in Figure 4.3. The stability ranges for AB_2 and AB_{13} are similar to those reported by Frenkel and coworkers,[16,17] but Monson and coworkers also determined a stable radius ratio range for AB to be between 0.2 and 0.42. This work further indicates that these superlattice structures are thermodynamically stable structures and can easily be tuned by changing various parameters like the microparticle radius ratio. A more detailed study using a wide range of microparticle radius ratios agree fairly well with Frenkel and Monson's predicted results.[20]

4.3 NANOPARTICLE SUPERLATTICES

Pusey and coworkers used microparticles of hundreds of nanometers to prepare binary microparticle superlattices.[14,15] The term *nanoparticles* typically refers to particles in the size range of ~1–100 nm. Bawendi and coworkers demonstrated that nanoparticle superlattices could also be formed with very small inorganic nanoparticles.[11] The authors prepared a range of superlattices, using CdSe nanoparticles with similar sizes

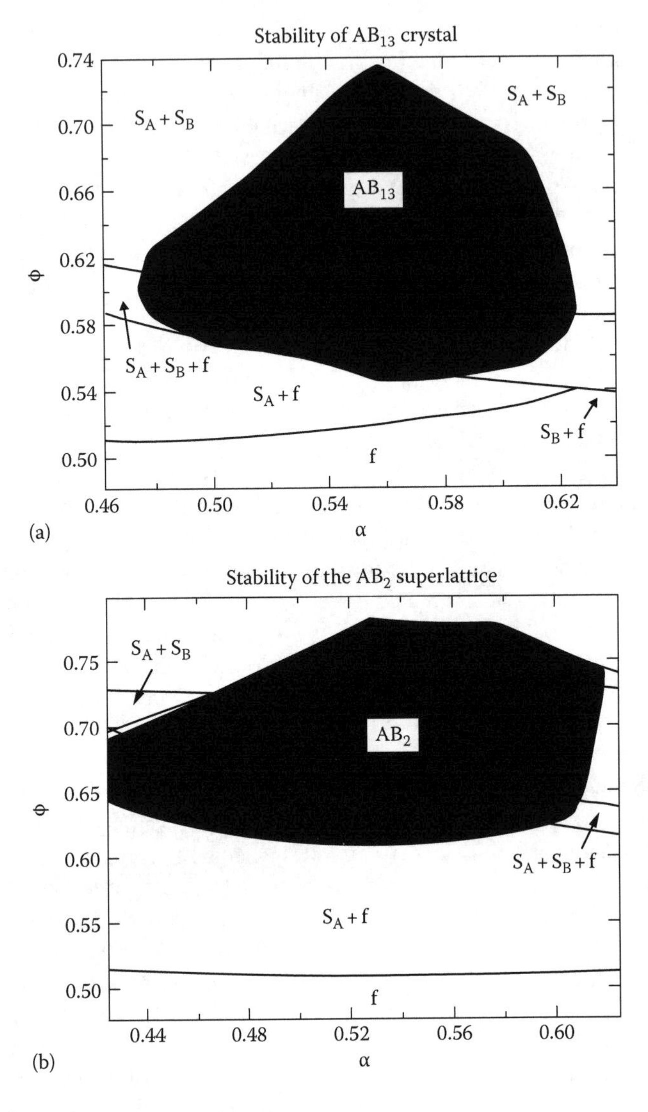

FIGURE 4.1 (a) Phase diagram of the computed stability of the AB_{13} superlattice as a function of packing fraction and microparticle radius ratio. (b) Phase diagram of the computed stability of the AB_2 superlattice as a function of packing fraction and microparticle radius ratio.[16,17] The shaded regions indicate thermodynamic stability of the superlattice. (a: Reprinted with permission from Eldridge, M.D. et al., *Mol. Phys.*, 79, 105. Copyright 1993, Taylor & Francis Ltd., http://www.tandfonline.com; b: Reprinted with permission from Eldridge, M.D. et al., *Mol. Phys.*, 80, 987. Copyright 1993, Taylor & Francis Ltd., http://www.tandfonline.com.)

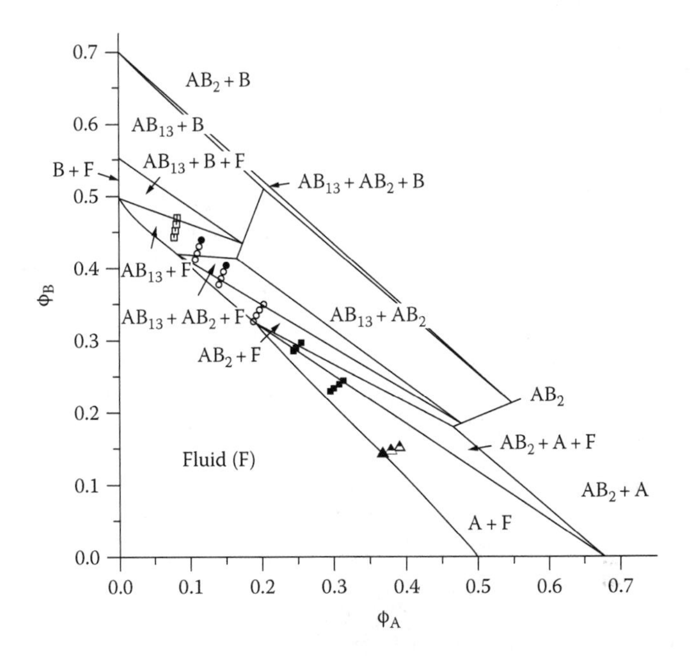

FIGURE 4.2 Phase diagram for a binary mixture of hard spherical microparticles with a radius ratio of 0.58 at constant volume. (Reprinted by permission from Macmillan Publishers Ltd. *Nature* Eldridge, M.D., Madden, P.A., and Frenkel, D., Entropy-driven formation of a superlattice in a hard-sphere binary mixture, 365, 35, copyright 1993.)

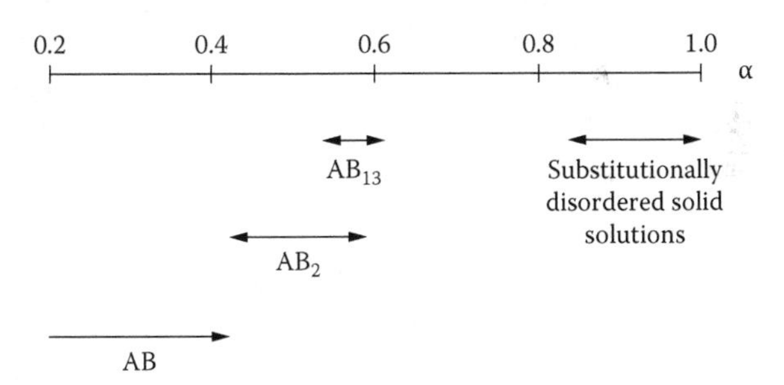

FIGURE 4.3 Regions of stability for the various binary lattices as a function of microparticle radius ratio. (Reprinted with permission from Cottin, X. and Monson, P.A., Substitutionally ordered solid-solutions of hard-spheres, *J. Chem. Phys.*, 102, 3354. Copyright 1995, American Institute of Physics.)

for each superlattice, with sizes ranging from 2.0–6.4 nm. Faceted colloidal crystals and colloidal superlattice thin films were also prepared. The authors show that the close-packed superlattice spacing can be tuned by tuning the nanoparticles diameter and by tuning the ligands size. The authors state "self-organization requires only a hard-sphere repulsion, a controlled size distribution, the inherent van der Waals attraction between particles, and a means of gently destabilizing the dispersion."[11]

This work demonstrates that even the use of nanoparticles as opposed to microparticles can lead to superlattices, and the authors imply that this strategy can be used with a wide variety of nanoparticle types and even binary nanoparticle superlattices should be possible from nanoparticles.

Building on previous work, Murray and coworkers demonstrated that binary nanoparticle superlattices with tunable structure can be formed using "small" inorganic nanoparticles. Some of the structures prepared are shown in Figure 4.4.[10] Elaborating from previous work, these structures were prepared with mixtures of semiconducting, metallic, and magnetic nanoparticles. The authors determined that the superlattice structure can be tuned by changing the surface charge of the nanoparticles using various additives in millimolar concentrations. When adding oleic acid to colloidal solutions of PbSe nanoparticles and Pd nanoparticles, AB and AB_2 superlattices were formed. Although when adding TOPO instead of oleic acid, AB_{13} superlattices were formed. This tunability was demonstrated for a range of binary nanoparticle mixtures of semiconducting, metallic, and magnetic nanoparticles. This work indicates that the assembly of these nanoparticle superlattices is a balance between van der Waals forces, dipolar interactions, and electrostatic interactions. Work by O'Brien and coworkers indicates that the binary nanoparticle superlattice structure, using semiconducting inorganic nanoparticles, can be tuned by tuning the radius ratio of the particles without having attractive electrostatic interactions.[21] Taking it a step further, the authors also assembled nanocrystals of different shapes to understand how they assemble. They discovered that the assembly of LaF triangular nanoplates with Au spherical nanocrystals can be tuned by the choice of the surface they are assembled on. This indicates that the surface can selectively orient anisotropic nanoparticles and that the nanoparticles shape can have a significant effect on the assembly of these superlattices.

It is difficult to predict the properties of bulk materials from the molecular or atomic properties of which it is composed. This is because as these atoms/molecules assemble into larger structures they typically do not retain the same properties and gain new properties. Talapin suggests that having a modular approach can solve this problem.[22] Atoms and molecules when assembled into various nanostructures can obtain some of these desirable bulk properties. The nanoparticles properties can be tuned during synthesis and by controlling the assembly of these modular nanoparticles into ordered mesoscale superlattices; it then gives us the opportunity to control the bulk materials properties. However, the random assembly of nanoparticles[23–26] has been demonstrated to prepare useful materials with desired properties. The preparation and control of ordered binary nanoparticle superlattices is a systematic approach to obtain tunable materials with desired properties. The driving force for assembly of ionic crystals is the attractive electrostatic interactions. When preparing nanoparticle assemblies one of the main driving forces for preparing these structures is entropy. Binary nanoparticle superlattices have been shown computationally to form thermodynamically stable structures and experimentally been prepared even when the attractive van der Waals forces are screened between particles. Although this may contradict common intuition of entropy relating to disorder, crystallization occurs because at high volume fractions of nanoparticles the accessible volume is larger in a crystalline lattice in the fluid phase and thus has more free-volume entropy.[16,17,27,28]

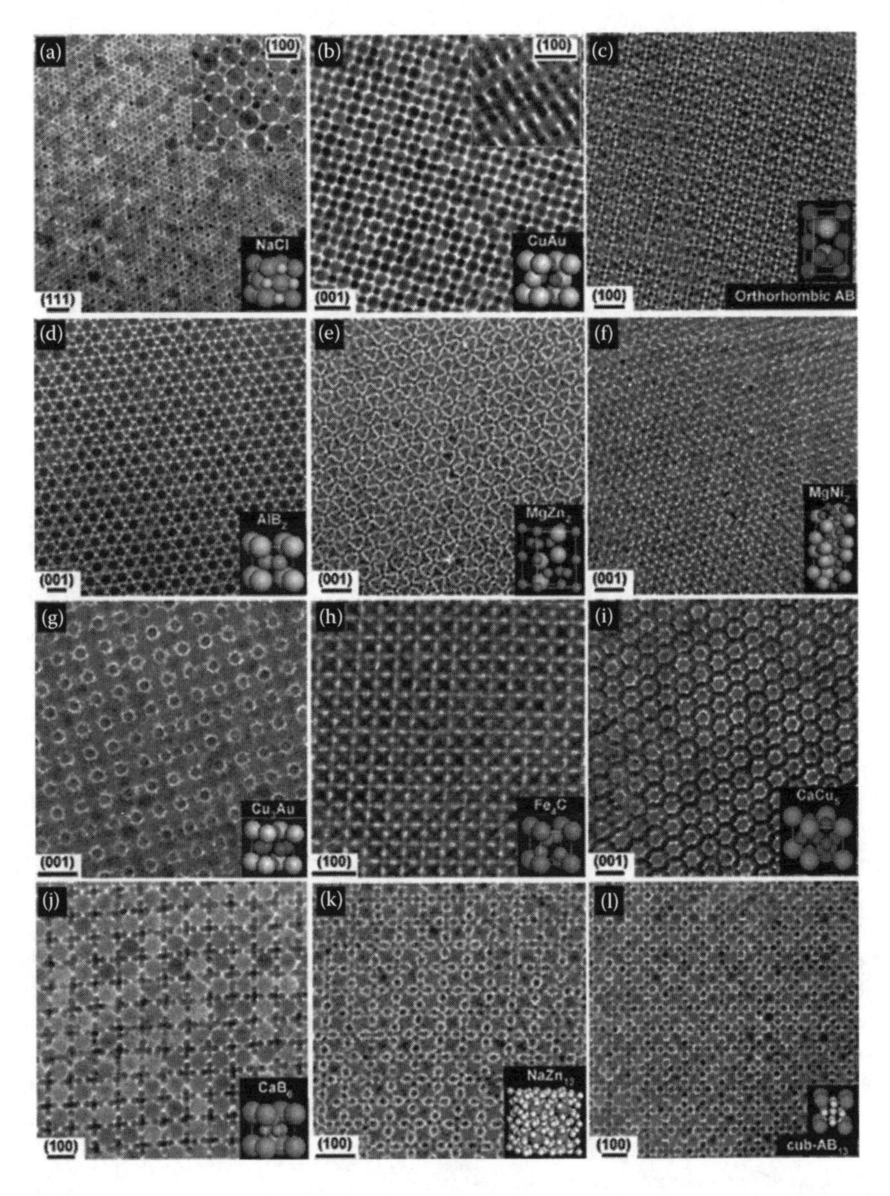

FIGURE 4.4 TEM images (a–l) of binary nanoparticle superlattices using inorganic nanoparticles (metallic, semiconducting, and magnetic) with a variety of radius ratios. Inserts are model projections of the corresponding atomic inorganic lattices that each binary nanoparticle superlattice emulates. (Reprinted by permission from Macmillan Publishers Ltd. *Nature* Shevchenko, E.V., Talapin, D.V., Kotov, N.A., O'Brien, S., and Murray, C.B., Structural diversity in binary nanoparticle superlattices, 439, 55, copyright 2006.)

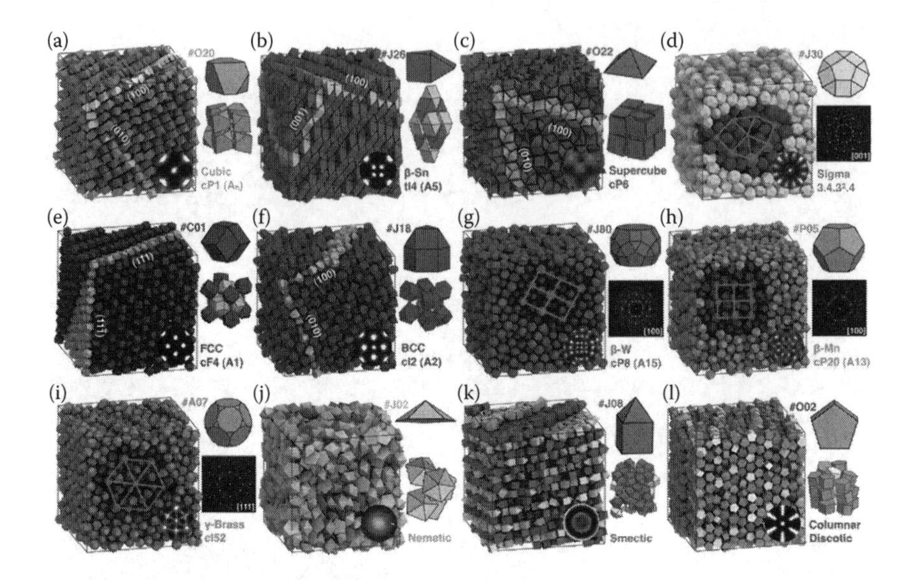

FIGURE 4.5 (a–l) Shows the simulated assembly of selected polyhedra that were simulated. A snapshot of the simulation is shown. The polyhedra simulated and its corresponding ID number is shown at the top right of each snapshot. An insert showing the nearest-neighbor particles for each assembly and the corresponding diffraction pattern is also shown for each assembly. (From Damasceno, P.F., Engel, M., and Glotzer, S.C., Predictive self-assembly of polyhedra into complex structures, *Science*, 337, 453, 2012. Reprinted with permission of AAAS.)

The next question then is how do other geometric shapes pack? Recent work from Glotzer's group has shown that the shape of a particle can directly affect and can be used to tune its assembly.[29] Glotzer and coworkers simulated the assembly of 145 different polyhedra and separated the packing of each shape into one of four categories of organization: disordered, plastic crystals, liquid crystals, and crystals. Interestingly although the simulations did not include attractive interactions between particles, 70% of the polyhedra used assembled on the timescale of their simulations. A variety of the simulated structures can be seen in Figure 4.5. The authors determined that by knowing the isoperimetric quotient of particles (shape) and the local order within the assembly, coordination number of the particles in the fluid phase before crystallization, the assembly of the particles could be predicted. This work demonstrates that a wide variety of assemblies can be prepared by changing the shape of the particles used. The experimental challenge now is to make nanoparticles of different and desired shapes.

4.4 ASSEMBLING NANOPARTICLES TO CONTROL THE MESOSCALE MORPHOLOGY IN OPVs

We provide here an example of how nanoparticle assemblies can be used to solve a longstanding problem in materials science. Controlling the morphology in organic solar cells is very difficult, and the current techniques to achieve the desired nanoscale

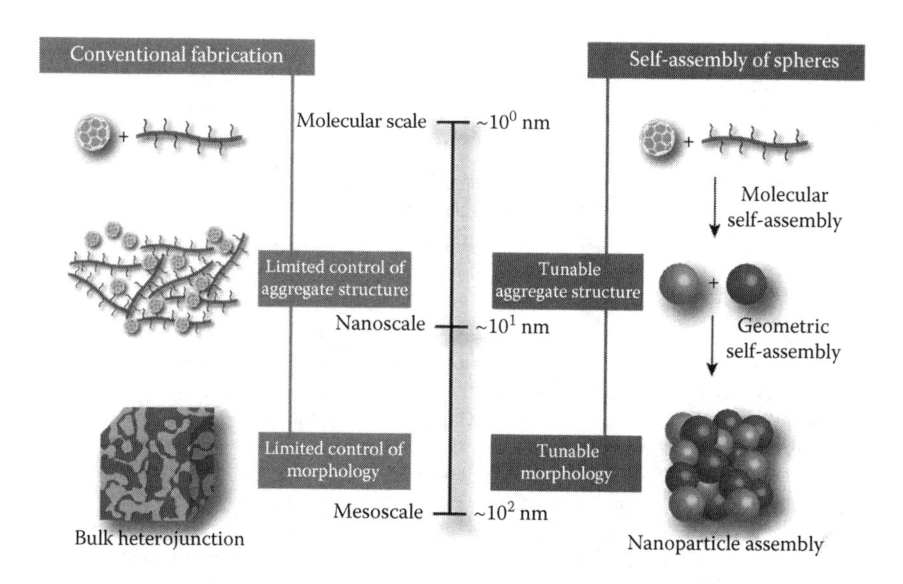

FIGURE 4.6 Comparison of nanoparticle-based method for organic photovoltaic morphology control with the conventional device fabrication method. (Reprinted with permission from Gehan, T.S., Bag, M., Renna, L.A., Shen, X.B., Algaier, D.D., Lahti, P.M., Russell, T.P., and Venkataraman, D., Multiscale active layer morphologies for organic photovoltaics through self-assembly of nanospheres, *Nano Lett.*, 14, 5238. Copyright 2014 American Chemical Society.)

packing are very trial and error based and can be very difficult to reproduce. In an attempt to solve this problem, Venkataraman and coworkers preformed the electron donor and electron acceptor as spherical nanoparticles and then assembled these nanoparticles to form the device active layer in an organic photovoltaic device.[23] A comparison of the nanoparticle assembly method and the conventional fabrication method is shown in Figure 4.6. The authors have shown that the internal structure of the nanoparticles can be tuned during the nanoparticle fabrication.[30] The authors also showed that they can tune the morphology by tuning the random assembly of electron-donating and electron-accepting nanoparticles by changing the nanoparticle size and the number ratio of donating to accepting particles. Devices prepared using this method have comparable efficiencies to bulk heterojunction devices, for the donor and acceptor used, and have the added advantage of being processed from benign aqueous solutions. This method offers the opportunity to systematically tune the nanosized domains directly by tuning the nanoparticles diameter. Although it is a random binary assembly of nanoparticles, the authors demonstrate there still are cocontinuous structures for each type of nanoparticle. In further work by Venkataraman and coworkers, the authors show that by changing the number ratio of two types of particles, the local connectivity can be systematically tuned and thus the connectivity within these "random" mesoscale structures can be tuned.[31] This method for preparing organic photovoltaics allows for the control of assembly from the molecular to the mesoscale. With this degree of control, it presents the opportunity to systematically elucidate the optimal morphology of organic photovoltaics.

Controlling the morphology of the active layer in organic photovoltaics has a direct impact on the device's efficiency. The previous example uses a random binary assembly of nanoparticles to systematically tune the active layer morphology. Using a random assembly has the added advantage of easily being scaled up. Although if one can create a binary nanoparticle superlattice, as described earlier, with electron-donating and electron-accepting nanoparticles, this would allow for more precise control of the morphology and the development of highly efficient devices.

REFERENCES

1. Hoeben, F. J. M.; Jonkheijm, P.; Meijer, E. W.; Schenning, A. About supramolecular assemblies of pi-conjugated systems. *Chem. Rev.* 2005, *105*, 1491.
2. Prins, L. J.; Reinhoudt, D. N.; Timmerman, P. Noncovalent synthesis using hydrogen bonding. *Angew. Chem. Int. Ed.* 2001, *40*, 2382.
3. Whitesides, G. M.; Grzybowski, B. Self-assembly at all scales. *Science* 2002, *295*, 2418.
4. Boncheva, M.; Whitesides, G. M. Making things by self-assembly. *MRS Bull.* 2005, *30*, 736.
5. Stupp, S. I.; Palmer, L. C. Supramolecular chemistry and self-assembly in organic materials design. *Chem. Mat.* 2014, *26*, 507.
6. Klug, A. The tobacco mosaic virus particle: Structure and assembly. *Philos. Trans. R. Soc. Lond. Ser. B Biol. Sci.* 1999, *354*, 531.
7. Rossmann, M. G.; Johnson, J. E. Icosahedral RNA virus structure. *Annu. Rev. Biochem.* 1989, *58*, 533.
8. Chandrasekar, V.; Johnson, J. E. The structure of tobacco ringspot virus: A link in the evolution of icosahedral capsids in the picornavirus superfamily. *Structure* 1998, *6*, 157.
9. Pauling, L. The principles determining the structure of complex ionic crystals. *J. Am. Chem. Soc.* 1929, *51*, 1010.
10. Shevchenko, E. V.; Talapin, D. V.; Kotov, N. A.; O'Brien, S.; Murray, C. B. Structural diversity in binary nanoparticle superlattices. *Nature* 2006, *439*, 55.
11. Murray, C. B.; Kagan, C. R.; Bawendi, M. G. Self-organization of CdSe nanocrystallites into 3-dimensional quantum-dot superlattices. *Science* 1995, *270*, 1335.
12. Talapin, D. V.; Shevchenko, E. V.; Bodnarchuk, M. I.; Ye, X. C.; Chen, J.; Murray, C. B. Quasicrystalline order in self-assembled binary nanoparticle superlattices. *Nature* 2009, *461*, 964.
13. Evers, W. H.; De Nijs, B.; Filion, L.; Castillo, S.; Dijkstra, M.; Vanmaekelbergh, D. Entropy-driven formation of binary semiconductor-nanocrystal superlattices. *Nano Lett.* 2010, *10*, 4235.
14. Bartlett, P.; Ottewill, R. H.; Pusey, P. N. Freezing of binary-mixtures of colloidal hard-spheres. *J. Chem. Phys.* 1990, *93*, 1299.
15. Bartlett, P.; Ottewill, R. H.; Pusey, P. N. Superlattice formation in binary-mixtures of hard-sphere colloids. *Phys. Rev. Lett.* 1992, *68*, 3801.
16. Eldridge, M. D.; Madden, P. A.; Frenkel, D. The stability of the AB_{13} crystal in a binary hard-sphere system. *Mol. Phys.* 1993, *79*, 105.
17. Eldridge, M. D.; Madden, P. A.; Frenkel, D. A computer-simulation investigation into the stability of the AB_2 superlattice in a binary hard-sphere system. *Mol. Phys.* 1993, *80*, 987.
18. Eldridge, M. D.; Madden, P. A.; Frenkel, D. Entropy-driven formation of a superlattice in a hard-sphere binary mixture. *Nature* 1993, *365*, 35.
19. Cottin, X.; Monson, P. A. Substitutionally ordered solid-solutions of hard-spheres. *J. Chem. Phys.* 1995, *102*, 3354.

20. Hunt, N.; Jardine, R.; Bartlett, P. Superlattice formation in mixtures of hard-sphere colloids. *Phys. Rev. E* 2000, *62*, 900.
21. Chen, Z.; O'Brien, S. Structure direction of II–VI semiconductor quantum dot binary nanoparticle superlattices by tuning radius ratio. *ACS Nano* 2008, *2*, 1219.
22. Talapin, D. V. LEGO materials. *ACS Nano* 2008, *2*, 1097.
23. Gehan, T. S.; Bag, M.; Renna, L. A.; Shen, X. B.; Algaier, D. D.; Lahti, P. M.; Russell, T. P.; Venkataraman, D. Multiscale active layer morphologies for organic photovoltaics through self-assembly of nanospheres. *Nano Lett.* 2014, *14*, 5238.
24. Bag, M.; Gehan, T. S.; Algaier, D. D.; Liu, F.; Nagarjuna, G.; Lahti, P. M.; Russell, T. P.; Venkataraman, D. Efficient charge transport in assemblies of surfactant-stabilized semiconducting nanoparticles. *Adv. Mater.* 2013, *25*, 6411.
25. Couto, R.; Chambon, S.; Aymonier, C.; Mignard, E.; Pavageau, B.; Erriguible, A.; Marre, S. Microfluidic supercritical antisolvent continuous processing and direct spray-coating of poly-(3-hexylthiophene) nanoparticles for OFET devices. *Chem. Commun.* 2015, *51*, 1008.
26. Zhou, X. J.; Belcher, W.; Dastoor, P. Solar paint: From synthesis to printing. *Polymers* 2014, *6*, 2832.
27. Frenkel, D. Order through disorder: Entropy strikes back. *Phys. World* 1993, *6*, 24.
28. Frenkel, D. Order through entropy. *Nat. Mater.* 2015, *14*, 9.
29. Damasceno, P. F.; Engel, M.; Glotzer, S. C. Predictive self-assembly of polyhedra into complex structures. *Science* 2012, *337*, 453.
30. Nagarjuna, G.; Baghgar, M.; Labastide, J. A.; Algaier, D. D.; Barnes, M. D.; Venkataraman, D. Tuning aggregation of poly(3-hexylthiophene) within nanoparticles. *ACS Nano* 2012, *6*, 10750.
31. Renna, L. A.; Bag, M.; Gehan, T. S.; Han, X.; Lahti, P. M.; Maroudas, D.; Venkataraman, D. Tunable percolation in semiconducting binary polymer nanoparticle glasses. *J. Phys. Chem. B* 2016, *120*, 2544.

5 Assemblies of Organic and Inorganic Molecular Systems on Solid Substrates

K. Chakrapani and S. Sampath

CONTENTS

This chapter focuses on the assembly of nanoparticles and certain organic molecules on various substrates. The preparation of spontaneous assembly of materials and their characteristics are reviewed with special reference to recent literature. Subsequently, gradients based on self-assembly like electrical properties and biosensors are briefly reviewed.

5.1 PREAMBLE

"Self-assembly" refers to the spontaneous association by which molecules, nanoparticles, or other discrete components assemble. It is one of the most important concepts developed in the last century, and the process itself is a mimic of natural phenomena. The organization could be due to direct specific interactions of the components and/or indirectly through their environment. The process is generally associated with thermodynamic equilibrium and the organized structures characterized by a minimum in the free energy of the total system. The essential requirement in the self-assembly process is that the primary building blocks organize into ordered, macroscopic structures, either through direct or indirect interactions by means of a template or an external stimulus. The technological applications based on organized structures require efficient scale-up and a high level of order, direction, and control. Various scaffolds such as solid surfaces of metals, semiconductors, and oxides with different topographies can be used for the construction of such supramolecular systems. Several review articles are available in the area of self-assembly, and this chapter focuses on assemblies and properties of organic and inorganic molecular systems on solid substrates, particularly with reference to recent studies. The section on the formation of assemblies of organic and inorganic molecules on different substrates is followed by few applications of self-assembled monolayers (SAMs) in the field of surface gradients, thin film transistors, and biosensors. Details on recent advances in nanoparticle ordering by self-assembly are given as well.

The two most common families of SAMs on solid substrates that have been well studied are alkylsilanes on oxides and sulfur-containing molecules on coinage metals such as gold (Figure 5.1).[1–4] Assembly of thiols has been very extensively studied due to their ease of preparation, spontaneous formation of densely packed monolayers, and amenability for functionalization.[3,4] The organosilane monolayers on oxides, particularly SiO_2, can easily be integrated into silicon technology and the covalent nature of the bonding of molecules with the substrate results in high stability.[2] The mechanism of the formation of SAMs has been extensively described elsewhere.[1,5] The terminal functional groups of the monolayers allow the flexibility of fine-tuning the interfacial surface properties in terms of, for example, chemical reactivity, conductivity, wettability, and (bio) compatibility.[6] The functionalization may be achieved through (1) the adsorption of prefunctionalized molecules or (2) the modification of the base monolayer with the functional group of interest. Complete synthesis is required in the first route while stepwise assembly is involved in the latter method. The advantages of the second method include the following among others: (1) it enables the incorporation of groups that are not compatible with the synthesis of the molecular system, (2) stepwise reaction does not affect the order of the underlying monolayer, and (3) the stepwise synthetic procedures are relatively simple.[1,5] The flip side of the process is that the surface reactions are never complete and further,

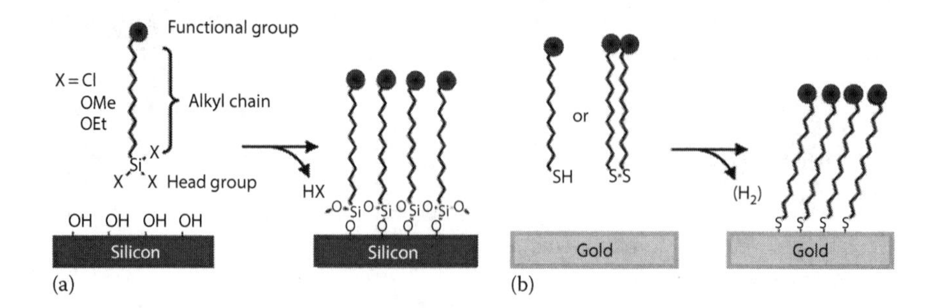

FIGURE 5.1 Formation of (a) functional silane on SiO_2 and (b) thiol monolayer on gold. (Adapted from Nicosia, C. and Huskins, J., *Mater. Horiz.*, 1, 32, 2014.)

purification of the functionalized systems is very difficult, and hence clean reactions such as click chemistry are essential.[2,4,7–9] Some of the recent literature includes that of Haensch and coworkers where chemical modification of silane-based monolayers involving nucleophilic substitution and Huisgen 1,3-dipolar cycloaddition of organic azides and acetylenes are described.[2] In another paper, Sullivan and Huck have illustrated nucleophilic substitution, esterification, acylation, and nucleophilic addition reactions on substrates, such as thiols on Au and silanes on SiO_2, to obtain functionalized systems with terminal amines, hydroxyls, carboxylic acids, aldehydes, and halogens.[7] Several other reactions based on nucleophilic substitutions, esterification, amidation, etc., are available to modify the surface chemistry of SAMs.[4,9] Reactive interfaces toward biochips have been reviewed by Jonkheijm and coworkers.[8]

5.2 ASSEMBLIES OF INORGANIC SYSTEMS

Inorganic systems basically involve integration of various nanocrystals and the strategies that have been reported for assemblies of colloidal nanocrystals on substrate, at an interface, and in solution are schematically shown in Figure 5.2. The thermodynamics that drive the self-assembly process may have to be modulated, either by

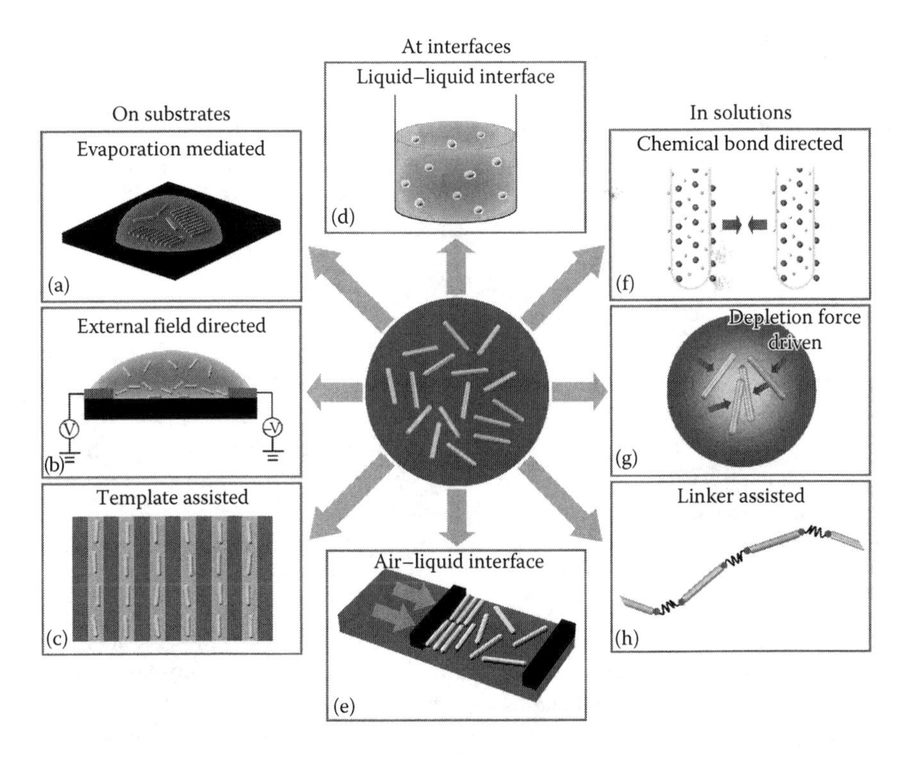

FIGURE 5.2 Schematics of self-assembly techniques for random colloidal inorganic nanocrystals to ordered structures (i) (a through c) on substrates; (ii) (d, e) at interfaces and (iii) (f through h) in solutions. (Adapted from Zhang, S.Y. et al., *Chem. Soc. Rev.*, 43, 2301, 2014.)

rational use of chemistry or templates or directed by external fields.[10,11] Techniques based on surface forces such as solvent evaporation–mediated methods have been shown to provide straightforward approaches to assemble highly ordered large-area nanoparticle structures on solid surfaces.[12–14] The rate of solvent evaporation is a critical parameter in these systems. Evaporation-mediated or drying-assisted assembly has been very widely used toward organizing spherical as well as anisotropic nanocrystals on solid substrates.[10,11,15,16] Electrostatic repulsive forces, hydrophobic interactions, capillary forces, and entropic considerations contribute substantially by mediating the assembly formation.[17,18] Single nanoparticle arrays and complex lattices of inorganic systems such as semiconductors, metals, and magnetic nanoparticles are shown to be formed by convective self-assembly, driven by solvent evaporation and assisted by electrostatic interactions, van der Waals forces, and dipolar interactions. It has further been proposed that particles are assembled either by a convective flow at the boundary of the nanoarray or by the attractive forces between particles due to surface tension at the interface.

Bentzon and coworkers in late 1980s observed the formation of ordered self-assembled nanostructures using transmission electron microscopy (TEM). This involves the dispersion of $Fe_{1-x}C_x$ particles in petroleum ether deposited onto a carbon-coated copper TEM grid and subsequently dried in air.[19] The $Fe_{1-x}C_x$ particles are found to be oxidized to iron oxide under ambient conditions and subsequently form three-dimensional close-packed structures of monodisperse spherical particles on amorphous carbon film upon drying. Murray and coworkers have demonstrated self-organization of ligand-stabilized CdSe quantum dots into two- and three-dimensional superlattices through nucleation at the liquid–gas interface, followed by controlled evaporation of the solvent.[20] Long-range ordering of quantum dots has been shown to possess different orientations of CdSe quantum dot superlattices.[21] Similar behavior has been reported with thiol-stabilized gold and silver nanoparticles and several other research groups followed this process to organize inorganic particles. Particle monodispersity is a crucial factor in the ordering process and an unusual and interesting assembly is reported for particles with bimodal distribution.[22,23]

The formation of ordered structures is quite influenced by several factors, including temperature, concentration, size of the primary particle, and nature of the solvent. Good and proper control of the critical parameters is necessary for a successful self-assembly.[24,25] Ryan and coworkers have investigated the effect of nanocrystal concentration and the nature of the solvent on the self-assembly of colloidal nanorods of cadmium chalcogenide semiconductors (CdS and CdSe).[14] When the concentration of nanocrystals is low, random deposition of the nanostructures on the substrate is observed and the assemblies are found to be inhomogeneous and not well ordered. The particles are far apart from each other that the attractive forces are possibly weak to trigger a good self-assembly process. When the concentration is high, only short-range ordering is observed suggesting that the crystals are too close that the repulsive forces become significant. Thus, for a highly ordered assembly, one should find an optimum concentration window such that the interparticle distances are small enough that the attractive forces between the nanoparticles overcome the repulsive forces.

5.2.1 External Field-Directed Assembly

External stimuli such as electric or magnetic fields, shear, or light offer a combination of speed and precision, as well as the ability to manipulate the nanoparticle assemblies.[26,27] The use of an external field to direct the alignment of nanocrystals is especially attractive where orientational ordering along with positional ordering of individual building blocks is required. Magnetic fields are particularly useful in the assembly of metal, metal oxide, and composite nanoparticles that are magnetic in nature. Ferromagnetic nanoparticles with sufficiently stable magnetic moments undergo spontaneous assembly, owing to dipole–dipole nanoparticle association, and the application of magnetic field particularly improves the organization. Superparamagnetic nanoparticles that possess randomly changing magnetic moment can be organized when torque exerted by the applied magnetic field exceeds nanoparticle thermal excitation energy.[28,29] Superlattices (one-dimensional as well as three-dimensional) are found to be formed when closely spaced magnetic nanoparticles are subjected to external magnetic field.

Electric field can be very useful in ordering nanoparticulate systems since the capping agents are generally terminated with some charges. Formation of electric field–assisted assembly[30] of CdSe nanorods on silicon substrate has been reported by Alivisatos and coworkers. The colloidal dispersion is deposited on the substrate and the solvent is slowly evaporated over a long time period (usually several hours) during which a DC voltage is applied. Binary cadmium chalcogenides possess noncentrosymmetric wurtzite lattice that results in permanent electric dipole moment. When placed under the influence of an external electric field, the nanorods experience a torque that rotates them and subsequently align their long axis in the field direction.[31] The rods can be aligned over large areas by following the field lines through the electrode gap. Alternating current (AC) electric field can also be employed to direct the formation of nanocrystal assemblies. The advantage of using AC signals is that it avoids the interference of electro-osmotic and electrochemical effects.[16] Mallouk and coworkers demonstrated the assembly of Au nanowires dispersed in a dielectric medium by the application of an AC field on coplanar interdigitated electrodes.[32] The metallic nanowires get easily polarized in the electric field due to charge separation at the surface and hence the alignment process can be moderated by the magnitude and frequency of the AC signal. The assemblies align in a short timescale when the magnitude and frequency of the AC voltage are increased.

Dielectrophoresis occurs due to the force exerted on the induced dipole moment of nanoparticles by an AC field.[33] The induced dipoles interact with the gradient of the nonuniform electric field and also with each other. Hence, dielectrophoresis can be used for the assembly of different types of nanoparticles in a variety of solvent environments without any undesired effect caused by electro-osmosis and electrolysis.[33] Velev and coworkers have shown the formation of chains of gold nanoparticles assembled from an aqueous medium under an AC field of 95–96 V cm^{-1}.[34] At high voltages, the chains grow on the electrode in the regions of high field intensity. Owing to strong attractive forces, the assembled structures remain stable even after the electric field is removed.

5.2.2 Template-Assisted Assembly

Another method to facilitate self-assembly of inorganic nanocrystals on substrates is to utilize templates.[35,36] Most templates are solids that are surface modified, possessing geometrically constrained sites and/or chemically functionalized regions that can nucleate selective deposition of nanocrystals, leading to the formation of assemblies. In creating templates, the solid substrate surface is often patterned with precise spatial and chemical control using techniques such as optical and electron beam lithography and/or microcontact printing. Hard templates such as chemically functionalized carbon nanotubes or inorganic nanowires offer well-defined shapes for the assembly but, in general, lack of control of interparticle spacing.[37,38] Liu Marzin and coworkers have reported the formation of chains of anionic poly(vinylpyrrolidone)-functionalized gold nanorods deposited on the surface of cationic poly(diallyldimethyl ammonium chloride)-coated nanotubes. The low surface potential at the end of the nanorods, as compared to the sides, favors end-to-end organization on the surface of coated carbon nanotubes.[39] Soft templates such as synthetic polymers, proteins, DNA molecules, or viruses possess distinct chemical features and provide multiple and well-defined binding sites for the attachment of nanoparticles.[40–42] Soft biological templates also allow nanoparticles to be organized in a hierarchical manner by exploiting strategies found in natural systems. In particular, DNA-based assemblies hold great promise, owing to its structural diversity, well-defined sequences, and a variety of functionalities. DNA scaffolds have been used to organize Au, Ag, CdSe, and CdSe/ZnS nanoparticles, CdS nanowires, and metal nanoparticles.[10,43]

5.2.3 Self-Assembly at Interfaces

Interfacial assembly of particles has been investigated for several decades. It is commonly observed in emulsions where solid particles spontaneously adsorb at the interface between two immiscible liquids to ensure that the emulsion droplets do not coalesce.[11,13,44] This phenomenon is exploited as a technique to spontaneously self-organize colloidal microscopic particles and nanocrystals. The high interfacial energy available when two immiscible fluids are in contact is reduced when particles adsorb at the interface. The driving force for the assembly is the decrease in interfacial energy, which is further supported by interfacial deformation-induced capillary forces holding the particles together. For relatively large micron-sized particles, the interfacial energy decrease is several orders of magnitude larger than the thermal energy (kBT) that causes fluctuations, leading to irreversible adsorption at the interface. However, for small nanometer-sized particles, the decrease in interfacial energy is comparable to the thermal excitation energy. Size-dependent particle exchange can occur at the interface where large nanoparticles displace the small ones at a rate that is consistent with the adsorption energies.[45,46] In addition to particle size and shape, wettability and interparticle interactions are important factors that determine the stability of the interfacial particle assembly.

The Langmuir–Blodgett (LB) technique is one of the well-studied methods to manipulate molecular ordering at the air–water interface and subsequently on the solid surface. The method involves the transfer of the monolayer present at the air–water (gas–liquid) interface to the substrate of choice. The LB apparatus includes a Langmuir trough with a dipping mechanism to lower or raise the substrate through the gas–liquid interface, an automated movable barrier that moves during the deposition process in order to maintain a controlled feedback surface pressure, and a surface pressure sensor. Traditional LB films of amphiphiles show poor mechanical stability. However, there has been progress reported on robust monolayer fabrication of ligand-stabilized gold nanoclusters, semiconducting quantum dots, and polymeric films.[47,48] Impressive long-range, hexagonal close-packed ordering of alkanethiol stabilized gold clusters that are devoid of defects, vacancies, or dislocations has been demonstrated.[49] It is reported that at low concentrations of the clusters, the level of order increases as observed from the fast Fourier transformation of images. By adopting a novel method of using bifunctional linkers as encapsulating ligands on the surface of gold nanoparticles, further enhancement in the mechanical stability of the monolayer film is shown.[50] However, the resulting particle assemblies lack long-term stability due to relatively weak interactions between neighboring particles as well as between particles and substrates. Fabrication of stand-alone, robust nanoparticle thin films are readily constructed by confining the particles and the cross-linking reagents to the water subphase, for instance, the cross-linking of aminoethanethiol-modified nanoparticles by glutaraldehyde in the subphase yield very ordered films.[51] Russell and coworkers have extended the assembly at the gas–liquid interface to a solid substrate by transferring the monolayer at constant surface pressure.[52] A schematic illustration of a typical assembly and the transfer experiment using an LB trough is shown in Figure 5.3. A dispersion of hydrophobically terminated Au nanorods in

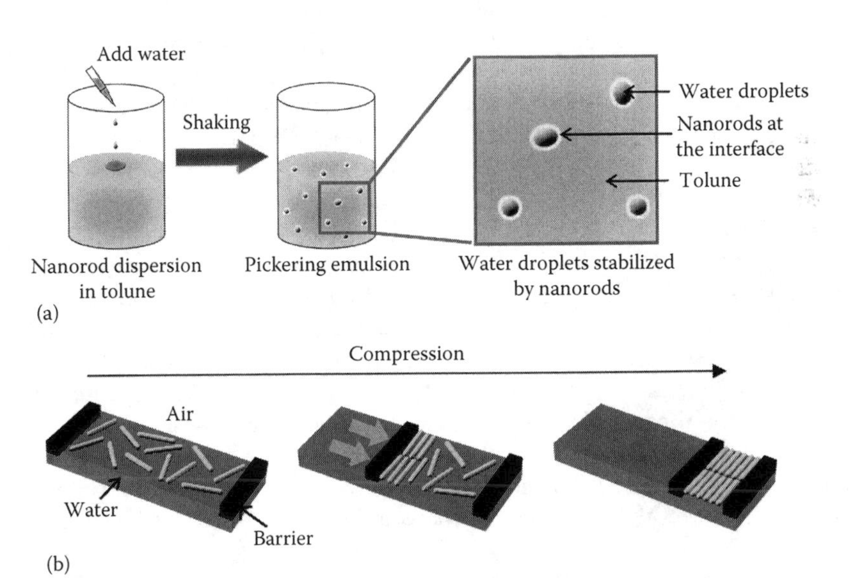

FIGURE 5.3 Assembly of colloidal nanocrystals (a) at liquid–liquid and (b) at gas–liquid interfaces. (a) Pickering emulsion droplets are stabilized through adsorption of nanorods at toluene–water interface. (b) Assembly of nanorods at air–water interface using the Langmuir–Blodgett (LB) technique. Compression using a set of barriers results in assembly of nanorods that are aligned parallel to the barriers. (Adapted from Zhang, S.Y. et al., *Chem. Soc. Rev.*, 43, 2301, 2014.)

an organic solvent is allowed to spread uniformly on the water surface of the LB trough, creating a monolayer of randomly oriented nanorods at the air–water interface. The hydrophobic coating is necessary to prevent aggregation and to make sure that the nanorods float on the water surface. Subsequently, the nanorods at the interface are compressed slowly, and the surface pressure monitored. The compression generates high-density assemblies of nanorods that are aligned parallel to the trough barriers. The structure of the final assembly is tuned by controlling the compression parameters. The assemblies are collected onto desired substrates like TEM grid, silicon wafer, etc., by using either vertical or horizontal lift-off.[53,54] A major advantage of the LB technique is its ability to produce 1D assemblies over large areas with ultrahigh packing density. It should be quickly pointed out that advantages of LB technique in yielding consistently near-perfect long-range ordering of monolayers and multilayers of species that self-assemble at the liquid surface are outweighed by the fact that the technique is cumbersome and time-consuming, and the required apparatus requires good maintenance.[55]

5.2.4 Electrostatic Self-Assembly

Self-assembly can also be directed through electrostatic interactions between nanoparticles or molecules.[10,11] This is different from the "field directed assembly" explained in an earlier section (5.2.1). Here, the opposite charges on the surfaces are allowed to interact spontaneously. Nanocomposite thin films composed of metal nanoparticles with amine-functionalized gold and the carboxylic acid-derivatized silver colloidal particles are good examples of deposition using the layer-by-layer technique where electrostatic interactions are used.[56] The charge on gold and silver colloidal particles control the self-assembly. Ionization of the terminal functional groups at suitable solution pH is essential and hence pKa/pKb values of the ionizable groups play a crucial role in the assembly. Liu and coworkers have demonstrated the layer-by-layer electrostatic self-assembly of nanoparticles with charged polymer containing Fe_3O_4 particles and poly (diallydimethyl ammonium chloride).[57] Mediation of the self-assembly process by hydrophilic and hydrophobic interactions, van der Waals forces, temperature control, and capillary forces is well studied in the literature.[17] Chen and Yu have demonstrated a low temperature, cost-effective, and simple method of constructing patterned SAM by exploiting hydrophobic and/or hydrophilic interactions using surfactant molecules coated on CdSe quantum dots. The wettability of nanocrystals is determined by the terminal functionality of capping molecules that are tethered to the nanocrystal surface; hydrophobic surfactant capped nanocrystals are obtained by derivatizing CdSe dots with tri-n-octylphosphine oxide, while hydrophilic counterparts are derived using 4-mercaptobenzoic acid.[58] The pattern on a gold substrate defines distinct hydrophilic and hydrophobic regions accomplished via microcontact printing of SAMs with different thiols.

5.3 SELF-ASSEMBLY OF POLYCYCLIC SYSTEMS

A majority of self-assembly experiments have been performed on metal substrates. However, due to their unique properties, the use of semiconductors and insulators as templates for molecular growth has recently been reported, although the presence of surface defects or dangling bonds makes the processes quite complicated. In the following section, the focus is on the adsorption and self-assembly of relatively large organic polycyclic molecules on metal oxide surfaces. Polycyclic systems show variation in electrical properties and hence their assemblies are useful in fabrication of electronic devices.

5.3.1 Adsorption of Planar Hydrocarbons on Oxide Surfaces

Among the various planar hydrocarbons, pentacene molecules are treated as model polyaromatic species, and their behavior has been extensively investigated on various substrates that exhibit metallic, semiconducting, and insulating characteristics.[59] High carrier mobility and high efficiency in photovoltaics make the assembly of pentacene molecules particularly useful in organic field-effect transistors.[60] Figure 5.4a and b shows the pentacene molecule on TiO_2 surface. Lanzilotto and coworkers have investigated the self-assembly of pentacene on TiO_2 surface using near-edge X-ray absorption fine structure (NEXAFS) and scanning tunneling microscopy (STM) measurements.[61] The molecules are observed to be physisorbed on the surface along the surface oxygen atoms. Increasing the coverage leads to the molecules forming molecular stripes that are perpendicular to surface oxygen. This is similar to the observations

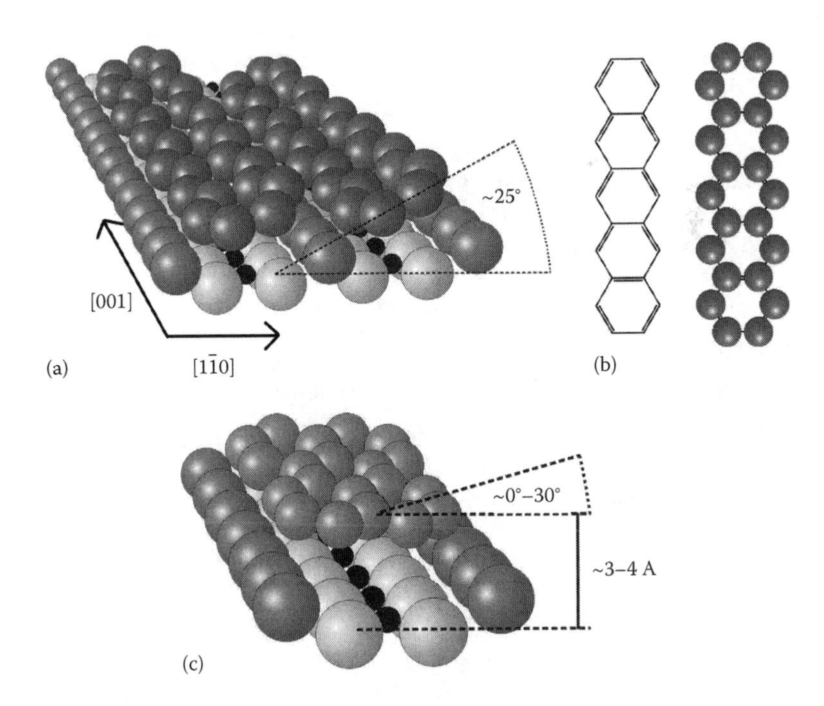

FIGURE 5.4 View of adsorbed (a) pentacene and (c) polycyclic hydrocarbon molecules on TiO_2 surface. (b) Structural model of pentacene molecule. (Adapted from Godlewski, S. and Szymonski, M., *Int. J. Mol. Sci.*, 14, 2946, 2013.)

on 3,4,9,10-perylene tetracarboxylic dianhydride (PTCDA) molecule. The arrangement of surface bridging oxygen along with repulsive forces among the adsorbates seems to influence the periodicity of the molecules on the surafce.[62] The weak binding with the substrate combined with relatively strong intermolecular interactions and tilting of neighboring molecules makes the structure of the film similar to pentacene layers in the bulk. This results in the development of bulk-like electronic properties of the molecular layer. Potapenko and coworkers have investigated the adsorption of anthracene using STM and temperature programmed desorption (TPD) techniques.[63] The multilayers of the molecule desorb around 270 K while desorption of the first layer occurs at approximately 360 K, and hence only a monolayer is feasible under ambient conditions of 25°C. The molecules are observed to follow flat-lying geometry with the long axis parallel to the rows of bridging oxygen atoms. The molecular lines run perpendicular to the surface rows of bridging oxygen atoms, but exhibit a wavy nature similar to the case of pentacene. Reiss and coworkers studied the adsorption characteristics of aromatic benzene derivatives using NEXAFS that reveal the plane of anthracene molecule tilted by about 28° with respect to the surface normal while for naphthalene molecules, the angle reaches close to 24°. The observed TPD spectra for benzene, naphthalene, and anthracene indicate that these molecules only weakly interact with the solid surface and the adsorption energy increases almost linearly as the number of carbon atoms increases.[64] These observations confirm that the molecules are all adsorbed in similar geometries. Simonson and subsequently Schuster have investigated the adsorption and self-assembly of perylene and perylene derivatives using x-ray photoelectron spectroscopy (XPS), ultraviolet photoelectron spectroscopy (UPS), and near edge x-ray absorption fine structure (NEXAFS) techniques and observed that the molecules weakly interact with the substrate and that no chemical bonds are formed.[65,66] According to these studies, the first layer does not exhibit any ordered structure. In summary, pure hydrocarbon polycyclic molecules only weakly interact with the oxide surface, primarily through van der Waals and electrostatic forces and consequently, the molecules are mobile on the surface and preferentially adsorb in a flat geometry with the plane of the molecule slightly tilted from the surface.

5.3.2 Self-Assembly of Polyaromatic Derivatives with Functional Groups and Phthalocyanines

Tekiel and coworkers have investigated the assembly of perylene derivative PTCDA on TiO_2 surfaces, as shown in Figure 5.5.[67] The PTCDA molecule has been examined on a variety of metallic, semiconducting, and insulating substrates.[68] It is observed that the molecules possess very intriguing behavior and get adsorbed in completely different configurations depending on the deposition conditions. When the molecules are deposited at temperatures below 75°C, the orientation is flat with the long molecular axis parallel to the surface rows forming small clusters, which are elongated and parallel to the direction. Deposition at high temperatures of 75°C leads to the molecules immobilized as single entities in a completely different geometry.[69]

Phthalocyanines are intensely colored macrocyclic organic compounds that are widely used in applications in areas like sensors, molecular electronics, dyeing, etc. The base phthalocyanine is a metal-free molecule (H_2Pc) that exhibits an intense

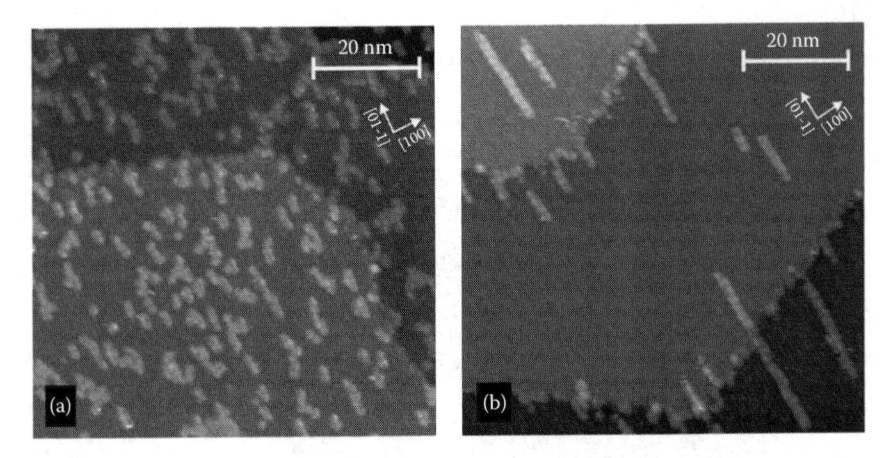

FIGURE 5.5 Micrograph of monolayer of 3,4,9,10-perylene tetracarboxylic dianhydride (PTCDA) on TiO_2 surface at (a) ambient room temperature and at (b) 100°C. (Adapted from Tekiel, A. et al., *Nanotechnology*, 19, 495304, 2008.)

blue-green color and forms complexes with many metals that substitute hydrogen in the central ring.[70] Substituted Pcs with metals are employed as organic dyes in dye-sensitized solar cells, in catalysis in the area of oxygen reduction reaction, and as donors in optoelectronic devices such as field effect transistors.[70] It has been reported that phthalocyanines strongly interact with the surface on which they are adsorbed and hence their electronic structure is significantly affected. This might be due to a charge transfer from the molecule to the surface or vice versa. Palmgren and coworkers have studied the adsorption and interface properties of metal-free phthalocyanines on TiO_2 surface using STM and high-resolution photoelectron spectroscopy. The molecules are found to lie flat on the surface and do not form any ordered structures up to 500 K (Figure 5.6). Further, they adsorb as single entities with the molecular arms pointing at an angle of 45° with respect to the surface normal.[71] Photoelectron spectra demonstrate that the molecules within the first layer bind strongly to the surface via chemisorption with charge transfer from the molecule to the substrate. The interactions are assumed to be between delocalized electrons from π orbitals of the phthalocyanines and the surface oxygen atoms. A striking difference in the interaction of the first layer and the second layer leads to bulk-like properties. The same authors have also investigated the assembly of iron phthalocyanines on TiO_2 surface using XPS and STM techniques and the behavior was found to be very similar to metal-free phthalocyanines.[72] The strong interaction leads to a significant alteration of the electronic structure of the molecule leading to the quenching of the highest occupied molecular orbital (HOMO) – lowest unoccupied molecular orbital (LUMO) shake-up transitions as observed from the carbon and nitrogen core-level spectra. The second layer grows in an unordered fashion in the form of an island, and the spectroscopic data reveal that the electronic properties of the phthalocyanines are retained due to weak interactions with the underlying molecules. Wang and coworkers and later Godlewski and coworkers investigated the assembly of CuPc on rutile

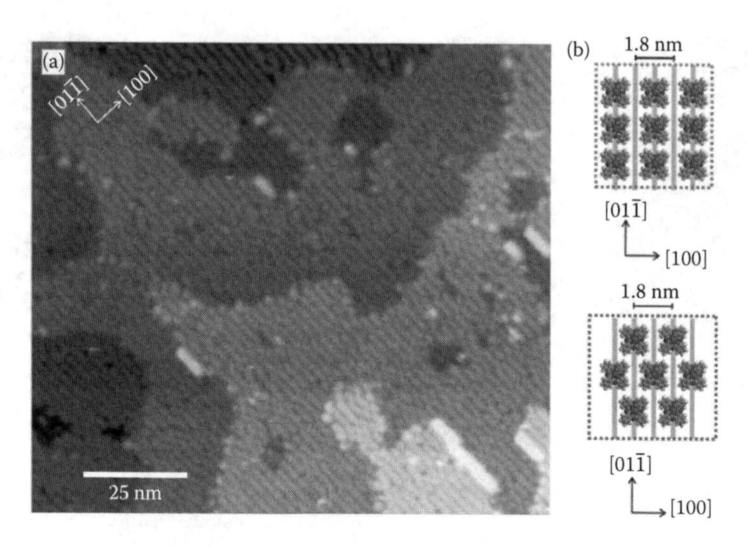

FIGURE 5.6 (a) STM image of CuPc molecules revealing quasi-ordered phase on TiO_2 surface. Bias voltage is +3.0 V, tunneling current used is 2 pA, black circle indicates single molecule within the layer; (b) structural models of two quasi-ordered phases, light gray lines indicate double zigzag oxygen rows. (Adapted from Godlewski, S. and Szymonski, M., *Int. J. Mol. Sci.*, 14, 2946, 2013.)

TiO_2 and found that the molecules strongly interact with the substrate.[73,74] However, the strength of the binding appears to depend on the crystallographic phase of titania involved in the adsorption process. The overall degree of ordering is rather poor. At low coverages, the single molecules adsorb on terraces, steps, and on defects.

Studies on cobalt-substituted phthalocyanines (CoPc) show that the adsorption characteristics strongly depend on the deposition process (Figure 5.7).[75,76] The evaporation at 25°C results in mobile species, and post-deposition annealing leads to the immobilization of molecules that are randomly distributed over the surface. The STS studies reveal that the molecules adsorbed on the surface introduce electronic states

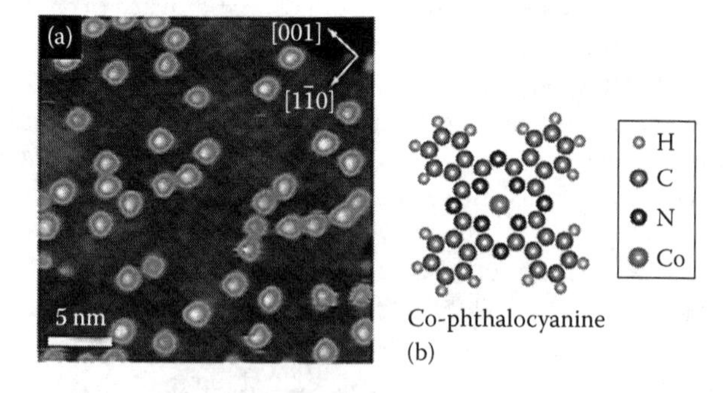

FIGURE 5.7 (a) High-resolution STM image of CoPc molecules on the TiO_2 surface (b) ball model of cobalt phthalocyanine molecule. (Adapted from Ishida, N. and Fujita, D., *J. Phys. Chem. C*, 116, 20300, 2012.)

within the band gap of TiO_2 and demonstrate that charge transfer has occurred between Pc molecules and the oxide surface.[77] The molecular orientation of silicon phthalocyanine thin films on three different substrates, highly oriented pyrolytic graphite (HOPG), indium thin oxide (ITO), and gold (Au) has been studied by NEXAFS using linearly polarized synchrotron radiation.[78] NEXAFS spectra show that the molecular planes of SiPc on the three substrates are found to be nearly parallel to the surface. The molecules are arranged almost parallel to the surface with a tilt angle of 2° on HOPG, while the tilt angle on the ITO surface is observed to be 26°. It is concluded that the morphology of the top surface layer of the substrate affects the molecular orientation.

5.4 GRADIENTS USING REACTIVE SAMs

Molecular layers are very amenable to form gradients of various properties in a single system. Click chemistry introduced by Sharpless and coworkers is quite useful in this direction since the reaction is quantitative, highly efficient, and selective.[79] Of course, one should make sure that the right functional groups are available. Various techniques based on the so-called "top-down" methods such as microcontact printing (mCP), scanning probe microscopy, UV, and e-beam lithography are employed to generate patterns of SAMs with sizes ranging from tens of nanometers to millimeters.[4,80,81] Microcontact printing, introduced by Whitesides and coworkers, offers a fast, flexible, simple, and inexpensive way to replicate patterns generated via photolithography.[82] In the conventional mCP method, a microstructured elastomeric poly-dimethylsiloxane stamp is employed to transfer molecules available in the "ink" (e.g., thiols) to the surface of the substrate by a conformal contact process. Patterns of alkanethiols on gold are conveniently used as etch-protecting layers for the fabrication of microstructures. Several potential applications in microelectronics and sensors are demonstrated.

Gradients on surfaces show gradual variation of at least one physicochemical property in space that may evolve in time (Figure 5.8). Achieving surface chemical gradients means gradual modulation of interfacial properties and can be exploited to generate smart materials and to investigate surface-driven transport phenomena.[83] This may include, for example, motion of water droplets on a wettability gradient, or the study of biological processes such as the directed migration and polarization of cells on

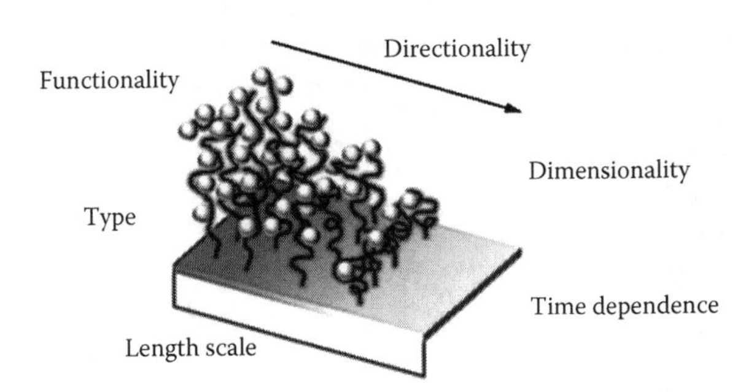

FIGURE 5.8 Surface chemical gradients—different attributes. (Adapted from Nicosia, C. and Huskins, J., *Mater. Horiz.*, 1, 32, 2014.)

biomolecular gradients. The surface gradients integrate a wide range of properties in a single sample, thus providing an invaluable tool for fast high-throughput analysis of several parameters. This avoids the problems associated with sample-to-sample variation, effect of distribution of properties in different samples, and a tedious analysis of multiple samples. The two most common methods employed for the development of monolayer-based surface chemical gradients are (1) controlled adsorption/desorption of SAMs on gold or silicon and (2) the chemical post-modification of reactive SAMs.[4]

Both thiols on gold and organosilanes on oxide surfaces produce stable, dense, monolayers and have been shown to be amenable for further modification using surface chemical reactions.[3,4] A common modification is via modular stepwise selective functionalization of preformed reactive SAMs. Given here are some of the recent examples in which reactive monolayers are employed in combination with click chemistry, either by reaction in solution or by lithography (e.g., mCP). For the present review, the most common and attractive click reactions based on the azide–alkyne cycloaddition, the thiol-ene reaction, Michael addition, the imine and oxime formation, and the Diels–Alder cycloaddition are considered.

One of the pioneering studies describing the fabrication of surface chemical gradient is that of Elwing and coworkers in 1987.[84] The gradient is the result of controlled silane diffusion in liquids. A hydrophilic silicon plate is placed in a cuvette filled with a biphasic solution of dimethyldichlorosilane in trichloroethylene covered with xylene. In this system, organosilane molecules diffuse to the xylene phase and deposit on the silicon substrate resulting in surface gradients. The obtained gradient is employed to study the wettability-driven adsorption and interaction of proteins and polymers at liquid–solid interfaces.[84] Since this study, a wide variety of methods and techniques are developed for the generation of surface gradients mainly based on the controlled adsorption/desorption of monolayers on substrates (Figure 5.9). Whitesides and coworkers prepared wettability surface gradients by vapor diffusion of decyltrichlorosilane on a silicon surface. This has become one of the most commonly used techniques to prepare silane-based gradients in the millimeter–centimeter scale.[85] These gradients are used to study the motion of water droplets based on the surface tension on liquid–solid interface. Certain dynamic methods that involve interfacial chemical reactions and interactions with the possibility of tailoring and controlling the functionalization of surfaces in space and time are given here.

A general strategy is based on the photochemical (de)protection of terminal photosensitive groups of SAMs.[86] Yousaf and coworkers have employed this methodology to pattern ligands and cells as gradients on inert surfaces. A nitroveratryloxycarbonyl (NVOC)-protected hydroquinone ethylene glycol-terminated thiol monolayer on gold surfaces undergoes photochemical (de)protection upon UV illumination, thus revealing the electrochemically active hydroquinone. The surface gradient of hydroquinone moieties is obtained as the irradiation is performed in the presence of a grayscale photomask. An aminooxy reactive quinone is obtained upon electrochemical oxidation of the hydroquinone, while the NVOC-protected units remain completely redox inactive. The obtained quinone gradient is reacted with soluble aminooxy-tagged ligands to form stable oxime conjugates via chemoselective ligation. A rhodamine–oxyamine is then used to visualize the surface gradient while an RGD–oxyamine peptide is immobilized to study cell migration and proliferation

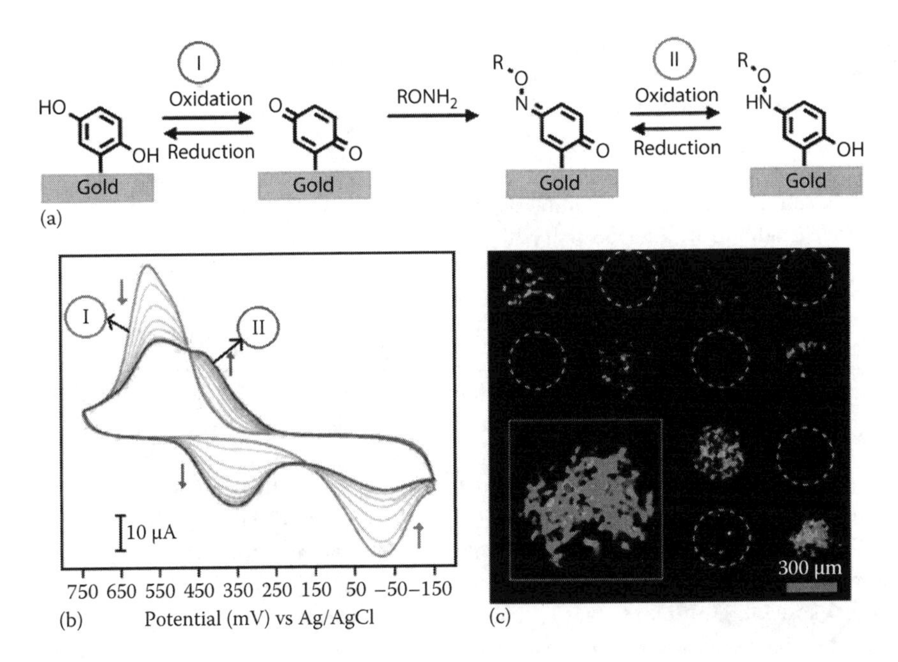

FIGURE 5.9 (a) Formation of redox-active hydroquinone/benzoquinone monolayer. (b) Cyclic voltammograms showing the extent of the interfacial reaction between soluble aminooxy acetic acid and a quinone monolayer. (c) Fluorescence microscopy image of cells attached to the surface. (Adapted from Nicosia, C. and Huskins, J., *Mater. Horiz.*, 1, 32, 2014.)

along the gradient. The dynamic character of the monolayers allows the electrochemical release of ligands by means of the reduction of the oxyamine bond and the restoration of the surface for further immobilization of the ligand. Ito and coworkers prepared a surface chemical gradient via photo degradation of octadecylsilane (ODS) monolayers on Si by vacuum UV (VUV) light at an excitation wavelength of 172 nm.[87] The VUV light is absorbed by the ODS layer forming radicals due to dissociation of C–C, C–H, and C–Si bonds and in turn react with oxygen and water to form surface-oxidized species. A surface gradient of oxidized groups (e.g., carboxy, aldehyde, and hydroxy) is thus obtained by moving the substrate, positioned on a sample holder, at a controlled speed of 50–100 mm and the resulting gradient is in micron-sized features. The formation of the gradient is followed by water contact angle measurements and fluorescence microscopy after labeling the carboxylate groups with fluoresceinamine. Further, the obtained surface gradient is used to investigate the motion of water microdroplets from the hydrophobic to the hydrophilic side. A similar approach described by Gallant and coworkers uses ODS (octadecylsilane) monolayer on Si that is gradually oxidized by exposing the modified substrate to the slit aperture of a UV lamp using a motorized stage.[88] The exposure time-dependent ozone-derived oxidation of the monolayer yields surface gradients of oxidized species based on various functional groups, alcohols, aldehydes, and carboxylic acids. A bifunctional propargyl-derivatized amino linker is attached to the acid gradient by using standard amidation procedures yielding a surface with varying coverages of alkyne groups. This is an essential platform for "click" modification to produce gradients. Electrochemically mediated reactions offer control over length scale, shape, and functionality on the surface. This is achieved by using electrochemically activated copper(I) azide–alkyne cycloaddition and atom transfer radical polymerization.[89] Surface gradients of covalently bound alkyne moieties are created on azide-functionalized conductive polymers. Hansen and coworkers have also fabricated surface gradients of fluorine-rich and bioactive alkyne-modified molecules using the electro-click process.[90] The amount of Cu(I) is tuned electrochemically, which leads to excellent spatial confinement of the active surface (Figure 5.10). The geometry or cell configuration is exploited in obtaining different sizes/shapes of the gradient on the conductive polymer (poly-3,4-(1-azidomethylethylene)-dioxythiophene (PEDOT-N3).[90] The distance between the counter electrode and the reactive surface dictates the speed of generation of the catalyst for the click process.

Huskens and coworkers described a method to investigate the reactivity of interfacial reactions in space and time.[91] Electrochemically assisted solution gradients of proton concentration (pH) and of a catalyst [Cu(I)] have been used to fabricate micron-scale surface chemical gradients and also to study the kinetics of surface-confined imine hydrolysis. The same group has also analyzed the directional spreading of multivalent ligands along the growing gradients on a receptor.[92] Fluorescence microscopy is used to understand surface diffusion of the ligands driven by the concentration gradient of free surface receptors. Dynamic control

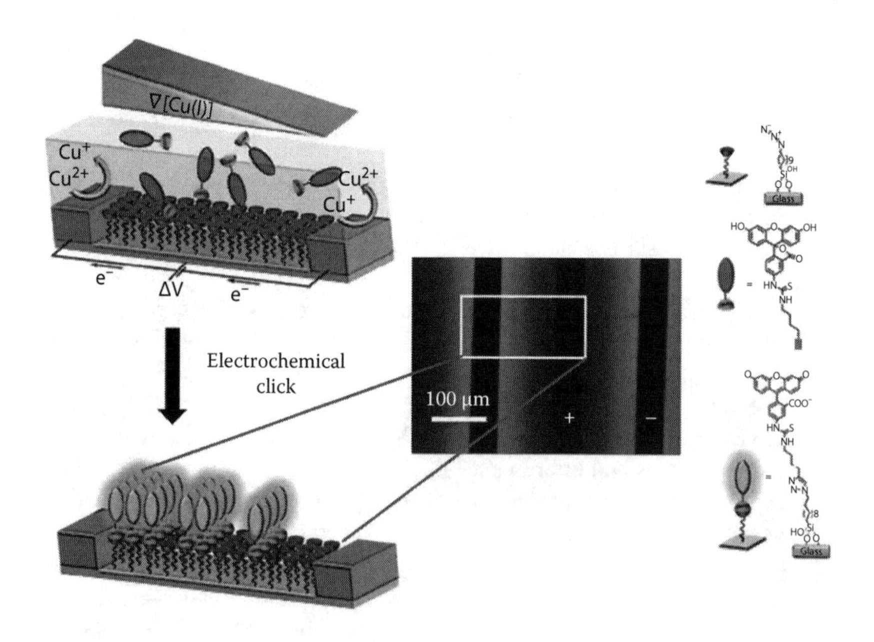

FIGURE 5.10 Surface chemical gradients via electrochemically promoted click reaction involving CuAAC of an alkyne-modified fluorescein on azide monolayer on glass. Platinum microarray electrodes are used. (Adapted from Krabbenborg, S.O. et al., *J. Nat. Commun.*, 4, 1667, 2013.)

using monolayers to fabricate gradients has been reported by Giuseppone and coworkers.[93] Aldehyde-terminated molecules on silicon in the presence of various amine-functionalized fluorophores with different pKa values [benzylamine (9.5) and alkylamine (10.5)] with simultaneous modulation of the pH and withdrawal of the sample at constant speed result in the selective functionalization of small molecules and proteins.

The conversion rate of reactions on the surface is a limiting factor, particularly by steric hindrance and slow diffusion. Development of strategies to control and switch the surface chemistry and properties by means of integration of dynamic monolayers has allowed the study of the behavior of biological systems at interfaces, for example, cell adhesion and migration.[4] In this direction, control of local surface composition of monolayers is particularly important for systematic investigation.

5.5 ELECTRICAL PROPERTIES OF SELF-ASSEMBLED SYSTEMS

The preparation and electrical properties of monolayers of different chain lengths have been studied by Mann and Kuhn.[94] The ordered monolayers via formation of chemical bonds between the precursor molecules and the substrate/conductive electrodes are the ones that rekindled the interest in the use of monolayers for organic electronics.[95,96] The use of organosilane precursors to form monolayers requires substrates with hydroxyl groups, including the technologically important surfaces of SiO_2, Al_2O_3, and tin-doped indium oxide (ITO).[97,98] The driving force for self-assembly is generally the *in situ* formation of siloxanes, via strong Si–O–Si bonds.[98] Since most the substrate surfaces are amorphous, the packing and ordering of the chemisorbed organosilanes are dictated by the underlying siloxane network, the extent of inter-chain interactions, and the temperature at which deposition is carried out. Vapor phase deposition of short-chain length molecules, in particular, has attracted wide attention.[99] Other classes of materials deposited on oxide surfaces include n-alkanoic acids and phosphonic acids on metal oxide surfaces such as Al_2O_3 and ITO that can compete with the robust SAMs of thiols on Au. In the case of organophosphonate SAMs, both vapor and solution deposition have been demonstrated.[100]

Conduction through monolayers derived from substituted aromatic or heteroaromatic alkanethiols has been investigated using Au–SAM–Au junctions (Figure 5.11).[101] The top Au electrode is deposited by cathodic deposition directly on top of the monolayer. This is very tricky since the deposition may alter the order and arrangement. In another related study, edge molecular junctions are used to investigate the electrical properties of conjugated 4,4′-biphenyldithiol (BPDT) SAMs and compared to aliphatic C9 dithiol alkyl SAMs.[102] It is concluded that the electrical conduction through the conjugated BPDT monolayer is via tunneling, despite the large currents and small HOMO–LUMO gap observed as compared to the C9 SAM. Conduction in SAMs depends on the nature of the chemical linker that binds the molecule to the bottom metal electrode and the overlap of orbitals with the energy levels of the metal surface.[103]

A correlation between molecular structure and electrical resistance (R) in the nonresonant tunneling regime is demonstrated by systematically varying the conjugation.[104] Mono- versus di-thiol substituted bridging molecules are studied to

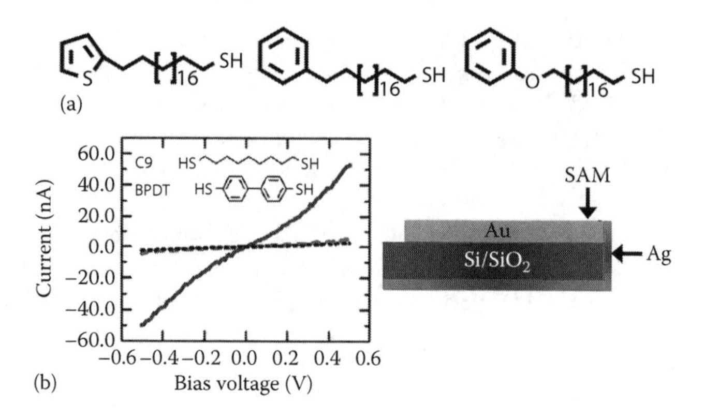

FIGURE 5.11 (a) Substituted aromatic or heteroaromatic alkanethiol as molecular architecture for Au–SAM–Au measurements and (b) illustration of an Au/SAM/Ag device and corresponding I–V curves for Au–C9–Ag and Au–BPDT–Ag device. (Adapted from Shamai, T. et al., *Appl. Phys. Lett.*, 91, 102108, 2007.)

exemplify this aspect. Interestingly, the contact resistance of alkane dithiols and their conjugated analogs are found to be similar and independent of the HOMO–LUMO gap. The difference in conductance values within a series of 1,4-butylene alkanes terminated with dimethyl phosphine, methyl sulfide, or amine is compared and nonresonant tunneling is assumed as the dominant conduction mechanism.[105] It has been found that phosphine termination provides the lowest contact resistance (highest conductance) in the series. As compared to thiol-based systems, the conduction mechanism of SAMs on oxide surfaces is less understood. It is shown to be independent of monolayer thickness indicating that tunneling is not the dominant process.[106] However, no alternate mechanism has been suggested. Waldeck and coworkers have performed photocurrent measurements on Si/SiO_2 deposited with different alkylsilanes and observed a weak dependence of conduction on monolayer thickness.[107] This is contrary to the expected and they have suggested hopping through traps in the film as a plausible mechanism to explain their observations. Cahen and coworkers measured conductance through alkanes in Hg/alkylsilane/SiO_2/Si–metal insulator semiconductor (MIS) devices and mentioned that the results, at first glance, would appear as if there is a clear dependence of current density on the chain length.[108] However, it is concluded that the curves are essentially identical within the experimental uncertainty. Subsequently, it is observed that the results are found to be in agreement with the results from earlier studies, where the conduction in alkylsilanes on native SiO_2 does not depend on chain length.

The notion of high capacitance SAMs and their use as gate dielectric in organic thin film transistors (OTFTs) seems to be attracting wide attention in the electronics community.[109] Vuillaume and coworkers pioneered this study where it is established that SAMs of n-ODTS grafted on Si native oxide are good insulators despite a very small thickness of ~2.8 nm. It is demonstrated that the interface state density of the OTS SAMs is reduced by one order of magnitude when the film is annealed at slightly elevated temperatures. The investigations based on the electrical properties of alkyl monolayers (versus varying chain lengths from 1.9 to 2.6 nm) reveal that the leakage currents are suppressed and low conductivities are observed for well-ordered SAMs whereas and large conductivities are observed in disordered monolayers.[110] One of the first attempts of utilizing SAMs of alkylsilanes with a –COOH end group as the gate insulator is in an OTFT device with p-type sexithiophene.[111] TFTs with aromatic terminated alkylsilane, 18-phenoxyoctadecyltrichlorosilane (PhO-OTS), and SAM gate dielectric are investigated by a different group, and drastically enhanced P5 TFT performance is observed. It is suggested that the OTS-OPh SAM is tightly packed due to intra-SAM interactions that lead to very low leakage currents and large breakdown fields.[112]

Thickness dependence of mobility for various semiconductors on SAMs has been studied and this opens up the possibility of using them as the active channels in thin film transistors.[113] The ODTS-based SAMs with different end groups such as methyl, thiol, thiophene, phenoxy, and biphenyl have been reported and of particular interest are the ones with good insulating properties. The insulating properties of alkyl chain along with lateral (in-plane) conductance of the doped biphenyl end group suggest that the SAMs could be used as TFTs. Further studies carried out on the dependence of pentacene with film thickness reveal that the mobility saturates at nearly 0.45 $cm^2 V^{-1} s^{-1}$ after about six layers. It is concluded that SAMs with different molecular dipole moments on conductive electrodes induce surface potential changes that possibly shift the work function of the metal or the electron affinity of the semiconductor electrode. As a consequence of the energy level shift, the transport/injection characteristics are influenced thus affecting both charge accumulation and conduction in the semiconducting channel of the OTFT.[114]

The monolayer-based systems have also been used to immobilize biomolecules to develop electrochemical biosensors. Itamar Willner and coworkers have pioneered this area several years ago wherein monolayers were used to immobilize enzymes, cofactors, and prothestic groups and subsequently used them in the sensing of various analytes.[115,116] This field has matured and has been widely researched.[117,118] Sturdy enzymes such as glucose oxidase (GO_x) has been the choice due to its availability, cost, and stability.[119] The method of immobilization determines sensor performance since electrical sensing signals are to be transported through the linker between the enzyme and the electrode. Various strategies, including electrostatic adsorption, covalent binding, electrochemical deposition, and the entrapment method have been adopted to immobilize different biomolecules.[120–122] Izumi and coworkers (Figure 5.12) have reported amperometric glucose biosensors, fabricated on gold substrates using SAMs with carboxylic acid end groups (COOH).[123,124] SAM of organic and inorganic systems is prepared as follows. The surface COOH groups of the SAM are activated by immersing the substrate in PBS containing 30 mM EDC and 15 mM NHS for 1.5 h followed by immersing the substrate in 5 mg/mL GOx solution for 12 h at room temperature. The sample at this stage is denoted as Au/SAM/GO_x. A Teflon LB trough is used for the LB film deposition of Prussian blue (PB) and nanometer-sized clusters are obtained at a fixed surface pressure to get a monolayer of PB on gold substrate. Glucose oxidase (GO_x) is then covalently immobilized on the SAMs and covered with Langmuir–Blodgett films, including nanometer-sized clusters of PB that play a role of mediator for

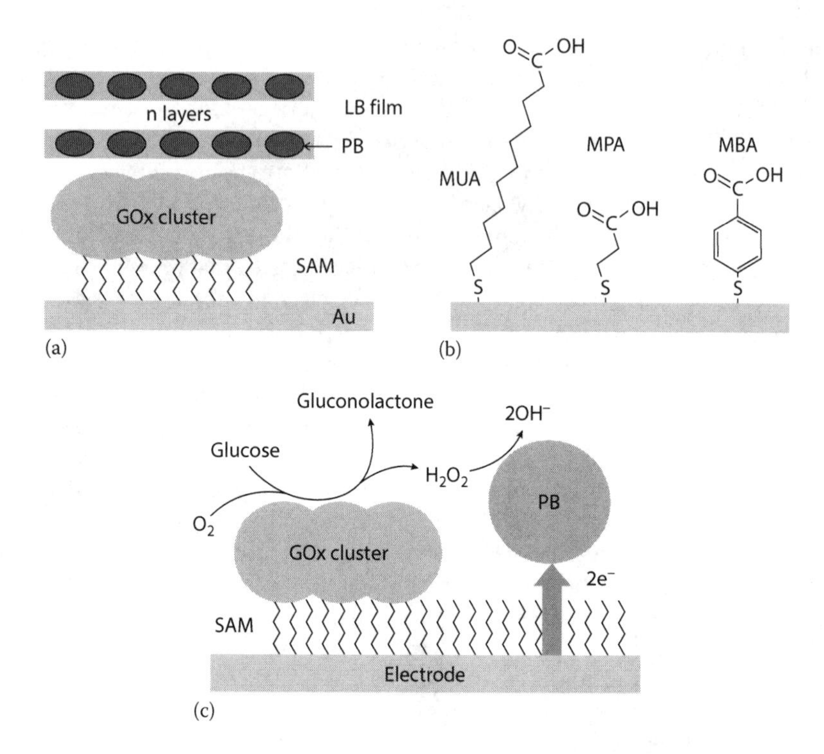

FIGURE 5.12 (a) Schematics of biosensor with combined SAM and LB films. (b) 3-mercaptopropionic acid (MPA), 11-mercaptoundecanoic acid (MUA) and 4-mercaptobenzoic acid (MBA) monolayers. (c) Schematic representation of redox cycle at the electrode surface. (Adapted from Wang, H. et al., *Sens. Actuators B Chem.*, 168, 249, 2012.)

glucose detection. An amperometric biosensor comprising a SAM of 4-mercaptobenzoic acid (MBA) exhibit a fast response current of 3 seconds with a detection limit of 12.5 μM and high sensitivity of 50 nA/(cm^2 mM).[124] The linearity ranges from 12.5 μM to 70 mM. The sensitivity is found to be significantly affected by the conductivity of the SAM layer. An important aspect in these studies is that the SAM used comprises short-chain alkanethiol. Hence, reduced concentration of the mediator produced during the enzyme reaction is still able to be detected at the underlying electrode.[125] Willner and coworkers, however, have shown the molecular level control over fabrication of the biorecognition interface and achieved control over the reaction.[126] They developed a method of linking enzymes to alkanethiol SAMs using a homo-bifunctional molecule that linked free amines from the surface lysine residues on enzymes to an amine-terminated SAM. They have shown that the same linker could be used to form multilayers of enzymes or enzyme combinations.[127]

Direct electron transfer to redox enzymes has been extensively studied where electron transfer between redox enzymes such as cytochrome c has been achieved. Various examples include horseradish peroxidase, azurin, hemoglobin, myoglobin, and laccase. SAMs act as a platform to incorporate enzymes by adsorption or covalent attachment and the electron transfer kinetics is found to be altered.[128,129] Enzyme active center in most of the cases is embedded deep within the glycoprotein and direct electron transfer is quite difficult. Willner and coworkers have shortened the distance with an electrically wired enzyme electrode using a mediator, pyrroloquinoline quinone (PQQ). The mediator is covalently attached to a cysteamine-modified gold surface using a coupling agent, 1-ethyl-3-(3-dimethylaminopropyl) carbodiimide). The PQQ-terminated SAM is then used to attach flavin adenine dinucleotide (FAD), a cofactor for GOx. The PQQ/FAD conjugate is then used to reconstitute apo-glucose oxidase to provide a GOx enzyme electrode mediated by PQQ.[130] Hess and coworkers reported another approach to achieve direct electron transfer to enzymes with a redox active center embedded deep within the glycoprotein where direct electron transfer to amine oxidase in its native configuration is achieved by the formation of SAMs on gold substrates.[131]

5.6 SUMMARY AND WAY FORWARD

The self-assembly process has given rise to a plethora of structures and several laboratory-scale applications have been demonstrated. However, still there are several issues that one can delve based on these systems. For example, intramolecular electron transfer in constrained systems is one of the outstanding problems that is of interest in photosynthesis, sensors, and molecular electronics. Photochemical and photoelectrochemical phenomenon are hardly exploited with self-assembled systems. There is no doubt that research in this direction will see positive contributions in the coming years.

REFERENCES

1. Love, J. C.; Estroff, L. A.; Kriebel, J. K.; Nuzzo, R. G.; Whitesides, G. M. *Chem. Rev.* 2005, *105*, 1103.
2. Haensch, C.; Hoeppener, S.; Schubert, U. S. *Chem. Soc. Rev.* 2010, *39*, 2323.
3. Vericat, C.; Vela, M. E.; Benitez, G.; Carro, P.; Salvarezza, R. C. *Chem. Soc. Rev.* 2010, *39*, 1805.
4. Nicosia, C.; Huskins, J. *Mater. Horiz.* 2014, *1*, 32.
5. Vericat, C.; Vela, M. E.; Corthey, G.; Pensa, E.; Cortes, E.; Fonticelli, M. H.; Ibanez, F.; Benitez, G. E.; Salvarezza, R. C. *RSC Adv.* 2014, *4*, 27730.
6. Onclin, S.; Ravoo, B. J.; Reinhoudt, D. N. *Angew. Chem. Int. Ed.* 2005, *44*, 6282.
7. Sullivan, T. P.; Huck, W. T. S. *Eur. J. Org. Chem.* 2003, 17.
8. Jonkheijm, P.; Weinrich, D.; Schroder, H.; Niemeyer, C. M.; Waldmann, H. *Angew. Chem. Int. Ed.* 2008, *47*, 9618.
9. Chechik, V.; Crooks, R. M.; Stirling, C. J. M. *Adv. Mater.* 2000, *12*, 1161.
10. Zhang, S. Y.; Regulacio, M. D.; Han, M.-H. *Chem. Soc. Rev.* 2014, *43*, 2301.
11. Grzelczak, M.; Vermant, J.; Furst, E. M.; Liz-Marzan, L. M. *ACS Nano* 2010, *4*, 3591.
12. Deegan, R. D.; Bakajin, O.; Dupont, T. F.; Huber, G.; Nagel, S. R.; Witten, T. A. *Nature* 1997, *389*, 427.
13. Boker, A.; He, J.; Emrick, T.; Russell, T. P. *Soft Matter* 2007, *3*, 1231.
14. Singh, A.; Gunning, R. D.; Ahmed, S.; Barrett, C. A.; English, N. J.; Garate J.-A.; Ryan, K. M. *J. Mater. Chem.* 2012, *22*, 1562.
15. Jana, N. R. *Angew. Chem. Int. Ed.* 2004, *43*, 1536.
16. Kinge, S.; Calama, M. C.; Reinhoudt, D. N. *ChemPhysChem* 2008, *9*, 20.

17. Min, Y.; Akbulut, M.; Kristiansen, K.; Golan, Y.; Israelachvili, J. *Nat. Mater.* 2008, *7*, 527.
18. Bishop, K. J. M.; Wilmer, C. E.; Soh, S.; Grzybowski, B. A. *Small* 2009, *5*, 1600.
19. Bentzon, M. D.; Wonterghem, J.; Mørup, S.; Tholen, A.; Koch, C. J. W. *Philos. Mag. B* 1989, *60*, 169.
20. Murray, C. B.; Kagan, C. R.; Bawendi, M. G. *Annu. Rev. Mater. Sci.* 2000, *30*, 545.
21. Zhao, N.; Liu, K.; Greener, J.; Nie, Z.; Kumacheva, E. *Nano Lett.* 2009, *9*, 3077.
22. Ristenpart, W. D.; Aksay, I. A.; Saville, D. A. *Phys. Rev. Lett.* 2003, *90*, 128303.
23. Guerrero-Martınez, A.; Perez-Juste, J.; Carbo-Argibay, E.; Tardajos, G.; Liz-Marzan, L. M. *Angew. Chem. Int. Ed.* 2009, *48*, 9484.
24. Zhou, Z.-Y.; Tian, N.; Li, J.-T.; Broadwell, I.; Sun, S.-G. *Chem. Soc. Rev.* 2011, *40*, 4167.
25. Baker, J. L.; Widmer-Cooper, A.; Toney, M. F.; Geissler, P. L.; Alivisatos, A. P. *Nano Lett.* 2010, *10*, 195.
26. Gast, A. P.; Zukoski, C. F. *Adv. Colloid Interface Sci.* 1989, *30*, 153.
27. Hu, Z.; Fischbein, M. D.; Querner, C.; Drndic, M., *Nano Lett.* 2006, *6*, 2585.
28. Ding, T.; Song, K.; Clays, K.; Tung, C. H. *Adv. Mater.* 2009, *21*, 1936.
29. Isojima, T.; Lattuada, M.; Vander Sande, J. B.; Hatton, T. A. *ACS Nano* 2008, *2*, 1799.
30. Ryan, K. M.; Mastroianni, A.; Stancil, K. A.; Liu, H.; Alivisatos, A. P. *Nano Lett.* 2006, *6*, 1479.
31. Mittal, M.; Lele, P. P.; Kaler, E. W.; Furst, E. M. *J. Chem. Phys.* 2008, *129*, 064513.
32. Smith, P. A.; Nordquist, C. D.; Jackson, T. N.; Mayer, T. S.; Martin, B. R.; Mbindyo, J.; Mallouk, T. E. *Appl. Phys. Lett.* 2000, *77*, 1399.
33. O'Brien, R. W.; White, L. R. *J. Chem. Soc., Faraday Trans.* 1978, *74*, 1607.
34. Hermanson, K. D.; Lumsdon, S. O.; Williams, J. P.; Kaler, E. W.; Velev, O. D. *Science* 2001, *294*, 1082.
35. Akey, A.; Lu, C.; Yang, L.; Herman, I. P. *Nano Lett.* 2010, *10*, 1517.
36. Jones, M. R.; Osberg, K. D.; Macfarlane, R. J.; Langille, M. R.; Mirkin, C. A. *Chem. Rev.* 2011, *111*, 3736.
37. Nepal, D.; Onses, M. S.; Park, K.; Jespersen, M.; Thode, C. J.; Nealey, P. F.; Vaia, R. A. *ACS Nano* 2012, *6*, 5693.
38. Hamon, C.; Postic, M.; Mazari, E.; Bizien, T.; Dupuis, C.; Jimenez, A.; Courbin, L.; Gosse, C.; Artzner, F.; Marchi-Artzner, V. *ACS Nano* 2012, *6*, 4137.
39. Correa-Duarte, M. A.; Perez-Juste, J.; Sanchez-Iglesias, A.; Giersig, M.; Liz-Marzan, L. M. *Angew. Chem. Int. Ed.* 2005, *44*, 4375.
40. Berry, V.; Saraf, R. F. *Angew. Chem. Int. Ed.* 2005, *44*, 6668.
41. Kuemin, C.; Nowack, L.; Bozano, L.; Spencer, N. D.; Wolf, H. *Adv. Funct. Mater.* 2012, *22*, 702.
42. Modestino, M. A.; Chan, E. R.; Hexemer, A.; Urban, J. J.; Segalman, R. A. *Macromolecules* 2011, *44*, 7364.
43. Thorkelsson, K.; Mastroianni, A. J.; Ercius, P.; Xu, T. *Nano Lett.* 2011, *12*, 498.
44. Binks, B. P. *Curr. Opin. Colloid Interface Sci.* 2002, *7*, 21.
45. Lin, Y.; Skaff, H.; Emrick, T.; Dinsmore, A. D.; Russell, T. P. *Science* 2003, *299*, 226.
46. Botto, L.; Lewandowski, E. P.; Cavallaro, M.; Stebe, K. *J. Soft Matter* 2012, *8*, 9957.
47. Wang, J.; Li, Q.; Knoll, W.; Jonas, U. *J. Am. Chem. Soc.* 2006, *128*, 15606.
48. Lee, J. A.; Meng, L.; Norris, D. J.; Scriven, L. E.; Tsapatsis, M. *Langmuir* 2006, *22*, 5217.
49. Jiang, L.; Chen, X.; Lu, N.; Chi, L. *Acc. Chem. Res.* 2014, *47*, 3009.
50. Vidoni, O.; Reuter, T.; Torma, V.; Meyer-Zaika, W.; Schmid, G. *J. Mater. Chem.* 2001, *11*, 3188.
51. Lundgren, A. O.; Bjorefors, F.; Olofsson, L. G. M.; Elwing, H. *Nano Lett.* 2008, *8*, 3989.
52. He, J.; Zhang, Q.; Gupta, S.; Emrick, T.; Russell, T. P. *Small* 2007, *3*, 1214.

53. Kim, F.; Kwan, S.; Akana, J.; Yang, P. *J. Am. Chem. Soc.* 2001, *123*, 4360.
54. Tao, A. R.; Huang, J. X.; Yang, P. D. *Acc. Chem. Res.* 2008, *41*, 1662.
55. Ariga, K.; Yamauchi, Y.; Mori, T.; Hill, J. P. *Adv. Mater.* 2013, *25*, 6477.
56. Gong, J.; Li, G.; Tang, Z. *Nano Today* 2012, *7*, 564.
57. Liu, K.; Zhao, N.; Kumacheva, E. *Chem. Soc. Rev.* 2011, *40*, 656.
58. Yu, K.; Chen, J. *Nanoscale Res. Lett.* 2009, *4*, 1.
59. Dimitrakopoulos, C. D.; Malenfant, P. R. L. *Adv. Mater.* 2002, *14*, 99.
60. Godlewski, S.; Szymonski, M. *Int. J. Mol. Sci.* 2013, *14*, 2946.
61. Lanzilotto, V.; Sanchez-Sanchez, C.; Bavdek, G.; Cvetko, D.; Lopez, M. F.; Martin-Gago, J. A.; Floreano, L. *J. Phys. Chem. C* 2011, *115*, 4664.
62. Godlewski, S.; Tekiel, A.; Piskorz, W.; Zasada, F.; Prauzner-Bechcicki, J. S.; Sojka, Z.; Szymonski, M. *ACS Nano* 2012, *6*, 8536.
63. Potapenko, D. V.; Choi, N. J.; Osgood, R. M. *J. Phys. Chem. C* 2010, *114*, 19419.
64. Reiss, S.; Krumm, H.; Niklewski, A.; Staemmler, V.; Wöll, C. *J. Chem. Phys.* 2002, *116*, 7704.
65. Simonsen, J. B.; Handke, B.; Li, Z.; Møller, P. J. *Surf. Sci.* 2009, *603*, 1270.
66. Schuster, B. E.; Casu, M. B.; Biswas, I.; Hinderhofer, A.; Gerlach, A.; Schreiberb, F.; Chasse, T. *Phys. Chem. Chem. Phys.* 2009, *11*, 9000.
67. Tekiel, A.; Godlewski, S.; Budzioch, J.; Szymonski, M. *Nanotechnology* 2008, *19*, 495304.
68. Tautz, F. S. *Prog. Surf. Sci.* 2007, *82*, 479.
69. Simonsen, J. B. *Surf. Sci.* 2010, *604*, 1300.
70. O'Regan, B.; Grätzel, M. *Nature* 1991, *353*, 737.
71. Palmgren, P.; Priya, B. R.; Niraj, N. P. P.; Göthelid, M. *Sol. Energy Mater. Sol. Cells* 2006, *90*, 3602.
72. Palmgren, P.; Yu, S.; Hennies, F.; Nilson, K.; Åkermark, B.; Göthelid, M. *J. Chem. Phys.* 2008, *129*, 074707.
73. Wang, Y.; Ye, Y.; Wu, K. *J. Phys. Chem. B* 2006, *110*, 17960.
74. Godlewski, S.; Tekiel, A.; Prauzner-Bechcicki, J. S.; Budzioch, J.; Szymonski, M. *ChemPhysChem* 2010, *11*, 1863.
75. Ishida, N.; Fujita, D. *J. Phys. Chem. C* 2012, *116*, 20300.
76. Ino, D.; Watanabe, K.; Takagi, N.; Matsumoto, Y. *J. Phys. Chem. B* 2005, *109*, 18018.
77. Peisert, H.; Biswas, I.; Knupfer, M.; Chasse, T. *Phys. Stat. Solid* 2009, *7*, 1529.
78. Deng, J.; Baba, Y.; Honda, M. *J. Phys. Condens. Matter* 2007, *19*, 196205.
79. Rostovtsev, V. V.; Green, L. G.; Fokin, V. V.; Sharpless, K. B. *Angew. Chem. Int. Ed.* 2002, *41*, 2596.
80. Nebhani, L.; Kowollik, C. B. *Adv. Mater.* 2009, *21*, 3442.
81. Iha, R. K.; Wooley, K. L.; Nystrom, A. M.; Burke, D. J.; Kade, M. J.; Hawker, C. *J. Chem. Rev.* 2009, *109*, 5620.
82. Xia, Y.; Whitesides, G. M. *Angew. Chem. Int. Ed.* 1998, *37*, 550.
83. Pulsipher, A.; Yousaf, M. N. *ChemBioChem* 2010, *11*, 745.
84. Elwing, H.; Welin, S.; Askendal, A.; Nilsson, U.; Lundstrom, I. *J. Colloid Interface Sci.* 1987, *119*, 203.
85. Chaudhury, M. K.; Whitesides, G. M. *Science* 1992, *256*, 1539.
86. Lee, E. J.; Chan, E. W. L.; Yousaf, M. N. *ChemBioChem* 2009, *10*, 1648.
87. Ito, Y.; Heydari, M.; Hashimoto, A.; Konno, T.; Hirasawa, A.; Hori, S.; Kurita, K. Nakajima, A. *Langmuir* 2007, *23*, 1845.
88. Gallant, N. D.; Lavery, K. A.; Amis, E. J.; Becker, M. L. *Adv. Mater.* 2007, *19*, 965.
89. Shida, N.; Ishiguro, Y.; Atobe, M.; Fuchigami, T.; Inagi, S. *ACS Macro Lett.* 2012, *1*, 656.
90. Hansen, T. S.; Lind, J. U.; Daugaard, A. E.; Hvilsted, S.; Andresen, T. L. Larsen, N. B. *Langmuir* 2010, *26*, 16171.
91. Krabbenborg, S. O.; Nicosia, C.; Chen, P.; Huskens, J. *Nat. Commun.* 2013, *4*, 1667.

92. Perl, A.; Gomez-Casado, A.; Thompson, D.; Dam, H. H.; Jonkheijm, P.; Reinhoudt, D. N.; Huskens, J. *Nat. Chem.* 2011, *3*, 317.
93. Tauk, L.; Schroeder, A. P.; Decher, G. Giuseppone, N. *Nat. Chem.* 2009, *1*, 649.
94. Mann, B.; Kuhn, H.; *J. Appl. Phys.* 1971, *42*, 4398.
95. Tran, E.; Duati, M.; Ferri, V.; Mullen, K.; Zharnikov, M.; Whitesides, G. M.; Rampi, M. A. *Adv. Mater.* 2006, *18*, 1323.
96. Sarkar, S.; Sampath, S. *Langmuir* 2006, *22*, 3388.
97. Li, J.; Lui, J.; Evmenenko, G. A.; Dutta, P.; Marks, T. J. *Lancet* 2008, *24*, 5755.
98. Liu, S.; Maoz, R.; Schmid, G.; Sagiv, J. *Nano Lett.* 2002, *2*, 1055.
99. Spori, D. M.; Venkatarman, N. V.; Tosatti, S. G. P.; Durmaz, F.; Spencer, N. D. *Langmuir* 2007, *23*, 8053.
100. Folkers, J. P.; Gorman, C. B.; Laibinis, P. E.; Buchholz, S.; Whitesides, G. M.; Nuzzo, R. G. *Langmuir* 1994, *11*, 813.
101. Maisch, S.; Effenberger, F. *J. Am. Chem. Soc.* 2005, *127*, 17315.
102. Shamai, T.; Ophir, A.; Selzer, Y. *Appl. Phys. Lett.* 2007, *91*, 102108.
103. Kim, B.; Beebe, J. M.; Jun, Y.; Zhu, X.-Y.; Frisbie, C. D. *J. Am. Chem. Soc.* 2006, *128*, 4970.
104. Wang, G.; Kim, T.-W.; Lee, H.; Lee, T. *Phys. Rev. B* 2007, *76*, 205320.
105. Park, Y. S.; Whalley, A. C.; Kamenetska, M.; Steigerwald, L. L.; Hybersten, M. S.; Nuckolls, C.; Venkataraman, L. *J. Am. Chem. Soc.* 2007, *129*, 15768.
106. Boulas, C.; Davidovits, J. V.; Rondelez, F.; Vuillaume, D. *Phys. Rev. Lett.* 1996, *76*, 4797.
107. Gu, Y.; Akhremitchev, B.; Walker, G. C.; Waldeck, D. H. *J. Phys. Chem. B* 1999, *103*, 5220.
108. Selzer, Y.; Salomon, A.; Cahen, D. *J. Phys. Chem. B* 2002, *106*, 10432.
109. Vuillaume, D.; Boulas, C.; Collet, J.; Davidovits, J. V.; Rondelez, F. *Appl. Phys. Lett.* 1996, *69*, 1646.
110. Fontaine, P.; Goguenheim, D.; Deresmes, D.; Vuillaume, D.; Garet, M.; Rondelez, F. *Appl. Phys. Lett.* 1993, *62*, 2256.
111. Collet, J.; Tharaud, O.; Chapoton, A.; Vuillaume, D. *Appl. Phys. Lett.* 2000, *76*, 1941.
112. Halik, M.; Klauk, H.; Zschieschang, U.; Schmid, G.; Dehm, C.; Schutz, M.; Maisch, S.; Effenberger, F.; Brunnbauser, M.; Stellacci, F. *Nature* 2004, *431*, 963.
113. Ruiz, R.; Papadimitratos, A.; Mayer, A. C.; Malliaras, G. C. *Adv. Mater.* 2005, *17*, 1795.
114. Mottaghi, M.; Lang, P.; Rodriguez, F.; Rumyantseva, A.; Yassar, A.; Horowitz, G.; Lefant, S.; Tondelier, D.; Vuillaume, D. *Adv. Funct. Mater.* 2007, *17*, 597.
115. Willner, I.; Riklin, A. *Anal. Chem.* 2004, *66*, 1535.
116. Zayats, M.; Baron, R.; Popov, I.; Willner, I. *Nano Lett.* 2005, *5*, 21.
117. Gooding, J. J.; Erokhin, P.; Hibbert, D. B. *Biosens. Bioelectron.* 2000, *15*, 229.
118. Turner, A. P. F. *Science* 2000, *290*, 1315.
119. Chaki, N. K.; Vijayamohanan, P. *Biosens. Bioelectron.* 2002, *17*, 1.
120. Wang, Z.; Liu, S.; Wu, P.; Cai, C. *Anal. Chem.* 2009, *81*, 1638.
121. Ammam, M.; Fransaer, J. *Biosens. Bioelectron.* 2009, *25*, 191.
122. Hecht, H. J.; Schomburg, D.; Kalisz, H.; Schmid, R. D. *Biosens. Bioelectron.* 1993, *8*, 197.
123. Ohnuki, H.; Saiki, T.; Kusakari, A.; Endo, H.; Ichihara, M.; Izumi, M. *Langmuir* 2007, *23*, 4675.
124. Wang, H.; Ohnuki, H.; Endo, H.; Izumi, M. *Sens. Actuators B Chem.* 2012, *168*, 249.
125. Gooding, J. J.; Ciampi, S. *Chem. Soc. Rev.* 2011, *40*, 2704.
126. Willner, I.; Riklin, A.; Shoham, B.; Rivenzon, D.; Katz, E. *Adv. Mater.* 1993, *5*, 912.
127. Willner, I.; Lion-Dagan, M.; Marx-Tibbon, S.; Katz, E. *J. Am. Chem. Soc.* 1995, *117*, 6581.

128. Shleev, S.; Tkac, J.; Christenson, A.; Ruzgas, T.; Yaropolov, A. I.; Whittaker, J. W.; Gorton, L. *Biosens. Bioelectron.* 2005, *20*, 2517.
129. Wei, J. J.; Liu, H. Y.; Dick, A. R.; Yamamoto, H.; He, Y. F.; Waldeck, D. H. *J. Am. Chem. Soc.* 2002, *124*, 9591.
130. Willner, I.; Heleg-Shabtai, V.; Blonder, R.; Katz, E.; Tao, G.; Bückmann, A. F.; Heller, A. *J. Am. Chem. Soc.* 1996, *118*, 10321.
131. Hess, C. R.; Juda, G. A.; Dooley, D. M.; Amii, R. N.; Hill, M. G.; Winkler, J. R.; Gray, H. B. *J. Am. Chem. Soc.* 2003, *125*, 7156.

6 Nanoparticles of Organic Conductors

Dominique de Caro, Christophe Faulmann, and Lydie Valade

CONTENTS

6.1 INTRODUCTION

As we approach the physical limits of conventional silicon-based electronics, there is a clear need for conducting molecular materials that can deliver smaller and smaller devices. However, reducing the size of the device usually requires a prior step: the reduction of the size of the molecule-based conductor itself. In this chapter, we will focus on organic conductors that belong to two families: the charge-transfer salts and the mixed-valence salts. Organic, electrically conducting polymers will not be discussed here.

Organic or metallo-organic molecules exhibiting a good planarity and containing large atoms (e.g., chalcogens, Figure 6.1) can stack in one direction to provide a framework for possible band structure formation. The partial filling of the conduction band (i.e., the metallic state) can be ensured either by partial oxidation (mixed-valence salts) or by partial charge transfer (charge-transfer salts). Organic conductors are commonly prepared and studied as macroscopic single crystals by using slow diffusion in organic solution or electrocrystallization techniques. Micrometer-sized crystals have nevertheless been grown by confined electrocrystallization [1], in an aqueous solution containing a copolymer block [2], or by pouring the organic compound on microfabricated Au electrodes on a Si substrate with a thermally grown oxide layer [3]. As mentioned at the beginning of this paragraph, intermolecular interactions between the planar or quasi-planar molecules are the strongest along the chain direction. Even orbital overlaps transverse to the stacking direction (but much weaker than those along the stacking direction) do exist, charge-transfer- or mixed-valence-based compounds are usually called quasi-one-dimensional conductors. Therefore, obtaining them as nanowires is now well established [4–12]. Nanorod arrays [13] and individual nanorods grown perpendicular to the surface of platinum

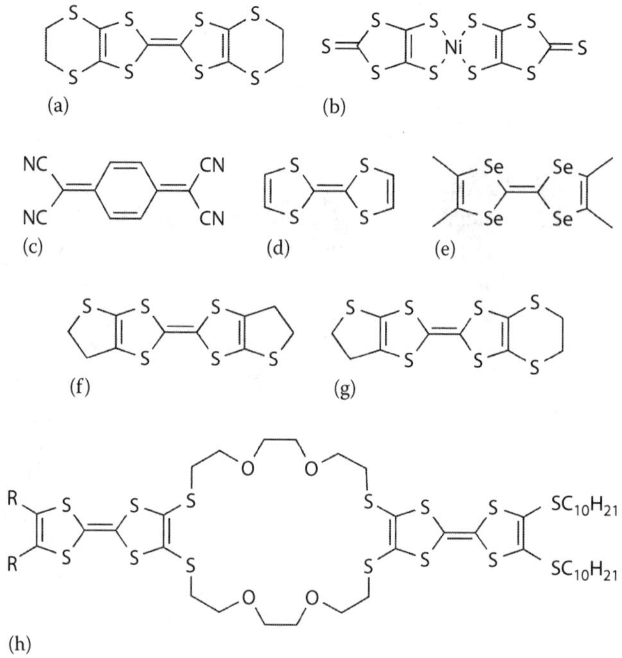

FIGURE 6.1 Molecular structures of various donors and acceptors: (a) BEDT-TTF or ET, (b) Ni(dmit)$_2$, (c) TCNQ, (d) TTF, (e) TMTSF, (f) (*E*)-BET-TTF, (g) ETEDT-TTF, and (h) macro-TTF.

nanoparticles [14] have also been reported. However, contrary to molecular magnets or spin crossover complexes, which are based on a coordination network, greatly helping for the growth as spherical nanoparticles [15], quasi 1-D molecular conductors do not show a natural tendency to grow as spherical objects. However, some attempts to prepare nanocrystals have been reported. For instance, $(BEDT\text{-}TTF)_2I_3$ (BEDT-TTF: bis(ethylenedithio)tetrathiafulvalene) and (perylene)(TCNQ) (TCNQ: tetracyanoquinodimethane) nanocrystals exhibiting irregular shapes have been grown by reprecipitation techniques (sizes in the 80–450 nm range) [16,17]. As a second example, bilayer composite films (polycarbonate/molecular conductor) have been generated by vapor exposure to iodine or bromine of the swollen polymer surface containing molecularly dispersed BET-TTF (BET-TTF: bis(ethylenethio)tetrathiafulvalene) [18] or ETEDT-TTF (ETEDT-TTF: (ethylenethio)(ethylenedithio)tetrathiafulvalene) [19]. The resulting conducting and transparent bilayers contained irregularly shaped nanocrystals of $\theta\text{-}(BET\text{-}TTF)_2Br\cdot 3H_2O$ or $(ETEDT\text{-}TTF)_xI_3/(ETEDT\text{-}TTF)_xI_yBr_{3-y}$ (sizes in the 200–300 nm range). Finally, the brief exposure of (111)-oriented Au surfaces to TTF·TCNQ (TTF: tetrathiafulvalene) resulted in the formation of TTF·TCNQ nanoclusters (irregular in shape) as evidenced by scanning tunneling microscopy (average thickness of 0.15 nm) [20,21]. A prolongated exposure of the Au surface resulted in the formation of highly oriented TTF·TCNQ needles, with their *b* axes (the molecular stacking direction) perpendicular to the substrate.

The first part of this chapter will describe the different synthetic strategies leading to nanoparticles of molecule-based conductors. The second part will be devoted both to their spectroscopic studies and to the description of some of their physical properties.

6.2 SYNTHETIC STRATEGIES TOWARD NANOPARTICLES OF ORGANIC OR METALLO-ORGANIC CONDUCTORS

6.2.1 Template Chemistry

Until now, the only example of roughly spherical nanoparticles prepared by template chemistry has been reported by Xu and coworkers [22]. They described the electrochemical growth of $[(CH_3)_4N][Ni(dmit)_2]_2$ nanoparticle chains within the pores of an anodic aluminum oxide film ($dmit^{2-}$: 2-thioxo-1,3-dithiole-4,5-dithiolato). Individual nanoparticles exhibited a diameter of about 30 nm (Figure 6.2). During the electrochemical procedure (oxidation of the $[Ni(dmit)_2]^-$ species into a partially charged anionic species, $[Ni(dmit)_2]^{\delta-}$), the authors evidenced oscillations of the voltage. Due to the anisotropic growth of the $[(CH_3)_4N][Ni(dmit)_2]_2$ material (1-D mixed-valence complex), the consequence of these voltage oscillations was that the particles did not match the columned pore's shape completely and had a spherical shape.

6.2.2 Nanoparticles Embedded in a Polymer Matrix

Polymers are widely used to stabilize and/or organize metal nanoparticles [23–25]. Moreover, polyethyleneglycol (PEG, Figure 6.3) and polyvinylpyrrolidone (PVP, Figure 6.3) had shown their efficiency to protect from the aggregation of nanoparticles of Prussian blue or spin crossover complexes [26,27].

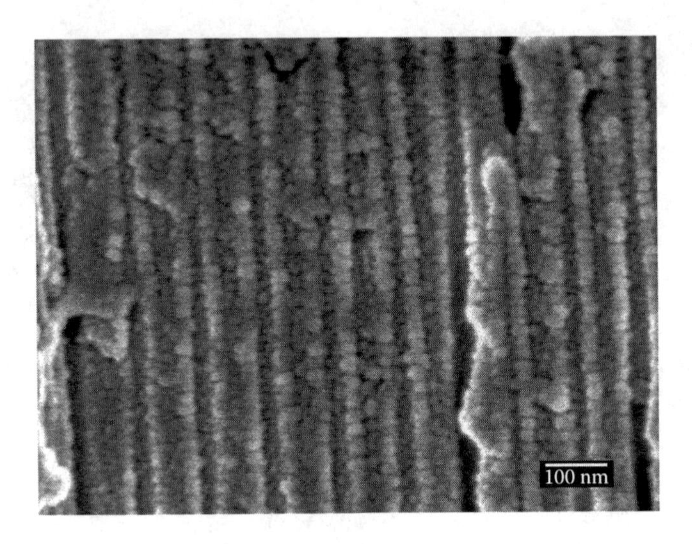

FIGURE 6.2 SEM images of the side view of $[(CH_3)_4N][Ni(dmit)_2]_2$ nanowire arrays within the pores of an anodic aluminum oxide film (pore diameter : 32 ± 4 nm) (scale bar 100 nm). (Reprinted with permission from Cui, G., Xu, W., Guo, C., Xiao, X., Xu, H., Zhang, D., Jiang, L., and Zhu, D., Conducting nanopearl chains based on the dmit salt, *J. Phys. Chem. B*, 108, 13638, 2004. Copyright 2014 American Chemical Society.)

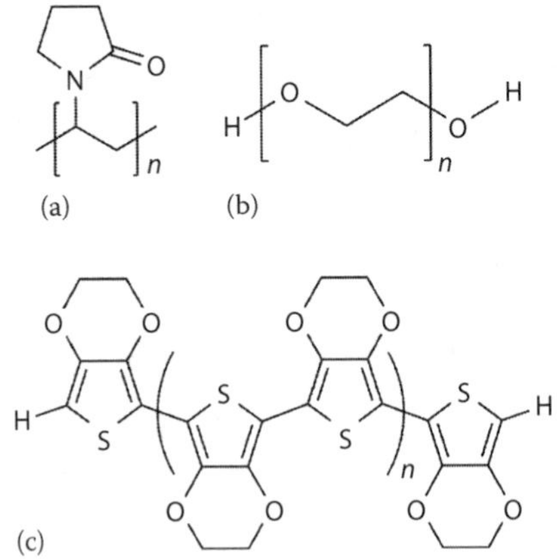

FIGURE 6.3 Molecular structures of selected polymers: (a) PVP, (b) PEG, and (c) PEDOT.

However, such polymers have been little used for controlling the growth and embedding molecule-based conductors. PVP was successfully used to prepare nanoparticles of $(BEDT\text{-}TTF)_2I_3$ and $TTF[Ni(dmit)_2]_2$ [28]. The resulting composites have been processed as thin films on SiO_2. Moreover, TTFBr and $TTFBr_{0.59}$ nanoparticles (diameter in the 5–40 nm range) have also been prepared and stabilized in poly(3,4-ethylenedioxythiophene) (PEDOT, Figure 6.3) [29]. Finally,

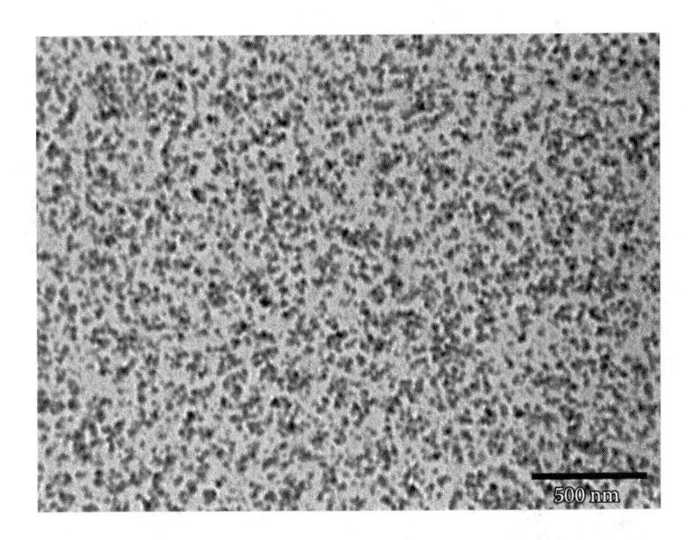

FIGURE 6.4 TEM image of {TTF-TCNQ}@PEG (scale bar 500 nm).

~40 nm-TTF·TCNQ nanoparticles have been grown in the presence of an acetone/acetonitrile solution of PEG (average molecular weight: 400; 3 eq. *vs.* 1 eq. TTF and 1 eq. TCNQ, Figure 6.4) [30]. In the latter case, the polymer in solution controlled the growth of the molecular conductor as nanoparticles but was not present at the particles' surface in the final nanopowdered material.

6.2.3 Nanoparticles Grown in the Presence of Imidazolium-Based Ionic Liquids or Long-Alkyl Chains Quaternary Ammonium Salts

6.2.3.1 Introduction

Ionic species such as ionic liquids containing 1-butyl-3-methylimidazolium, that is, $BMIM^+$ cation (Figure 6.5) or long-alkyl chains quaternary ammonium salts were most often used as growth controlling agents for the preparation of molecule-based conductors as nanoparticles. Originally, imidazolium-based ionic liquids ([BMIM][X], where X^- stands for BF_4^-, PF_6^-, $CF_3SO_3^-$, $(CF_3SO_2)_2N^-$, …), were very good candidates to stabilize metallic [31] or Prussian blue [32] nanoparticles. In these two cases, [BMIM][X] usually acted as both the solvent and the stabilizing agent. Mainly at low temperatures, the ionic liquid organized into nonpolar microdomains in which the confinement of the growing species was more efficient, affording to very small particles [33]. Macroscopic single crystals of $(TMTSF)_2NbF_6$ (TMTSF:

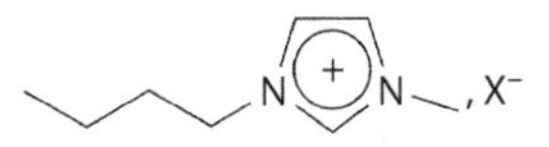

1-butyl-3-methylimidazolium
$BMIM^+$; $X^- = BF_4^-$, $[N(CF_3SO_2)_2]^-$

FIGURE 6.5 Molecular structure of $BMIM^+$.

tetramethyltetraselenafulvalene) [34] and $(BEDT-TTF)_2H_2F_3$ [35] have been grown by electrocrystallization using $[EMIM][NbF_6]$ or $[EMIM][H_2F_3]$ ionic liquids as supporting electrolytes (EMIM: 1-ethyl-3-methylimidazolium). Moreover, electronically and ionically conductive gels of various ionic liquids and TTF·TCNQ have been reported by Ouyang and coworkers [36]. Gels were prepared by the mechanical grinding of neutral TTF and neutral TCNQ in ionic liquids. They contained up to 50 wt% of the conducting TTF·TCNQ material. The morphology of TTF·TCNQ incorporated into the ionic liquid was similar to that observed for the bulk material, that is, needle-like crystallites. The great majority of molecule-based conductors are prepared using the electrocrystallization method, conducted under galvanostatic or potentiostatic conditions. Quaternary ammonium salts such as $[R_4N]X$ (R: methyl, ethyl, propyl, or butyl group) were the supporting electrolytes of choice [37]. For mixed-valence compounds, the counterion, X^- (present in the ammonium salt), also entered the composition of the conductor, in association with the partially oxidized TTF-based molecule.

6.2.3.2 Ionic Liquids for the Growth of Nanoparticles of Molecular Conductors

The preparation of nanoparticles of molecule-based conductors in the presence of an imidazolium-based ionic liquid was conducted either by a chemical or by an electrochemical process.

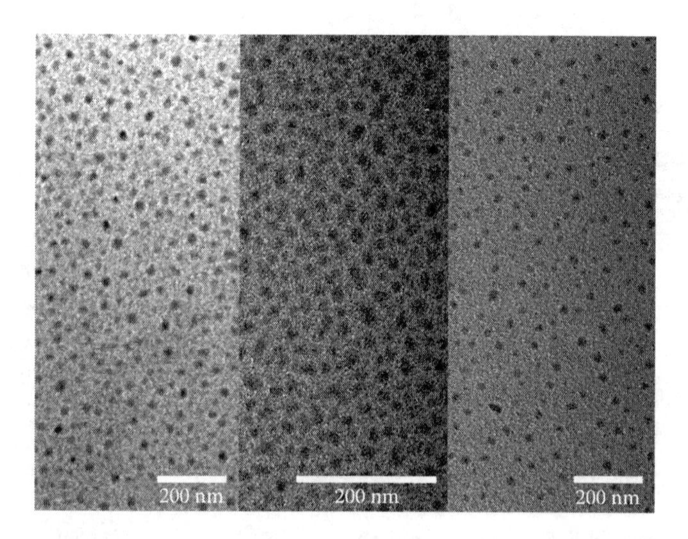

FIGURE 6.6 TEM image of TTF-TCNQ, $TTF[Ni(dmit)_2]_2$, and $(BEDT-TTF)_2I_3$ obtained as nanoparticles by chemical oxidation in the presence of $BMIM^+$ (scale bar 200 nm). (de Caro, D., Jacob, K., Hahioui, H., Faulmann, C., Valade, L., Kadoya, T., Mori, T., Fraxedas, J., and Viau, L., Nanoparticles of organic conductors: Synthesis and application as electrode material in organic field effect transistors, *New J. Chem.*, 35, 1315–1319, 2011. Reproduced by permission of The Royal Society of Chemistry and the Centre National de la Recherche Scientifique; de Caro, D., Faulmann, C., Valade, L., Jacob, K., Chtioui, I., Foulal, S., de Caro, P. et al.: Four molecular superconductors isolated as nanoparticles. *Eur. J. Inorg. Chem.* 2014. 4010–4016. Copyright Wiley-VCH Verlag GmbH & Co. KGaA. Reproduced with permission.)

The three systems processed as nanoparticles using a chemical procedure were two charge-transfer salts, that is, TTF·TCNQ and TTF$[Ni(dmit)_2]_2$, and one mixed-valence salt, that is, $(BEDT\text{-}TTF)_2I_3$. The bulk materials of these compounds were obtained by adding an acetonitrile solution of TTF (or $(TTF)_3(BF_4)_2$ or BEDT-TTF) to an acetonitrile solution of TCNQ (or $[(n\text{-}C_4H_9)_4N][Ni(dmit)_2]$ or I_2) at 25°C for TTF·TCNQ and TTF$[Ni(dmit)_2]_2$ or 85°C for $(BEDT\text{-}TTF)_2I_3$ [37]. When the TTF-based molecule (neutral TTF, monocationic TTF in $(TTF)_3(BF_4)_2$, or BEDT-TTF) in the presence of [BMIM][X]/S (S: acetonitrile or acetone) was added to a solution of the second species, nanoparticles of TTF·TCNQ [30,38,39], TTF$[Ni(dmit)_2]_2$ [30,38,40], or $(BEDT\text{-}TTF)_2I_3$ [41] were observed (Figure 6.6).

The ionic liquid should contain an unreactive counterion X^- versus the TTF-based cation. Thus, the two ionic liquids used were [BMIM]$[BF_4]$ and [BMIM]$[(CF_3SO_2)_2N]$. Reactions were conducted in a mixture of ionic liquid and conventional solvent (acetonitrile or acetone) because reactants were not soluble in pure ionic liquid. However, the observation of nanoparticles (vs. needles in the absence of ionic liquid) confirmed that such a mixture was not deleterious to a growth control. To support this point, Gonfa and coworkers reported that imidazolium-based ionic liquids were able to self-organize in a solvent [42]. This nanoscale self-organization was the result of the interplay between Coulomb and van der Waals interactions, which led to the formation of high charge density–permeated nonpolar regions. Size, morphology, and state of dispersion for TTF·TCNQ nanoparticles mainly depended on the temperature, the amount of ionic liquid, the nature of the solvents used, and finally the stirring intensity and duration [30,39] (see Table 6.1 for the influence of the temperature). TTF$[Ni(dmit)_2]_2$ nanoparticles were roughly spherical in shape at low temperatures (see Table 6.1), whereas they were irregular in shape at room temperature [30,38,40]. $(BEDT\text{-}TTF)_2I_3$ nanoparticles were prepared by the chemical oxidation of BEDT-TTF by iodine in the presence of [BMIM]$[(CF_3SO_2)_2N]$ [41]. This procedure afforded well-dispersed nanoparticles showing an average size of 35 nm (see Table 6.1).

The two systems processed as nanoparticles using an electrochemical procedure were the charge-transfer salt, TTF$[Ni(dmit)_2]_2$, and the mixed-valence salt, $(TMTSF)_2PF_6$. They were commonly prepared as needle-shaped single crystals by electro-oxidation of TTF or TMTSF in the presence of $[(n\text{-}C_4H_9)_4N][Ni(dmit)_2]$ or $[(n\text{-}C_4H_9)_4N]PF_6$ under current densities higher than 50 μA cm^{-2} [37,43]. In the case of TTF$[Ni(dmit)_2]_2$ nanoparticles, [BMIM]$[BF_4]$ or [BMIM]$[(CF_3SO_2)_2N]$ was used as growth controlling agents. Nanoparticles exhibiting mean diameters in the 10–20 nm range were observed [41] (Figure 6.7). In the second case, [BMIM]$[PF_6]$ acted as a supporting electrolyte, growth controlling agents, and finally as PF_6^- counteranion providers [41,44]. $(TMTSF)_2PF_6$ nanoparticle sizes were current density dependent (see Table 6.1 and Figure 6.7). Electrochemical syntheses took place in an H-shaped electrocrystallization cell [37] in which the anodic compartment was vigorously stirred. Stirring the medium prevented the formation of large single crystals, increased the amount of ionic liquid at the electrode vicinity, and favored the growth control of the material as nanoparticles.

TABLE 6.1
Synthesis Conditions for Nanoparticles of Molecular Conductors Grown in the Presence of an Ionic Liquid

Precursor Molecules	Molar Ratio	Solvent (Temperature, If Applicable: Applied Current Density)	TEM Observation
TTF; TCNQ; [BMIM][BF_4]	1/1/20	AN (25°C)	20 nm
TTF; TCNQ; [BMIM][BF_4]	1/1/20	AN (−80°C)	20 nm
TTF; TCNQ; [BMIM][$(CF_3SO_2)_2N$]	1/1/20	AN (25°C)	Aggregates of nanoparticles
TTF; TCNQ; [BMIM][$(CF_3SO_2)_2N$]	1/1/20	AN (−80°C)	Mixture of spherical and elongated nanoparticles
$(TTF)_3(BF_4)_2$; [$(n\text{-}C_4H_9)_4N$][$Ni(dmit)_2$]; [BMIM][BF_4]	1/2/20	AN/AC: 1/1 vol. (−80°C)	25 nm
$(TTF)_3(BF_4)_2$; [$(n\text{-}C_4H_9)_4N$][$Ni(dmit)_2$]; [BMIM][$(CF_3SO_2)_2N$]	1/2/20	AN/AC: 1/1 vol. (−80°C)	30 nm
BEDT-TTF; I_2; [BMIM][$(CF_3SO_2)_2N$]	1/1/20	THF (85°C)	35 nm
TMTSF; [BMIM][PF_6]	1/10	CH_2Cl_2 (25°C, 10^2 μA cm^{-2})	30 nm
TMTSF; [BMIM][PF_6]	1/10	CH_2Cl_2 (25°C, 2×10^3 μA cm^{-2})	50 nm
TMTSF; [BMIM][PF_6]	1/10	CH_2Cl_2 (25°C, 5×10^3 μA cm^{-2})	55 nm

TEM, transmission electron microscopy; AN, acetonitrile; AC, acetone; THF, tetrahydrofurane.

6.2.3.3 Long-Alkyl Chains Quaternary Ammonium Salts for the Growth of Nanoparticles of Molecular Conductors

As for [BMIM][PF_6], used for the preparation of $(TMTSF)_2PF_6$ nanoparticles, long-alkyl chains quaternary ammonium salts, that is, [$(CH_3)(n\text{-}C_8H_{17})_3N$]X or [$(n\text{-}C_8H_{17})_4N$]X played the same "triple" role: supporting electrolyte, growth controlling agent, and X^- counteranion provider. Various mixed-valence salts have been synthesized at 25°C as roughly spherical nanoparticles or irregularly shaped nanocrystals using an electrochemical procedure under an applied current density of about 10^2 μA cm^{-2} [30,45] (see Table 6.2). For TTF or BEDT-TTF-based compounds, well-dispersed spherical nanoparticles were observed, whereas nanocrystals (irregular in

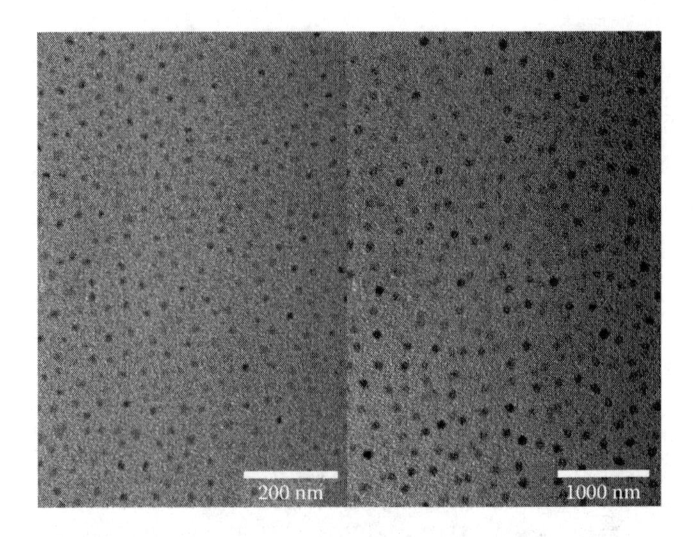

FIGURE 6.7 TEM image of $TTF[Ni(dmit)_2]_2$ (scale bar 200 nm) and $(TMTSF)_2PF_6$ (scale bar 1000 nm) obtained as nanoparticles by electrochemical oxidation in the presence of $BMIM^+$. (de Caro, D., Faulmann, C., Valade, L., Jacob, K., Chtioui, I., Foulal, S., de Caro, P. et al.: Four molecular superconductors isolated as nanoparticles. *Eur. J. Inorg. Chem.* 2014. 4010–4016. Copyright Wiley-VCH Verlag GmbH & Co. KGaA. Reproduced with permission.)

TABLE 6.2
Synthesis Conditions for Nanoparticles of Molecular Conductors Grown in the Presence of a Long-Alkyl Chains Quaternary Ammonium Salt at 10^2 μA cm^{-2}

Precursor Molecules	Molar Ratio (Solvent)	Compound	TEM Observation (nm)
TTF; $[(CH_3)(n\text{-}C_8H_{17})_3N]Cl$	1/10 (AN)	$TTFCl_{0.77}$	20
TTF; $[(n\text{-}C_8H_{17})_4N]Br$	1/10 (AN)	$TTFBr_{0.59}$	30
BEDT-TTF; $[(CH_3)(n\text{-}C_8H_{17})_3N]Cl$	1/10 (THF)	$(BEDT\text{-}TTF)Cl_{0.66}$	50
BEDT-TTF; $[(n\text{-}C_8H_{17})_4N]Br$	1/10 (THF)	$(BEDT\text{-}TTF)Br_{0.50}$	40
TMTSF; $[(CH_3)(n\text{-}C_8H_{17})_3N]ClO_4$	1/10 (CH_2Cl_2)	$(TMTSF)_2ClO_4$	20–60 (irregular in shape)

TEM, transmission electron microscopy; AN, acetonitrile; THF, tetrahydrofurane.

shape) were obtained for the $(TMTSF)_2ClO_4$ phase (Figure 6.8). These nanocrystals were actually made of agglomerates of smaller nanoparticles with diameters in the 3–5 nm range [41]. The observed interplanar spacings were indexed in a vast majority of cases according to the well-known $(TMTSF)_2ClO_4$ triclinic structure (Figure 6.9).

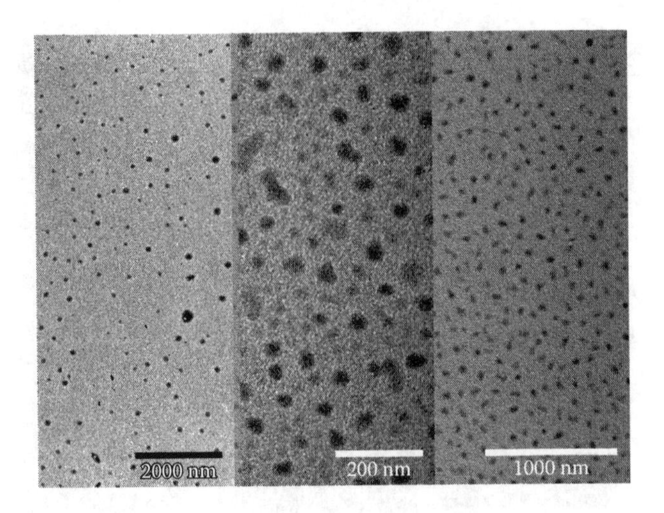

FIGURE 6.8 TEM image of $TTFCl_{0.77}$, $(BEDT\text{-}TTF)_2Br$, and $(TMTSF)_2ClO_4$ obtained as nanoparticles in the presence of long-alkyl chains quaternary ammonium salts. (de Caro, D., Valade, L., Faulmann, C., Jacob, K., Van Dorsselaer, D., Chtioui, I., Salmon, L. et al., Nanoparticles of molecule-based conductors, *New J. Chem.*, 37, 3331–3336, 2013. Reproduced by permission of The Royal Society of Chemistry and the Centre National de la Recherche Scientifique.)

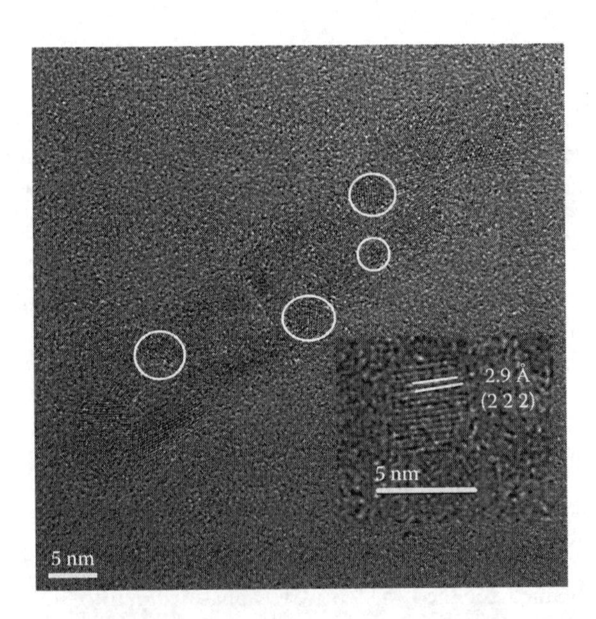

FIGURE 6.9 HRTEM image taken at 200 kV of aggregated nanoparticles of $(TMTSF)_2ClO_4$ with $[(CH_3)(n\text{-}C_8H_{17})_3N]ClO_4$ exhibiting diameters in the 3–5 nm range (scale bar 5 nm) (insert: one indexed 4.5 nm nanoparticle). (de Caro, D., Faulmann, C., Valade, L., Jacob, K., Chtioui, I., Foulal, S., de Caro, P. et al.: Four molecular superconductors isolated as nanoparticles. *Eur. J. Inorg. Chem.* 2014. 4010–4016. Copyright Wiley-VCH Verlag GmbH & Co. KGaA. Reproduced with permission.)

6.2.4 Nanoparticles Grown in the Presence of a Neutral Amphiphilic Molecule

6.2.4.1 Introduction

An amphiphilic molecule is related to a compound having a polar water-soluble group attached to a water-insoluble hydrocarbon chain. Neutral amphiphilic molecules have been widely used to prepare colloidal solutions of conventional metals, metal oxides, metal sulfides, etc. The amphiphilic molecule, acting as a stabilizing agent, controlled the particle growth through coordination to the metal center and was responsible for their solubility [46]. The most frequently encountered were saturated or unsaturated long-alkyl chain amines, carboxylic acids or thiols, such as octylamine (OA), hexadecylamine, octanethiol, oleic acid, or oleylamine. For the growth of molecule-based conductors as nanoparticles, three cases can arise: the amphiphilic molecule is itself one of the precursors, the amphiphilic molecule reacts with one of the precursors, or its presence in solution controls the growth as nanoparticles but is not present in the final material.

6.2.4.2 Precursors Acting as the Amphiphilic Molecule

Until now, the only example of spherical nanoparticles (nanodots, as called by the authors) prepared with an amphiphilic TTF-based precursor has been reported by W. Akutagawa and coworkers [4,47]. The precursor was a bis-TTF annulated macrocyclic compound bearing two $S(CH_2)_9CH_3$ groups, simplified as macro-TTF in the following (Figure 6.1). Langmuir–Blodgett films of macro-TTF on mica evidenced nanodot array and nanodot ring structures. This organization was favored by the terminal ethylenedithio group, which increased the intermolecular interactions through lateral S–S contacts, in addition to the π–π interaction. The reaction of macro-TTF in the presence of two equivalents of I_2 led to the mixed-valence conductor (macro-TTF)$(I_2)_2$. A chloroform/acetonitrile solution of (macro-TTF)$(I_2)_2$ was spin coated on freshly cleaved mica. Isolated nanodots were observed in spin-coated films (Figure 6.10). Their height (measured by AFM) depended on the rotation speed of the spinner: from 20 nm at 6000 rpm to 80 nm at 500 rpm (see Table 6.3).

6.2.4.3 Precursors Reacting with the Amphiphilic Molecule

Long-alkyl chain amines ($C_nH_{2n+1}NH_2$) are excellent stabilizing agents for controlling the growth of metal particles exhibiting diameters as low as 1 nm [46]. They can prevent agglomeration of the nanoparticles by an efficient coordination on surface metal atoms. TTF·TCNQ nanoparticles stabilized by OA ($C_8H_{17}NH_2$) have also been reported (Figure 6.11). The reaction of TTF, TCNQ, and OA (1:1:1 in molar proportions) led to TTF·TCNQ nanoparticles as a black stable dispersion in tetrahydrofurane [48]. AFM images evidenced nanoparticles with heights in the 10–35 nm range (see Table 6.3). The mechanism of their stabilization in organic solvents (such as tetrahydrofurane, acetone, ethanol, diethyl ether) was elucidated. During the growth of the TTF·TCNQ adduct, some TCNQ molecules reacted with OA leading to a TCNQ-OA compound (TCNQ in which a cyano group was substituted by the amino group of OA). Nanoparticles were considered as a TTF·TCNQ core surrounded by TCNQ-OA molecules (bound via π–π stacking to the core). The remaining free OA

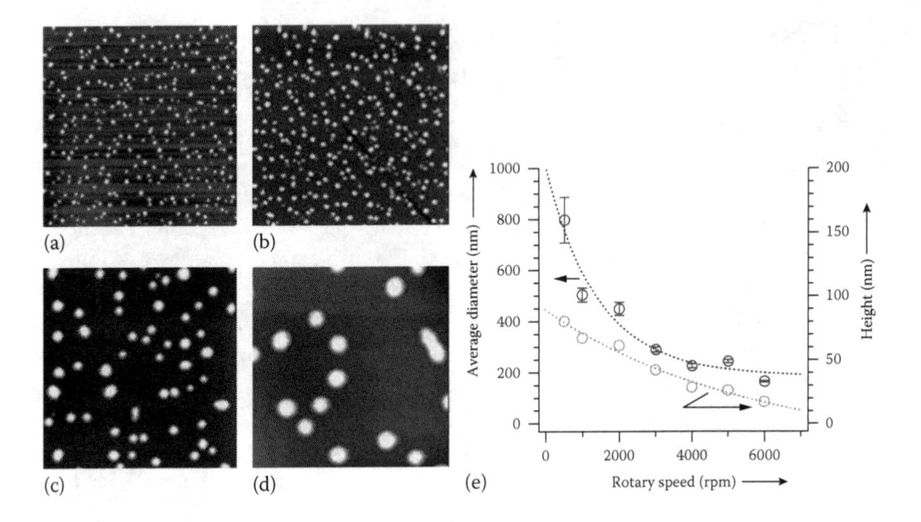

FIGURE 6.10 (a–d) AFM images of size-controllable nanodots of the (macro-TTF)$(I_2)_2$ complex with an open-shell electronic structure on mica. The rotation speed of the substrate on the spinner was 6000, 4000, 2000, and 500 rpm for (a–d), respectively. The scale of all AFM images is 10×10 μm². (e) Rotational-speed dependence of the average diameter and height of the nanodots on mica. The average diameter and height of the nanodots were obtained by a statistical analysis of the AFM images (10×10 μm²). (Akutagawa, T., Kakuichi, K., Hagegawa, T., Noro, S., Nakamura, T., Hasegawa, H., Mashiko, S., and Becher, J.: Molecularly assembled nanostructures of a redox-active organogelator. *Angew. Chem. Int. Ed.* 2005. 44. 7283–7287. Copyright Wiley-VCH Verlag GmbH & Co. KGaA. Reproduced with permission.)

was then associated with TCNQ-OA surface molecules via hydrophobic interactions. This OA-based protecting shell prevented nanoparticles from the aggregation and explained their solubility in common organic solvents. Smaller nanoparticles (heights in the 3–5 nm range according to AFM images, see Table 6.3) were isolated by adding a well-defined amount of pure TCNQ-OA to the reaction medium, also containing TTF, TCNQ, and free OA [48].

6.2.4.4 The Amphiphilic Molecule Simply as a Growth Controlling Agent

The neutral amphiphilic molecules explored for nanoparticles growth control (see Table 6.3) did not react with the TTF-based precursor. However, its presence in solution (as for ionic species described in Section 6.2.3) allowed the growth of $(TMTSF)_2ClO_4$ as nanoparticles [41]. They were obtained by the oxidation of neutral TMTSF in the presence of amphiphilic molecules gathered in Table 6.3, $[(n\text{-}C_4H_9)_4N]ClO_4$ being the perchlorate source. Except for hexadecylamine, electron micrographs exhibited irregularly shaped nanocrystals with sizes in the 35–70 nm range (Figure 6.12). In the presence of hexadecylamine, spherical, 35 nm diameter nanoparticles were observed (Figure 6.12). When hexadecylamine was used in combination with $[(CH_3)(n\text{-}C_8H_{17})_3N]ClO_4$, also bearing long alkyl chains, smaller spherical nanoparticles (diameter of about 20 nm) were obtained in addition with few remaining nanocrystals (Figure 6.12).

TABLE 6.3
Synthesis Conditions for Nanoparticles of Molecular Conductors Grown in the Presence of an Amphiphilic Precursor or Growth Controlling Agent

Precursor Molecules	Molar Ratio (Solvent, If Applicable: Applied Current Density)	Compound	AFM or TEM Observation
Bis-TTF annulated macrocycle (macro-TTF); I_2	1/2 ($CHCl_3$/AN: 8/2 vol.)	(macro-TTF)$(I_2)_2$	AFM: 20–80 nm (depending on the rotation speed of the substrate)
TTF; TCNQ; octylamine	1/1/1 (THF)	TTF·TCNQ	AFM: 10–35 nm
TCNQ-OA; TTF; TCNQ; octylamine	0.5/1/1/0.5 (THF)	TTF·TCNQ	AFM: 3–5 nm
TMTSF; octylamine	1/10 (THF, 10^2 μA cm^{-2})	$(TMTSF)_2ClO_4$	TEM: 35–70 nm
TMTSF; dodecylamine	1/10 (THF, 10^2 μA cm^{-2})	$(TMTSF)_2ClO_4$	TEM: 40–65 nm
TMTSF; hexadecylamine	1/10 (THF, 10^2 μA cm^{-2})	$(TMTSF)_2ClO_4$	TEM: 35 nm
TMTSF; methyloleate	1/10 (THF, 10^2 μA cm^{-2})	$(TMTSF)_2ClO_4$	TEM: 35–70 nm
TMTSF; N-octylfurfurylimine	1/10 (THF, 10^2 μA cm^{-2})	$(TMTSF)_2ClO_4$	TEM: 35–70 nm

AFM, atomic force microscopy; TEM, transmission electron microscopy; AN, acetonitrile; THF, tetrahydrofurane.

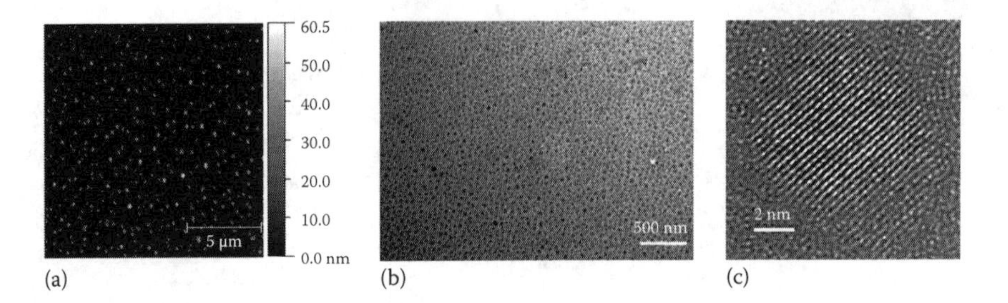

FIGURE 6.11 (a) AFM image of TTF·TCNQ nanoparticles deposited on freshly cleaved mica. (b) TEM image of TTF–TCNQ nanoparticles (scale bar = 500 nm). (c) HRTEM of a 10 nm single TTF·TCNQ nanoparticle. (Reprinted with permission from de Caro, D., Souque, M., Faulmann, C., Coppel, Y., Valade, L., Fraxedas, J., Vendier, O., and Courtade, F., Colloidal solutions of organic conductive nanoparticles, *Langmuir*, 29, 8983, 2013. Copyright 2014 American Chemical Society.)

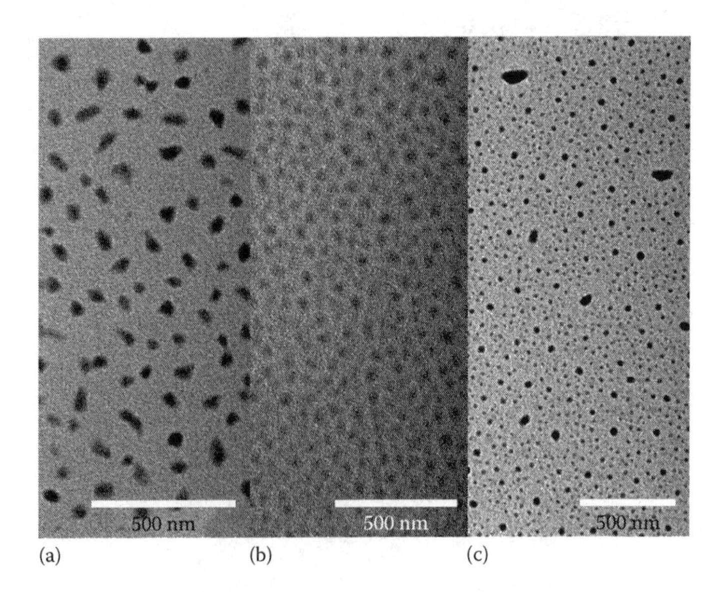

FIGURE 6.12 Electron micrographs for $(TMTSF)_2ClO_4$ with (a) $[(CH_3)(n\text{-}C_8H_{17})_3N]ClO_4$, (b) hexadecylamine and $[(n\text{-}C_4H_9)_4N]ClO_4$, (c) hexadecylamine and $[(CH_3)(n\text{-}C_8H_{17})_3N]ClO_4$ (all scale bars are 500 nm). (de Caro, D., Faulmann, C., Valade, L., Jacob, K., Chtioui, I., Foulal, S., de Caro, P. et al.: Four molecular superconductors isolated as nanoparticles. *Eur. J. Inorg. Chem.* 2014. 4010–4016. Copyright Wiley-VCH Verlag GmbH & Co. KGaA. Reproduced with permission.)

6.3 SPECTROSCOPIC STUDIES AND SOME PHYSICAL PROPERTIES OF NANOPARTICLES OF ORGANIC OR METALLO-ORGANIC CONDUCTORS

6.3.1 Spectroscopic Studies

Infrared, Raman, and x-ray photoelectron spectroscopy (XPS) were the most widely used techniques to characterize nanoparticles of molecule-based conductors. The first two techniques allowed the determination of the degree of charge transfer. X-ray photoelectron spectroscopy provided in some cases the stoichiometry, that is, a confirmation of the chemical composition of the particles.

Infrared spectroscopy is particularly suitable for molecules bearing polar groups such as cyano groups in TCNQ for example. The vibration frequency of the CN groups of TCNQ is characteristic of the formal charge of TCNQ, thus of the charge transfer between the TTF-based molecule and the TCNQ entity. Except for those stabilized by OA, the infrared spectra of TTF·TCNQ nanoparticles [49] were very similar to those reported for bulk single crystals or nanowires [5]. The strong CN stretching mode at 2204 cm^{-1} indicated a charge transfer of 0.59 electron from the donor TTF to the acceptor TCNQ [39], a value identical to that observed on TTF·TCNQ single crystals [50]. For TTF·TCNQ particles capped by TCNQ-OA, four absorptions were observed in the CN region: 2104, 2126, 2175, and 2205 cm^{-1} [48]. The infrared absorption at 2205 cm^{-1} was due to the TTF·TCNQ core, whereas the others were assigned to both TCNQ-OA and TCNQ surface molecules disrupted by the stacking with TCNQ-OA.

Raman spectroscopy is a powerful technique to investigate polarizable molecules. It is a technique of choice for studying molecules exhibiting large delocalized π systems, such as those listed in Figure 6.1. Except for those stabilized by OA, the Raman spectra of TTF·TCNQ nanoparticles [49] were very similar to those reported for bulk single crystals [51]. In the C=C region, signals at 1418, 1461, 1516, and 1604 cm^{-1} were observed. The ν_4 a_g mode for TCNQ (located at 1418 cm^{-1}) led to the determination of a −0.55 partial charge borne by the TCNQ, in relatively good agreement with that for TTF·TCNQ single crystals (−0.59). Solid-state UV-visible spectroscopy also confirmed the existence of charge transfer within the TTF·TCNQ nanoparticles and their conducting character [49]. The plasma reflection was observed at 10700 cm^{-1}. From this value, the conduction-band width was evaluated to about 0.59 eV, in good agreement with that calculated for TTF·TCNQ single crystals (0.62 eV) [51]. A charge-transfer band (12,600 cm^{-1}) and two intermolecular transitions of TCNQ (21,700 and 24,000 cm^{-1}) were also observed in the solid-state UV-visible spectrum [49]. For TTF·TCNQ nanoparticles, an additional confirmation of the charge transfer was given by XPS. N 1s signals at 397.5 and 399.0 eV corresponded to charged and neutral TCNQ species, respectively, while S 2p lines at 163.5 and 164.5 eV arose from neutral and charged TTF species, respectively. Both dynamical configurations, neutral and charged, were observed by XPS because photoemission is an intrinsic rapid process (about 10^{-15} s). Raman spectra for OA-stabilized TTF·TCNQ nanoparticles evidenced C=C stretching modes at 1420, 1460, 1513, 1567, and 1604 cm^{-1}. The additional signal

TABLE 6.4
Degree of Charge Transfer and Chemical Composition for Chemically or Electrochemically Grown Nanoparticles

TTF-Based Molecule	Raman Shift for the Central C=C Band (cm^{-1})	Degree of Charge Transfer	Composition
TTF	1447	0.77	$TTFCl_{0.77}$
TTF	1457	0.59	$TTFBr_{0.59}$
BEDT-TTF	1464	0.66	$(BEDT\text{-}TTF)Cl_{0.66}$
BEDT-TTF	1471	0.50	$(BEDT\text{-}TTF^{0.50+})_2(Br^-)$
BEDT-TTF	1471	0.50	$(BEDT\text{-}TTF^{0.50+})_2(I_3^-)$
TMTSF	1462	0.50	$(TMTSF^{0.50+})_2(ClO_4^-)$
TMTSF	1460	0.50	$(TMTSF^{0.50+})_2(PF_6^-)$

at 1567 cm^{-1} could be due to TCNQ-OA surface molecules [52]. Raman spectroscopy was also performed to determine the degree of charge transfer, and then the chemical composition, for nanoparticles of mixed-valence salts (see Table 6.4). Indeed, the position of the more intense C=C mode (mainly due to the central Csp^2–Csp^2 bond in the TTF-based molecule) linearly depends on the degree of charge transfer.

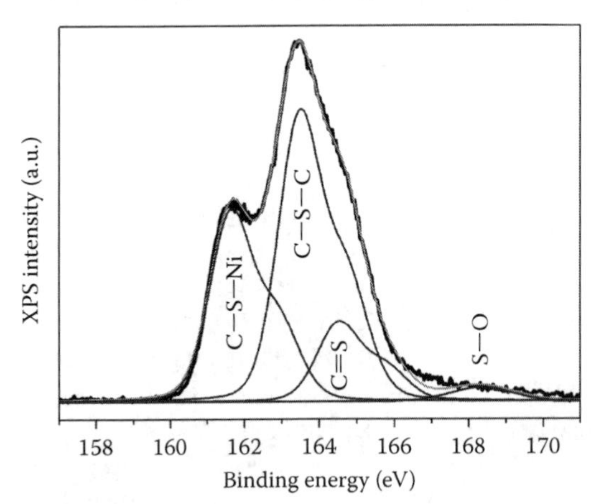

FIGURE 6.13 XPS S 2p line measured ex situ at room temperature for $TTF[Ni(dmit)_2]_2$ nanocrystalline powders (continuous black line). A least-square fit using a combination of Gaussians and Lorentzians after a Shirley-type background subtraction is also shown (continuous grey lines). Dark gray lines correspond to the different contributions associated with different chemical environments: C–S–Ni, C–S–C, and C=S. The small contribution from S–O bonding, due to contamination, is also taken into account. Each line contains two components, $2p_{3/2}$ and $2p_{1/2}$, separated by 1.2 eV, with an intensity ratio 2:1, imposed by the chemical composition of the $Ni(dmit)_2$ molecule. The light gray line corresponds to the sum of all contributions, which satisfactorily fits the experimental data. (Reprinted from *Synth. Met.*, 160, de Caro, D., Jacob, K., Faulmann, C., Legros, J.P., Senocq, F., Fraxedas, J., and Valade, L., Ionic liquid-stabilized nanoparticles of charge transfer-based conductors, 1223–1227, 2010, Copyright 2014, with permission from Elsevier.)

The XPS method also allowed the determination of the stoichiometry for $TTF[Ni(dmit)_2]_2$ nanoparticles grown in the presence of an ionic liquid, acting as a growth controlling agent. This compound contains sulfur atoms in the donor part (TTF) and in the acceptor part ($Ni(dmit)_2$), these atoms being in different chemical environments (simply bound to carbon, to nickel, or doubly bound to carbon). The S 2p line was satisfactorily decomposed into three contributions: 163.5 eV (C–S–C), 161.7 eV (C–S–Ni), and 164.5 eV (C=S). The ratios of peak areas gave ~3:2:1 in excellent agreement with the nominal 12 C–S–C, 8 C–S–Ni, and 4 C=S bonds, thus confirming the 1 TTF/2 [$Ni(dmit)_2$] stoichiometry (Figure 6.13).

Finally, X-ray diffraction patterns for nanoparticle powders or nanoparticle films were usually in agreement with simulations calculated using single crystal data: monoclinic for TTF·TCNQ [38], for $TTF[Ni(dmit)_2]_2$ [40], and for $[(CH_3)_4N][Ni(dmit)_2]_2$ [22]; triclinic for $(BEDT\text{-}TTF)_2I_3$ [41]; orthorhombic for $TTFCl_{0.77}$ [45]. However, in some cases, preferential orientations [38,48] or dominant textures [22] were observed.

6.3.2 Physical Properties

Transport measurements were performed on nanoparticle powders (prepared in the presence of ionic liquids or ammonium salts), on nanopearl chains of $[(CH_3)_4N][Ni(dmit)_2]_2$, and on isolated nanodots of (macro-TTF)$(I_2)_2$. The conductivity for nanoparticle powders was measured on a layer deposited from a dispersion of the nanoparticles in ether on a support containing four connecting gold lines. Conductivity values were in the 1–10 S cm^{-1} range [30]. These values were one or two orders of magnitude lower than those of single crystals but were very close to values on compressed pellets of single crystals. The conductivity of a nanoparticle powder resulted from the contribution of nanograin boundaries in addition to that of each nanograin. For OA-stabilized nanoparticles, conductivity values (0.01–0.1 S cm^{-1}) were one or two orders of magnitude lower than those of nanoparticles grown in the presence of [BMIM][X] or $[R_4N]X$ [48]. In this case, lower values were due to the resistive contribution of the neutral OA layer around each nanoparticle. The conductive properties of nanopearl chains of $[(CH_3)_4N][Ni(dmit)_2]_2$ were studied by conductive atomic force microscopy (C-AFM) [22]. Values of about 1 S cm^{-1} along a single chain of nanoparticles were reported (Figure 6.14). The conductance of single nanodots of (macro-TTF)$(I_2)_2$ fabricated on highly ordered pyrolytic graphite substrates was measured by C-AFM [47]. The conductance of a single nanodot was obtained from the slope of the current–voltage curve (Figure 6.15). A value of 20 nS (at 5×10^{-3} Pa) was observed for a nanodot exhibiting a height of 23 nm according to AFM images. This value increased to about 85 nS under ambient pressure conditions.

TTF·TCNQ nanoparticles grown in the presence of [BMIM][X] were used as source and drain electrode in pentacene-based organic field effect transistors (OFETs) [39]. Organic-based conductors were of interest for this application because the reduced contact resistance at the organic/organic (TTF·TCNQ/pentacene) interface improved the performance of bottom-contact transistors. Mobility values of

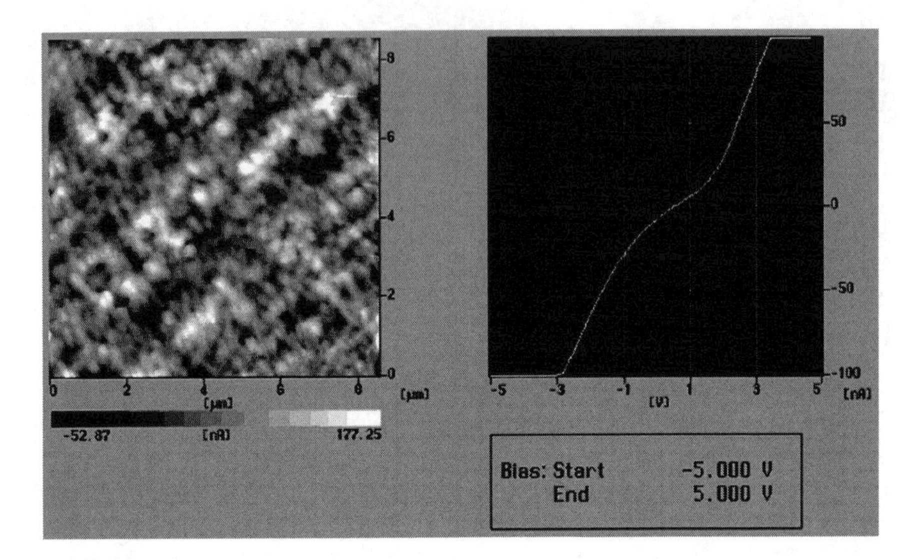

FIGURE 6.14 The C-AFM data of the $[(CH_3)_4N][Ni(dmit)_2]_2$ nanowire arrays. The left is a simultaneous current image of the nanowire arrays at a bias voltage of 5 V. After the template was partially dissolved, the C-AFM data were measured on the top of nanowire arrays with the average pore diameter of 49 ± 2 nm. (Reprinted with permission from Cui, G., Xu, W., Guo, C., Xiao, X., Xu, H., Zhang, D., Jiang, L., and Zhu, D., Conducting nanopearl chains based on the dmit salt., *J. Phys. Chem. B*, 108, 13638–13642, 2004. Copyright 2014 American Chemical Society.)

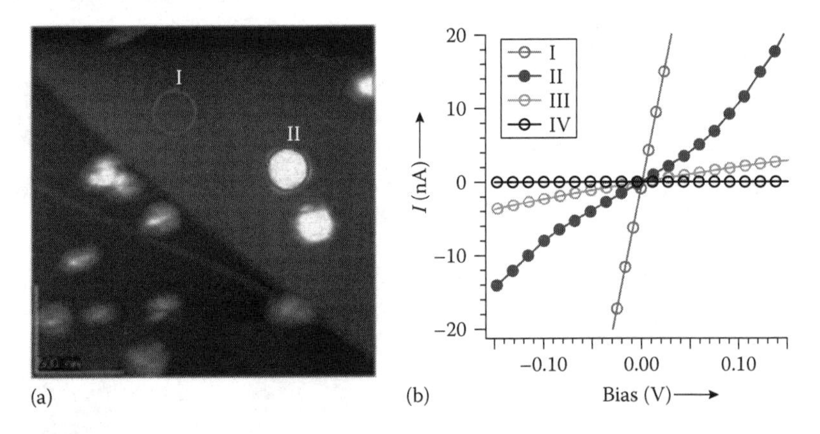

FIGURE 6.15 (a) AFM image of nanodots of (macro-TTF)$(I_2)_2$ on HOPG for conducting-AFM (C-AFM) measurements (2 × 2 μm²). After measurement of the topographical image, the Pt-coated AFM tip contacted the bare HOPG surface (point I) and a single nanodot (point II) to measure their *I–V* characteristics. The nanodot structures were retained after the conductivity measurement, as confirmed by AFM measurement. (b) *I–V* characteristics at points I (HOPG), II (single nanodot in air), III (single nanodot in vacuum at 5×10^{-3} Pa), and IV (bundle of fibers). DC bias was applied from −0.2 to 0.2 V. (Akutagawa, T., Kakuichi, K., Hagegawa, T., Noro, S., Nakamura, T., Hasegawa, H., Mashiko, S., and Becher, J.: Molecularly assembled nanostructures of a redox-active organogelator. *Angew. Chem. Int. Ed.* 2005. 44. 7283–7287. Copyright Wiley-VCH Verlag GmbH & Co. KGaA. Reproduced with permission.)

0.042 cm^2 V^{-1} s^{-1} were obtained, larger than those for bottom-contact devices with usual Au electrodes (0.011 cm^2 V^{-1} s^{-1}) because the TTF·TCNQ/pentacene interface removed the undesirable interfacial potential and morphological discontinuity [53]. However, values of 0.042 cm^2 V^{-1} s^{-1} were lower than those obtained with evaporated thin TTF·TCNQ films as electrodes (0.340 cm^2 V^{-1} s^{-1}) [53]. This was consistent with the larger amount of grain boundaries in nanoparticle-based films.

Superconductivity in $(TMTSF)_2X$ systems was first discovered by Jérome and coworkers [54] in $(TMTSF)_2PF_6$ in 1980 and by Bechgaard [55] in $(TMTSF)_2ClO_4$ in 1981. $(TMTSF)_2ClO_4$ is the only member of the $(TMTSF)_2X$ series displaying superconductivity at ambient pressure (critical temperature T_C = 1.3 K according to resistivity measurements). Gubser and coworkers evidenced the superconducting transition on a bundle of $(TMTSF)_2ClO_4$ single crystals by *ac* magnetic susceptibility measurements [56]. The critical temperature, defined as the midpoint of the relative susceptibility curve (plotted as a function of temperature), was 1.1 K. The superconducting properties of $(TMTSF)_2ClO_4$ nanoparticles (prepared in the presence of $[(CH_3)(n\text{-}C_8H_{17})_3N]ClO_4$) have been investigated by inductive susceptibility measurements using a tunnel diode oscillator [57] (Figure 6.16). The relative susceptibility curve exhibited the same feature as that reported by Gubser and coworkers [56]. The critical temperature was the same, about 1.2 K. Until now, it is the only example of organic conductor nanoparticles (3–5 nm in diameter organized as 20–60 nm nanocrystals) undergoing a superconducting transition. The observation of such a transition is not

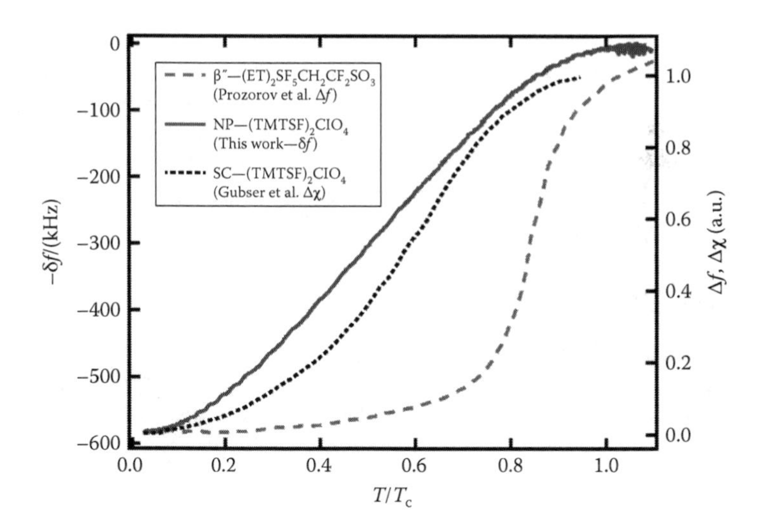

FIGURE 6.16 Observation of the superconducting transition of nanoparticles of $(TMTSF)_2ClO_4$ as a function of temperature using a TDO (top curve). The middle curve is that reported by Gubser et al. [56]. (Reprinted with permission from Winter, L.E., Steven, E., Brooks, J.S., Benjamin, S., Park, J.-H., de Caro, D., Faulmann, C. et al., Spin density wave and superconducting properties of nanoparticle organic conductor assemblies, *Phys. Rev. B*, 91, 035437–1/7, 2015. Copyright 2015 by the American Physical Society.)

too much surprising because the coherence length for TMTSF-based mixed-valence compounds is about 50 nm, value in agreement with nanocrystal sizes. However, it has been reported that λ-$(BETS)_2GaCl_4$ exhibited superconductivity in the minute size of four pairs of $(BETS)_2GaCl_4$ based on scanning tunneling microscopy study (BETS: bis(ethylenedithio)tetraselenafulvalene) [58]. The classical mechanism for superconductivity in organic superconductors (electron coupled in pairs, the so-called Cooper pairs, forming a Bose condensate flowing without resistance) [59] will perhaps be reviewed in light of these novel nano-objects. Tunneling Cooper pairs, when the size of the object is lower than the coherence length, is one possible answer.

6.4 CONCLUDING REMARKS

As shown in this chapter, the synthesis of nanoparticles of quasi-one-dimensional conductors is a recent field. It is largely a new scientific concept that has not so far been explored. The physical characteristics of the corresponding bulk conductors are well known. However, physical studies on nanoparticles are at a very early stage. Their use as active components for source and drain electrodes in OFETs is the only application known at the moment [39]. Their use for future electronic device applications could develop rapidly. Indeed, nanoparticles of organic conductors are prepared under mild conditions, whereas commonly used metals and semiconductors usually require metallurgical elaboration methods. Moreover, slight changes in the molecular structure on at least one of the molecules constituting the conductor can drastically influence its physical properties. Finally, an important aspect for applications is that, contrary to bulk organic conductors, nanoparticles exhibit significant solubility in common organic solvents [41,47,48]. Synthetic methods described in this chapter allowed the preparation of nanoparticle powders, except for nanodots of (macro-TTF)$(I_2)_2$. Ionic liquids, neutral or ionic species, bearing one or several long alkyl chains served as growth controllers. A future synthetic challenge is the organization of molecule-based nanoparticles on nano-structured or functionalized surfaces. Physical properties on regular arrangements of nano-objects or on individual nano-objects would then be more easily investigated.

ACKNOWLEDGMENTS

The authors are thankful to K. Jacob, J.P. Legros, M. Souque, I. Chtioui, D. Van Dorsselaer, Y. Coppel, M.T. Carayon, L. Salmon, L. Viau, P. de Caro, M. Bergez-Lacoste, O. Vendier, F. Courtade, J. Fraxedas, B. Ballesteros, T. Mori, T. Kadoya, J.S. Brooks, E. Steven, and L.E. Winter.

REFERENCES

1. Deluzet, A., Perruchas, S., Bengel, H., Batail, P., Molas, S., and Fraxedas, J. 2002. Thin single crystals of organic insulators, metals, and superconductors by confined electrocrystallization. *Adv. Funct. Mater.* 12:123–128.
2. Xiao, J., Yin, Z., Li, H., Zhang, Q., Boey, F., Zhang, H., and Zhang, Q. 2010. Postchemistry of organic particles: When TTF microparticles meet TCNQ microstructures in aqueous solution. *J. Am. Chem. Soc.* 132:6926–6928.

3. Mas-Torrent, M., Hadley, P., Bromley, S.T., Ribas, X., Tarrés, J., Mas, M., Molins, E., Veciana, J., and Rovira, C. 2004. Correlation between crystal structures and mobility in organic field-effect transistors based on single crystals of tetrathiafulvalene derivatives. *J. Am. Chem. Soc.* 126:8546–8553.
4. Gomar-Nadal, E., Puigmartí-Luis, J., and Amabilino, D.B. 2008. Assembly of functional molecular nanostructures on surfaces. *Chem. Soc. Rev.* 37:490–504.
5. de Caro, D. 2011. Nanowires of molecule-based charge-transfer salts. In *Nanowires, Fundamental Research*, Hashim, A. ed., pp. 509–526. Rijeka, Croatia: Intech.
6. Savy, J.P., de Caro, D., Faulmann, C., Valade, L., Almeida, M., Koike, T., Fujiwara, H., Sugimoto, T., Ondarçuhu, T., and Pasquier, C. 2007. Nanowires of molecule-based charge-transfer salts. *New J. Chem.* 31:519–527.
7. Ren, L., Xian, X., Yan, K., Fu, L., Liu, Y., Chen, S., and Liu, Z. 2010. A general electrochemical strategy for synthesizing charge-transfer complex micro/nanowires. *Adv. Funct. Mater.* 20:1209–1223.
8. Ji, H.X., Hu, J.S., Tang, Q.X., Hu, W.P., Song, W.P., and Wan, L.J. 2006. Bis(ethylenedithio)tetrathiafulvalene charge-transfer salt nanotube arrays. *Adv. Mater.* 18:2753–2757.
9. Jung, Y.J., Kim, Y., Kim, G.T., Kang, W., and Noh, D.Y. 2012. Electrochemical fabrication of $(TMTSF)_2X$ (X = PF_6, BF_4, ClO_4) nanowires. *J. Nanosci. Nanotechnol.* 12:5397–5401.
10. Puigmartí-Luis, J., Schaffhauser, D., Burg, B.R., and Dittrich, P.S. 2010. A microfluidic approach for the formation of conductive nanowires and hollow hybrid structures. *Adv. Mater.* 22:1–5.
11. Liu, H., Li, J., Lao, C., Huang, C., Li, Y., Wang, Z.L., and Zhu, D. 2007. Morphological tuning and conductivity of organic conductor nanowires. *Nanotechnology* 18:495704(7pp).
12. Lv, J., Liu, H., and Li, Y. 2008. Self-assembly and properties of low-dimensional nanomaterials based on π-conjugated organic molecules. *Pure Appl. Chem.* 80:639–658.
13. Huang, C., Zhang, Y., Liu, H., Cui, S., Wang, C., Jiang, L., Yu, D., Li, Y., and Zhu, D. 2007. Controlled growth and field-emission properties of the organic charge-transfer complex of κ-$(BEDT\text{-}TTF)_2Cu(SCN)_2$ nanorod arrays. *J. Phys. Chem. C* 111:3544–3547.
14. Favier, F., Liu, H., and Penner, R.M. 2001. Size-selective growth of nanoscale tetrathiafulvalene bromide crystallites on platinum particles. *Adv. Mater.* 13:1567–1570.
15. Catala, L., Volatron, F., Brinzei, D., and Mallah, T. 2009. Functional coordination nanoparticles. *Inorg. Chem.* 48:3360–3370.
16. Jeszka, J.K., Tracz, A., Wostek, D., Boiteux, G., and Kryszewski, M. 2001. Preparation of ET_2I_3 nanocrystals. *J. New Mater. Electrochem. Syst.* 4:149–153.
17. Van Keuren, E. and Nishida, M. 2009. Synthesis of nanocomposite materials using the reprecipitation method. *Comput. Mater. Con.* 14:61–77.
18. Mas-Torrent, M., Laukhina, E., Rovira, C., Veciana, J., Tkacheva, V., Zorina, L., and Khasanov, S. 2001. New transparent metal-like bilayer composite films with highly conducting layers of θ-$(BEDT\text{-}TTF)_2Br{\cdot}3H_2O$ nanocrystals. *Adv. Funct. Mater.* 11:299–303.
19. Mas-Torrent, M., Ribera, E., Tkacheva, V., Mata, I., Molins, E., Vidal-Gancedo, J., Khasanov, S. et al. 2002. New molecular conductors based on ETEDT-TTF trihalides: From single crystals to conducting layers of nanocrystals. *Chem. Mater.* 14:3295–3304.
20. Hossick Schott, J. and Ward, M.D. 1994. Snapshots of crystal growth: Nanoclusters of organic conductors on Au(111) surfaces. *J. Am. Chem. Soc.* 116:6806–6811.
21. Hillier, A.C., Hossick Schott, J., and Ward, M.D. 1995. Molecular nanoclusters as precursors to conductive thin films and crystals. *Adv. Mater.* 7:409–413.
22. Cui, G., Xu, W., Guo, C., Xiao, X., Xu, H., Zhang, D., Jiang, L., and Zhu, D. 2004. Conducting nanopearl chains based on the dmit salt. *J. Phys. Chem. B* 108:13638–13642.
23. Shenhar, R., Norsten, T.B., and Rotello, V.M. 2005. Polymer-mediated nanoparticle assembly: Structural control and applications. *Adv. Mater.* 17:657–669.

24. Zhang, A.Q., Cai, L.J., Sui, L., Qian, D.J., and Chen, M. 2013. Reducing properties of polymers in the synthesis of noble metal nanoparticles. *Polym. Rev.* 53:240–276.
25. Toshima, N. 2010. Inorganic nanoparticles for catalysis. In *Inorganic Nanoparticles: Synthesis, Applications and Perspectives*, Altavilla, C. and Ciliberto, E. eds., pp. 475–509. London, U.K.: CRC Press.
26. Uemura, T. and Kitagawa, S. 2003. Prussian blue nanoparticles protected by poly(vinylpyrrolidone). *J. Am. Chem. Soc.* 125:7814–7815.
27. Tissot, A. Rechignat, L., Bousseksou, A., and Boillot, M.L. 2012. Micro- and nanoparticles of iron(III) spin-transition material $[Fe^{III}(3\text{-MeO-SalEen})_2]PF_6$. *J. Mater. Chem.* 22:3411–3419.
28. Kadoya, T., de Caro, D., Jacob, K., Faulmann, C., Valade, L., and Mori, T. 2011. Charge injection from organic charge-transfer salts to organic semiconductors. *J. Mater. Chem.* 21:18421–18424.
29. Souque, M., de Caro, D., and Valade, L. 2011. Tetrathiafulvalenium-bromide systems as nanosticks and nanoparticles stabilized by PEDOT. *Synth. Met.* 161:1001–1004.
30. de Caro, D., Valade, L., Faulmann, C., Jacob, K., Van Dorsselaer, D., Chtioui, I., Salmon, L. et al. 2013. Nanoparticles of molecule-based conductors. *New J. Chem.* 37:3331–3336.
31. Bhatt, A., Mechler, Á., Martin, L.L., and Bond, A.M. 2007. Synthesis of Ag and Au nanostructures in an ionic liquid: Thermodynamic and kinetic effects underlying nanoparticle, cluster and nanowire formation. *J. Mater. Chem.* 17:2241–2250.
32. Clavel, G., Larionova, J., Guari, Y., and Guérin, C. 2006. Synthesis of cyano-bridged magnetic nanoparticles using room-temperature ionic liquids. *Chem. Eur. J.* 12:3798–3804.
33. Gutel, T., Garcia-Antõn, J., Pelzer, K., Philippot, K., Santini, C.C., Chauvin, Y., Chaudret, B., and Basset, J.M. 2007. Influence of the self-organization of ionic liquids on the size of ruthenium nanoparticles: Effect of the temperature and stirring. *J. Mater. Chem.* 17:3290–3292.
34. Sakata, M., Yoshida, Y., Maesato, M., Saito, G., Matsumoto, K., and Hagiwara, R. 2006. Preparation of superconducting $(TMTSF)_2NbF_6$ by electrooxidation of TMTSF using ionic liquid as electrolyte. *Mol. Cryst. Liq. Cryst.* 452:103–112.
35. Yoshida, Y., Sakata, M., Saito, G., Matsumoto, K., and Hagiwara, R. 2007. New α'-type ET salt $(ET)_2H_2F_3$ by electrocrystallization using ionic liquid. *Chem. Lett.* 36:226–227.
36. Mei, X. and Ouyang, J. 2011. Electronically and ionically conductive gels of ionic liquids and charge-transfer tetrathiafulvalene-tetracyanoquinodimethane. *Langmuir* 27:10953–10961.
37. Valade, L. and Tanaka, H. 2010. Molecular inorganic conductors and superconductors. In *Molecular Materials*, Bruce, D.W., O'Hare, D., and Walton, R.I. eds., pp. 211–280. John Wiley & Sons, Chichester.
38. de Caro, D., Jacob, K., Faulmann, C., Legros, J.P., Senocq, F., Fraxedas, J., and Valade, L. 2010. Ionic liquid-stabilized nanoparticles of charge transfer-based conductors. *Synth. Met.* 160:1223–1227.
39. de Caro, D., Jacob, K., Hahioui, H., Faulmann, C., Valade, L., Kadoya, T., Mori, T., Fraxedas, J., and Viau, L. 2011. Nanoparticles of organic conductors: Synthesis and application as electrode material in organic field effect transistors. *New J. Chem.* 35:1315–1319.
40. de Caro, D., Jacob, K., Faulmann, C., Valade, L., and Viau, L. 2012. $TTF[Ni(dmit)_2]_2$: Now as nanoparticles. *C. R. Chimie* 15:950–954.
41. de Caro, D., Faulmann, C., Valade, L., Jacob, K., Chtioui, I., Foulal, S., de Caro, P. et al. 2014. Four molecular superconductors isolated as nanoparticles. *Eur. J. Inorg. Chem.* 4010–4016.
42. Gonfa, G., Bustam, S.A., Man, Z., and Abdul Mutalib, M.I. 2011. Unique structure and solute-solvent interaction in imidazolium based ionic liquids: A review. *Asian Trans. Eng.* 1:24–34.

43. Bechgaard, K., Jacobsen, C.S., Mortensen, K., Pedersen, H.J., and Thorup, N. 1980. The properties of five highly conducting salts $(TMTSF)_2X$, $X^- = PF_6^-$, AsF_6^-, SbF_6^-, BF_4^- and NO_3^-, derived from tetramethyltetraselenafulvalene (TMTSF). *Solid State Commun.* 33:1119–1125.
44. de Caro, D., Jacob, K., Faulmann, C., and Valade, L. First nanoparticles of Bechgaard salts. *C. R. Chimie* 16:629–633.
45. de Caro, D., Jacob, K., Mazzi, S., Carayon, M.T., Faulmann, C., and Valade, L. 2012. First TTF-based conductor nanoparticles by electrocrystallization. *Synth. Met.* 162:805–807.
46. Philippot, K. and Chaudret, B. 2007. Organometallic derived metals, colloids and nanoparticles. In *Comprehensive Organometallic Chemistry III*, Vol. 12, Crabtree, R.H. and Mingos, M.P. eds., pp. 71–99. Amsterdam, the Netherlands: Elsevier.
47. Akutagawa, T., Kakuichi, K., Hagegawa, T., Noro, S., Nakamura, T., Hasegawa, H., Mashiko, S., and Becher, J. 2005. Molecularly assembled nanostructures of a redox-active organogelator. *Angew. Chem. Int. Ed.* 44:7283–7287.
48. de Caro, D., Souque, M., Faulmann, C., Coppel, Y., Valade, L., Fraxedas, J., Vendier, O., and Courtade, F. 2013. Colloidal solutions of organic conductive nanoparticles. *Langmuir* 29:8983–8988.
49. de Caro, D., Jacob, K., Souque, M., and Valade, L. 2012. Vibrational and optical studies of organic conductor nanoparticles. In *Vibrational Spectroscopy*, de Caro, D. ed., pp. 141–152. Rijeka, Croatia: Intech.
50. Chappell, J.S., Bloch, A.N., Bryden, W.A., Maxfield, M., Poelher, P.O., and Cowan, D.O. 1981. Degree of charge transfer in organic conductors by infrared absorption spectroscopy. *J. Am. Chem. Soc.* 103:2442–2443.
51. Graja, A. 1997. *Spectroscopy of Materials for Molecular Electronics*. Poznań, Poland: Scientific Publishers.
52. Souque, M. 2011. Nanostructured molecule-based conductors and metal oxides for space industry. PhD dissertation, Paul Sabatier (Toulouse 3) University, Toulouse, France.
53. Shibata, K., Wada, H., Ishikawa, K., Takezoe, H., and Mori, T. 2007. (Tetrathiafulvalene)(tetracyanoquinodimethane) as low-contact-resistance electrode for organic transistors. *Appl. Phys. Lett.* 90:193509–193511.
54. Jérome, D., Mazaud, A., Ribault, M., and Bechgaard, K. 1980. Superconductivity in a synthetic organic conductor $(TMTSF)_2PF_6$. *J. Phys. Lett.* 41:L95–L98.
55. Bechgaard, K., Carneiro, K., Olsen, M., and Rasmussen, F.B. 1981. Zero-pressure organic superconductor: Di-(Tetramethyltetraselenafulvalenium)-perchlorate $[(TMTSF)_2ClO_4]$. *Phys. Rev. Lett.* 46:852–855.
56. Gubser, D.U., Fuller, W.W., Poehler, T.O., Cowan, D.O., Lee, M., Potember, R.S., Chiang, L.Y., and Bloch, A.N. 1981. Magnetic susceptibility and resistive transition of superconducting $(TMTSF)_2ClO_4$: Critical magnetic films. *Phys. Rev. B* 24:478–480.
57. Winter, L.E., Steven, E., Brooks, J.S., Benjamin, S., Park, J.-H., de Caro, D., Faulmann, C. et al. 2015. Spin density wave and superconducting properties of nanoparticle organic conductor assemblies. *Phys. Rev. B* 91:035437-1/7.
58. Clark, K., Hassanien, A., Khan, S., Braun, K.F., Tanaka, H., and Hla, S.W. 2010. Superconductivity in just four pairs of $(BETS)_2GaCl_4$ molecules. *Nat. Nanotechnol.* 5:261–265.
59. Kanoda, K. 2008. Mott transition and superconductivity in Q2D organic conductors. In *The Physics of Organic Superconductors and Conductors*, Lebed, A. ed., pp. 623–642. Berlin, Germany: Springer.

Section III

Nanostructures/Metal Oxides

7 Metal Silicate and Phosphate Nanomaterials

Pratap Vishnoi and Ramaswamy Murugavel

CONTENTS

7.1 BACKGROUND

Silicates comprise a class of compounds that are often built up of anionic tetrahedral SiO_4^{4-} building units and are the major constituents of a large number of naturally occurring materials, from zeolites to clay, from minerals to glasses. In these materials, SiO_4^{4-} units are cross-linked through corners of shared oxygen, featuring open-framework structures. Utility of silicates as molecular sieves, ion exchangers, drying agents, and heterogeneous catalysts and catalytic support in a myriad of chemical transformations has led to the enormous research activities in this field. Metal phosphates, on the other hand, exist in nature as minerals or as constituents of living animals and play unique roles in biological processes. Although the building units of silicates and phosphates are chemically different, they are quite similar to each other in terms of structure and from the reactivity point of view (Chart 7.1). Thus, the self-assembly process of silicates, phosphates and their porous open-frameworks and other fascinating molecular architectures is a constantly evolving arena of materials chemistry. Ever since the first laboratory synthesis of aluminosilicate zeolite (ZSM-5) by Milton and coworkers in 1949,[1] and aluminophosphates (AlPOs) by Flanigen and coworkers in 1982,[2] a large number of inorganic silicates and phosphates have been designed and synthesized.

Ideally, $Si(OH)_4$ (orthosilicic acid) and $P(O)(OH)_3$ (orthophosphoric acid) could serve as the anion sources for silicates and phosphates, respectively. However, these acids are metastable in their free state and tend to form extended frameworks in an uncontrolled fashion. This ultimately limits the utility of these materials. Therefore, synthetic chemists have sought to use alternative starting materials for the incorporation of Si–O and P–O linkages. This has led to the use of a variety of organosilanols, esters of orthosilicic acid, phosphonic acids, and esters of orthophosphoric acid as the precursors (Chart 7.1). In this connection, studies carried out by Tilley et al.[3]

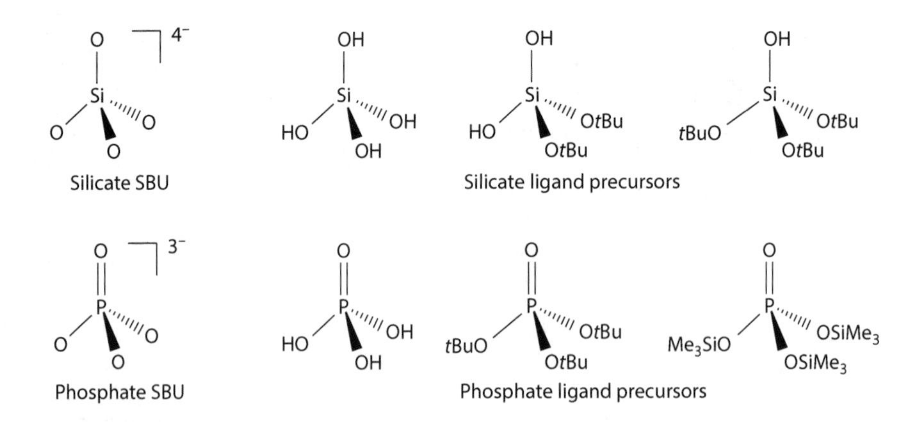

CHART 7.1 Tetrahedral silicate and phosphate building units and ligand precursors.

and Murugavel et al.[4] have established that alkyl esters of these acids can form molecular complexes with metal ions, known as *single source precursors* (SSPs). The SSPs essentially carry elemental combination of the material to be synthesized. Low-temperature thermal decomposition of the SSPs offers materials with tailored properties and structures such as homogeneous dispersion of elements, well-defined porosity, isolated catalytic sites, and particle size. This approach is often referred to as thermolytic molecular precursors (TMP) method.

7.2 SYNTHETIC APPROACH

Molecular precursors for the TMP methods are typically oxygen-rich complexes of $L_nM(O_xE(OR)_y]_z$, where L = alkoxide, amide, alkyl; E = Si, P, B; and R = *t*Bu or $Si(OtBu)_3$. Various molecular precursors based on oxygen-rich ligands have been reported that provide access to materials with three or more heteroelements (i.e., M, Si, P, and/or B) from a single source. Tri-*tert*-butoxysilanol [$HOSi(OtBu)_3$] and di-*tert*-butyl phosphate [$HOP(O)(OtBu)_2$] and their alkali metal derivatives are the most utilized ligands for silicates and phosphates, respectively. Some studies have been carried out with $(HO)_2Si(OtBu)_2$ and $(Me_3SiO)_3P{=}O$. These ligands have attracted interest mainly because of their ability to bind to a variety of metal ions and solubility in common organic solvents. Also, the generation of Si–O or P–O linkages is relatively easy and can be achieved by β-elimination of Si–O*t*Bu/P–O*t*Bu groups at low temperature or facile hydrolysis of Si–O*t*Bu/P–$OSiMe_3$ groups in solution. The thermolysis leads to the conversion of P–O*t*Bu moieties into P–OH, which subsequently undergo condensation to form metal phosphates, as shown in Scheme 7.1.

The molecular precursor species can be condensed with oxide support (i.e., aluminosilicate and silica gel) through covalent grafting. This approach provides well-defined catalytic sites. The first step involves condensation reaction between the precursors and the surface Si–OH groups to yield silica-bound species through $Si\text{–}O\text{–}_{(surface)}$ or $M\text{–}O\text{–}_{(surface)}$ linkages with the loss of HO*t*Bu or $HOSi(OtBu)_3$. Calcination at low temperature (<200 °C) converts these species into $MOx{\cdot}nSiO_2$ or $MOx(n-1)SiO_2$ surface-supported sites with the concomitant loss of isobutylene and water (Scheme 7.2). While several metal silicates and phosphates have been synthesized and have found wide applications, we shall limit our discussion in this chapter to the materials obtained mainly from $HOSi(OtBu)_3$ and HOP(O)$(OtBu)_2$.

M, O, OtBu, P, OtBu, M, O
Thermolysis
$-H_2C{=}CMe_2$
M, O, OH, P, OH, M, O
Condensation
$-H_2O$
Metal phosphate materials

SCHEME 7.1 Schematic presentation of the formation of phosphate materials.

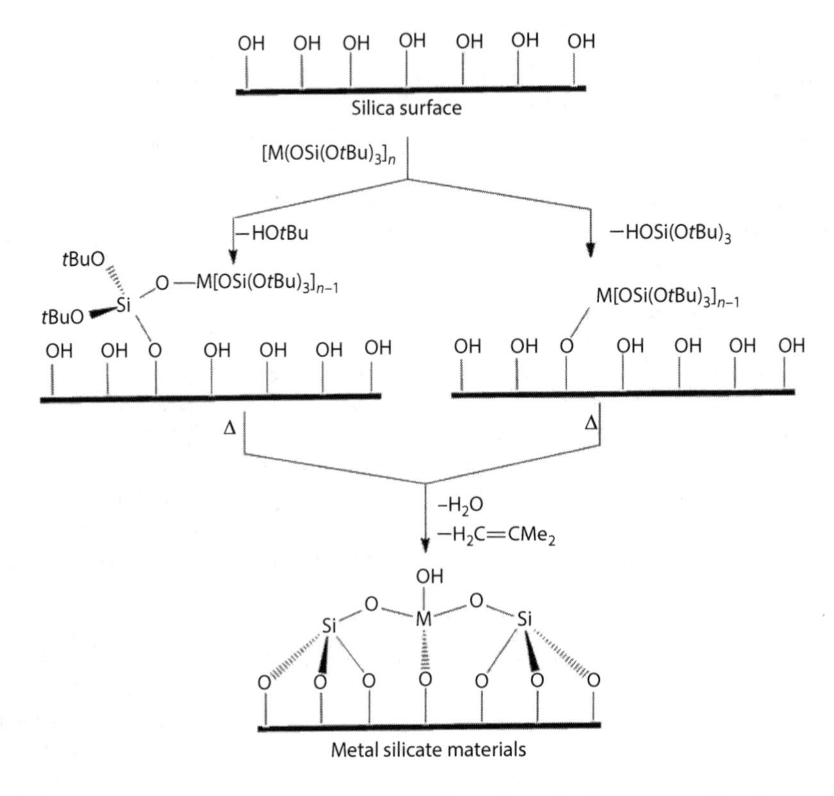

SCHEME 7.2 Schematic presentation of the formation of silicate materials via covalent grafting.

7.3 MOLECULAR METALLOSILICATES

7.3.1 Alkali and Alkaline Earth Metal Silicates

The primary use of alkali and alkaline earth metal siloxides lies in the synthesis of transition and main group metal siloxides. Alkali metal siloxides $MOSi(OtBu)_3$, where M = Li (**1**), Na (**2**), K (**3**), have been readily synthesized by the reaction of $HOSi(OtBu)_3$ with 1 equivalent of corresponding metals in hexanes followed by crystallization at low temperature (Scheme 7.3).[5] Alkaline earth metal siloxides such as magnesium siloxide $Mg[OSi(OtBu)_3]_2$ (**4**) have been synthesized by the reaction of $Mg(nBu)_2$ with $HOSi(OtBu)_3$ in a 1:2 molar ratio in THF. Complex **4** immediately hydrolyzes in moist air and hence could not be isolated as single crystals.[3,6] The stable form of this complex $Mg_4[\mu_2\text{-}OSi(OtBu)_3]_4[OSi(OtBu)_3]_2[\mu_3\text{-}OH]_2$ (**5**) has been isolated as single crystals from a concentrated benzene solution at ambient temperature under argon atmosphere.[7] Other isolated and structurally characterized siloxides include $Mg_5(O)[OSi(OtBu)_3]_5Me_3$ (**6**), obtained from the reaction of $Mg(Me)_2$ with 2 equivalents of $HOSi(OtBu)_3$ in hexane.[7]

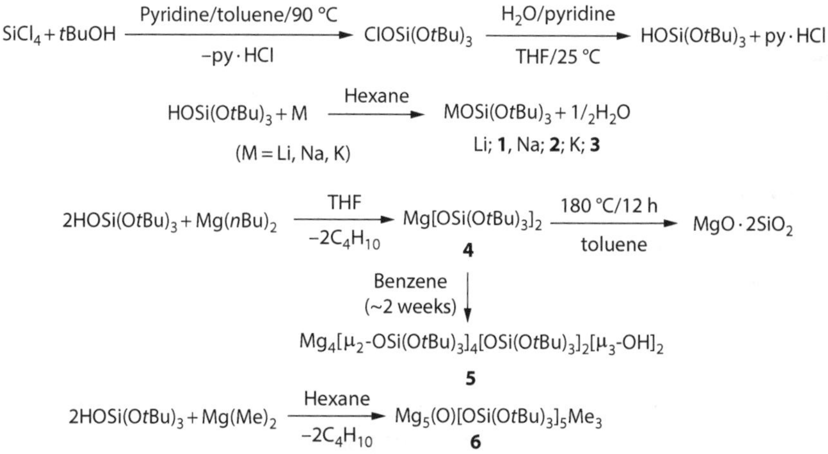

$$SiCl_4 + tBuOH \xrightarrow[-py\cdot HCl]{\text{Pyridine/toluene/90 °C}} ClOSi(OtBu)_3 \xrightarrow[\text{THF/25 °C}]{H_2O/\text{pyridine}} HOSi(OtBu)_3 + py\cdot HCl$$

$$\underset{(M = Li, Na, K)}{HOSi(OtBu)_3 + M} \xrightarrow{\text{Hexane}} \underset{Li;\ \mathbf{1},\ Na;\ \mathbf{2};\ K;\ \mathbf{3}}{MOSi(OtBu)_3 + 1/_2H_2O}$$

$$2HOSi(OtBu)_3 + Mg(nBu)_2 \xrightarrow[-2C_4H_{10}]{\text{THF}} \underset{\mathbf{4}}{Mg[OSi(OtBu)_3]_2} \xrightarrow[\text{toluene}]{180\ °C/12\ h} MgO\cdot 2SiO_2$$

$$\mathbf{4} \xrightarrow{\text{Benzene (~2 weeks)}} \underset{\mathbf{5}}{Mg_4[\mu_2\text{-}OSi(OtBu)_3]_4[OSi(OtBu)_3]_2[\mu_3\text{-}OH]_2}$$

$$2HOSi(OtBu)_3 + Mg(Me)_2 \xrightarrow[-2C_4H_{10}]{\text{Hexane}} \underset{\mathbf{6}}{Mg_5(O)[OSi(OtBu)_3]_5Me_3}$$

SCHEME 7.3 Synthesis of $HOSi(OtBu)_3$ and its conversion to alkali and alkaline earth metal silicates.

The thermolytic decomposition of **4** in toluene leads to the formation of magnesia–silica xerogel $MgO{\cdot}2SiO_2$ as a fine texture composed of nanoparticles (≤5 nm) with BET surface area of 245 m^2g^{-1}. The processing of the xerogel using supercritical CO_2 extraction results in the formation of aerogel with improved surface area of 640 m^2g^{-1} (Figure 7.1).[6]

FIGURE 7.1 TEM micrograph of the aerogel obtained from **4**. (Reproduced from Kriesel, J.W. and Tilley, T.D., *J. Mater. Chem.*, 11, 1081, 2001. With permission.)

7.3.2 Transition Metal Silicates

Molecular complexes incorporating TM–O–Si groups (TM = transition metal) are of great interest, particularly as models of metal complexes immobilized on silica and silicate surfaces. The metal silicates particularly derived from $(HO)Si(OtBu)_3$ and $(HO)_2Si(OtBu)_2$ have attracted great interest owing to their wide applicability in nanomaterials and catalysis.[8]

7.3.2.1 Group 4 Metal Silicates

Tilley and coworkers prepared titanium molecular silicate, $Ti[OSi(OtBu)_3]_4$ (**7**), from the reaction of $Ti(NEt_2)_4$ with 4 equivalents of $HOSi(OtBu)_3$ in nonpolar solvents such as toluene or pentane (Scheme 7.4).[9] This complex has been isolated as single crystals from pentane at −78 °C. The crystals are unstable, which turn opaque in vacuum and hydrolyze back to $HOSi(OtBu)_3$ and metal oxide gels. Bohra et al. have isolated a similar complex $Ti(OtBu)[OSi(OtBu)_3]_3$ (**8**) from dichloromethane, which is a distorted tetrahedral molecule with one-O*t*Bu and three-$OSi(OtBu)_3$ coordinated ligands.[10]

The solid-state pyrolysis of **7** at 500 °C affords low surface area (22 m^2g^{-1}) titania–silica nanoparticles of $TiO_2{\cdot}4SiO_2$ (**7a**) with an average diameter of *ca.* 25 nm (Figure 7.2).[11] The xerogels (**7b**) obtained from the solution-phase thermolysis of **7** show much higher surface area (552 m^2g^{-1} for samples heated to

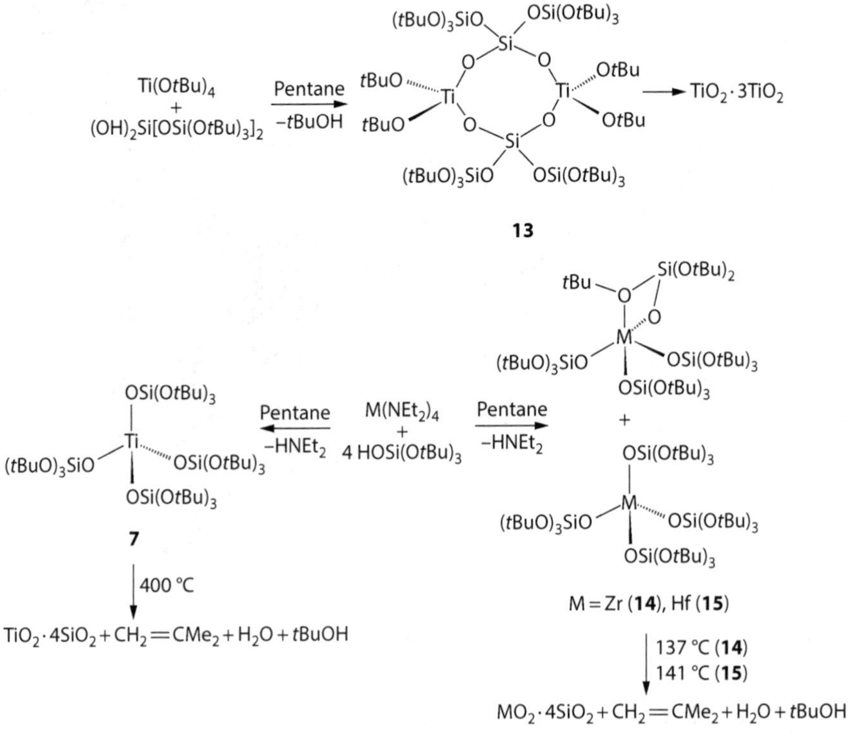

$Ti(OtBu)_4$ + $(OH)_2Si[OSi(OtBu)_3]_2$ —(Pentane, −*t*BuOH)→ **13** → $TiO_2{\cdot}3TiO_2$

7 ←(Pentane, $-HNEt_2$)— $M(NEt_2)_4$ + 4 $HOSi(OtBu)_3$ —(Pentane, $-HNEt_2$)→ $M(OtBu)[\kappa^2\text{-}O_2Si(OtBu)_2][OSi(OtBu)_3]_3$ + $M[OSi(OtBu)_3]_4$; M = Zr (**14**), Hf (**15**)

7 —(400 °C)→ $TiO_2{\cdot}4SiO_2 + CH_2{=}CMe_2 + H_2O + tBuOH$

14, **15** —(137 °C (**14**), 141 °C (**15**))→ $MO_2{\cdot}4SiO_2 + CH_2{=}CMe_2 + H_2O + tBuOH$

SCHEME 7.4 Synthesis of group 4 metal silicates.

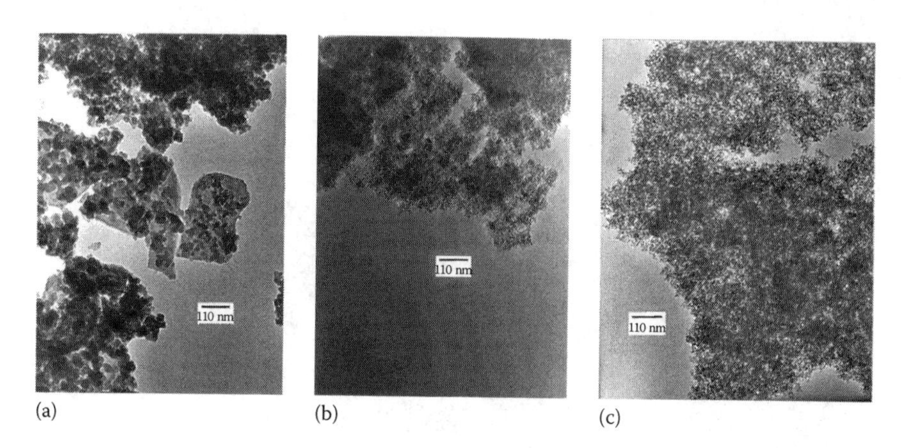

FIGURE 7.2 TEM micrographs of $TiO_2{\cdot}4SiO_2$ materials calcined at 500 °C for 2 h in oxygen. (a) **7a**, (b) **7b**, and (c) **7c**. (Reproduced from Coles, M.P. et al., *Chem. Mater.*, 12, 122, 1999. With permission.)

500 °C) and smaller primary particles (≤*ca.* 5 nm). The supercritical drying of the wet gel **7b** in CO_2 affords an aerogel (**7c**) morphologically similar to the xerogels and a slightly higher surface area (677 m^2g^{-1} for samples heated to 500 °C). Complex **7** serves as a homogeneous catalyst for the selective epoxidation of cyclohexene to cyclohexene oxide using cumene hydroperoxide (CHP). The aerogel **7c** shows superior catalytic performance in olefin epoxidation. Subsequently, various related complexes $(tBuO)_3TiOSi(OtBu)_3$ (**9**), $iPrOTi[OSi(OtBu)_3]_3$ (**10**), $(iPrO)_2Ti[OSi(OtBu)_3]_2$ (**11**), and $(tBuO)Ti[OSi(OtBu)_3]_3$ (**12**) have also been reported and their thermal decomposition behavior has been explored.[10,12,13] The related dinuclear complex $[(tBuO)_2Ti\{\mu\text{-}O_2Si[OSi(OtBu)_3]_2\}]_2$ (**13**) has been isolated via the reaction of $Ti(OtBu)_4$ with 1 equivalent of $(OH)_2Si[OSi(OtBu)_3]_2$ in pentane (Figure 7.3).[14] Solid-state pyrolysis of **9**, **10**, and **11** in air results in the formation of crystalline TiO_2–SiO_2 at 600 °C–650 °C, 700 °C–750 °C, and 800 °C–850 °C, respectively. The thermolysis of **13** yields $TiO_2{\cdot}3SiO_2$, which is highly active and selective for the epoxidation of cyclohexene into cyclohexene oxide using cumene hydroperoxide.

The covalent grafting of **7**, **9**, and **10** onto the inner walls of MCM-41 and SBA-15 silica materials provides catalysts with a large concentration of accessible, isolated, and structurally well-defined active sites.[14,15] The introduction of titanium species onto the silica surface eventually provides $Ti(SO)_4$, $Ti(SO)_3$, and $Ti(SO)_2$ environments in the single-site catalysts. Grafting of **7** is less selective as it gives a mixture of both $(\equiv SiO)Ti[OSi(OtBu)_3]_3$ and $(\equiv SiO)[(tBuO)_2SiO]Ti[OSi(OtBu)_3]_2$. Replacing one of the bulky $-OSi(OtBu)_3$ groups with $-OiPr$ yields mononuclear $(\equiv SiO)Ti(OiPr)[OSi(OtBu)_3]_2$. The SBA-15-supported catalysts are highly active for cyclohexene epoxidation at low Ti loading (0.25−1.77 wt%).

Zirconium and hafnium siloxides $Zr[OSi(OtBu)_3]_4$ (**14**) and $Hf[OSi(OtBu)_3]_4$ (**15**) have been synthesized by the reaction of $M(NEt_2)_4$ (M = Zr or Hf) and 4 equivalents of $HOSi(OtBu)_3$ in nonpolar solvents.[9] Compounds **14** and **15** are structurally isomorphous and each crystallizes as a mixture of two isomers.[16] One isomer contains

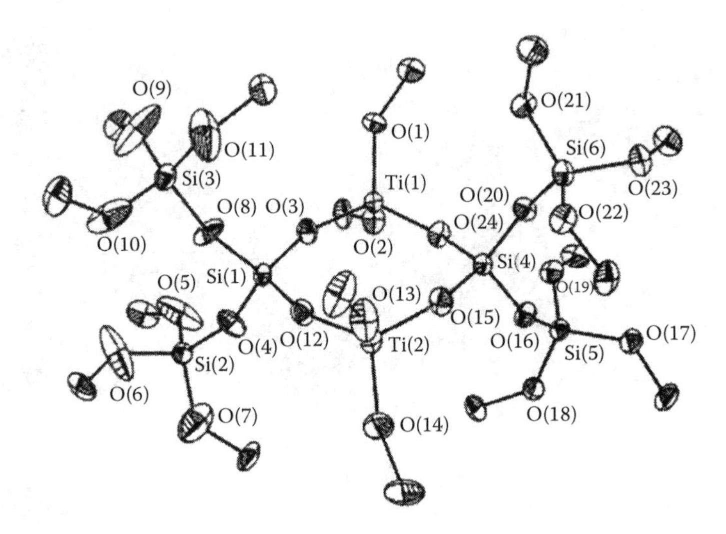

FIGURE 7.3 ORTEP diagram of **13** at the 50% probability level. Terminal methyl groups and hydrogen atoms have been omitted for clarity. (Reproduced from Brutchey, R.L. et al., *J. Mol. Cat. A*, 238, 1, 2005. With permission.)

a 4-coordinated metal center similar to titanium complex **7**, while the other one is a 5-coordinated complex with a chelating η^2-OSi(O*t*Bu)$_3$ ligand (Figure 7.4). Thermal decomposition of **14** and **15** gives ceramic materials $MO_2{\cdot}4SiO_2$ at *ca.* 137 °C and 141 °C, respectively. The $ZrO_2{\cdot}4SiO_2$ exhibits an open and fibrous structure, which is somewhat ordered with ~20 nm pores. Solution-phase thermolysis of **14** results in xerogels composed of small primary particles (≤5 nm), showing a high surface area (700 m^2g^{-1} for samples heated at 200 °C). The controlled hydrolysis of **14** and **15** with 1 and 2 equivalents of water in THF produces isolable monoaqua complexes, M[OSi(O*t*Bu)$_3$]$_4$(H_2O) [M = Zr (**16**) and Hf (**17**)] and diaqua M[OSi(O*t*Bu)$_3$]$_4$(H_2O)$_2$ [M = Zr (**18**) and

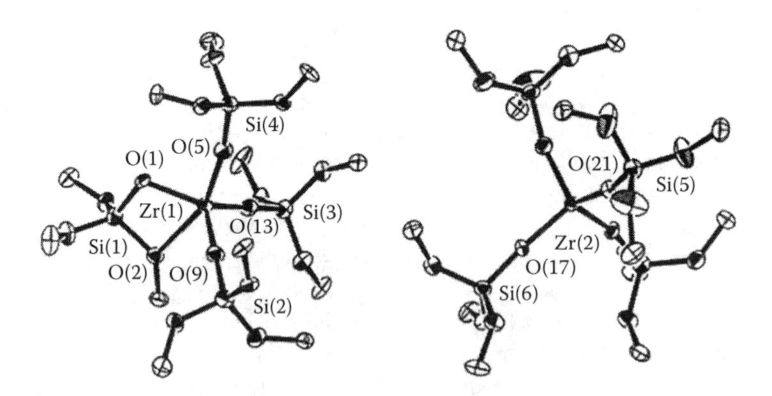

FIGURE 7.4 ORTEP diagrams of two structural isomers of **14**. Methyl groups have been removed for clarity. (Reproduced from Terry, K.W. et al., *J. Am. Chem. Soc.*, 119, 9745, 1997. With permission.)

Hf (**19**)].[17] Addition of excess water results in the formation of $MO_x(OH)_y(OH_2)_z$ gel and $HOSi(OtBu)_3$ through rapid hydrolysis. Other related zirconium siloxides including $(iPrO)_2Zr[OSi(OtBu)_3]_2$ (**20**) and $[Zr\{OSi(OtBu)_3\}_4(H_2O)_2]\cdot 2H_2O$ (**21**) have been isolated as a viscous liquid and single crystals, respectively.[10]

With the objective of synthesizing tricomponent oxide materials (i.e., Zr/Si/B), Tilley and coworkers isolated $Cp_2Zr(Me)OB[OSi(OtBu)_3]_2$ (**22**) via a room temperature reaction of Cp_2ZrMe_2 with $HOB[OSi(OtBu)_3]_2$ in pentane.[18] The complex crystallizes in the triclinic space group *P*-1, exhibiting 18 independent molecules in the asymmetric part of the unit cell. This complex converts into $Cp_2Zr[OSi(OtBu)_3]_2$ (**23**) upon thermolysis or by storing in ether at room temperature.

7.3.2.2 Group 5 Metal Silicates

Vanadia–silica materials have been widely studied as catalysts for the selective oxidation of hydrocarbons, including the conversion of methane to formaldehyde and the oxidative dehydrogenation (ODH) of ethane and propane. Tris(*tert*-butoxy)siloxide precursors for V(IV)and V(V) silicates can be prepared *via* simple silanolysis reactions. For example, $OV[OSi(OtBu)_3]_3$ (**24**) has been obtained by the reaction of $OVCl_3$ with excess $HOSi(OtBu)_3$ in pyridine.[19a] The calcination of **24** results in vanadia–silica xerogels, showing high porosity and high catalytic activity for propane oxidative dehydrogenation. Solution-phase cothermolysis of **24** and $Zr(OCMe_2Et)_4$ results in a ternary vanadia–zirconia–silica catalyst that exhibits impressive activities and selectivity for oxidative dehydrogenation of propane.[19b] The reaction of $V(OtBu)_4$ with 1 and 2 equivalents of $HOSi(OtBu)_3$ in toluene gives $(tBuO)_3OV[OSi(OtBu)_3]_3$ (**25**) and $(tBuO)_2OV[OSi(OtBu)_3]_2$ (**26**), respectively (Figure 7.5).[20] These precursors have been applied to generate materials of approximate formulas $VO_{2.5}\cdot SiO_2$ and $VO_{2.5}\cdot 2SiO_2$ *via* solid-state and solution thermolysis, respectively.

SSPs to tantala- and niobia-silica materials have also been reported. For example, $(iPrO)_2M[OSi(OtBu)_3]_3$ [M = Ta (**27**) and Nb (**28**)] have been obtained by the reaction of $M(OiPr)_5$ and excess $HOSi(OtBu)_3$ in pentane.[21] While these compounds have not been structurally characterized, an analogous tantalum complex $(EtO)_2Ta[OSi(OtBu)_3]_3$ (**29**) has been crystallographically characterized.[21] Immobilized **27** on TaSBA-15 exhibits excellent catalytic selectivity for cyclohexene epoxidation (>98% after 2 h) in aqueous H_2O_2 as the oxidant.[22] Wolczanski et al. have reported trivalent tantalum siloxide, $Ta[OSi(OtBu)_3]_3$.[23] Thermolytic decomposition of **27** and **29** yields tantala-silica $Ta_2O_5\cdot 6SiO_2$. The Ta_2O_5-containing materials are considered to be the potential heterogeneous catalysts for esterification, condensation, and hydration reactions.

7.3.2.3 Group 6 Metal Silicates

Invention of Phillip catalyst (CrO_3–SiO_2) in 1953 and its application in industrial polyethylene production has triggering an explosion of research efforts in developing chromia-silica materials.[24] In 1993, Tilley synthesized Cr(II) and Cr(III) siloxides $Cr[OSi(OtBu)_3]_2(NHEt_2)_2$ (**30**) and $Cr[OSi(OtBu)_3]_3(NHEt_2)_2$ (**31**) as precursors to chromium-supported catalysts.[25] These complexes have been obtained as a mixture of Cr(II) and Cr(III) siloxides by the reaction of $Cr(NEt_2)_4$ and 4 equivalents of $HOSi(OtBu)_3$ in toluene. The mixture was extracted with pentane and crystallized

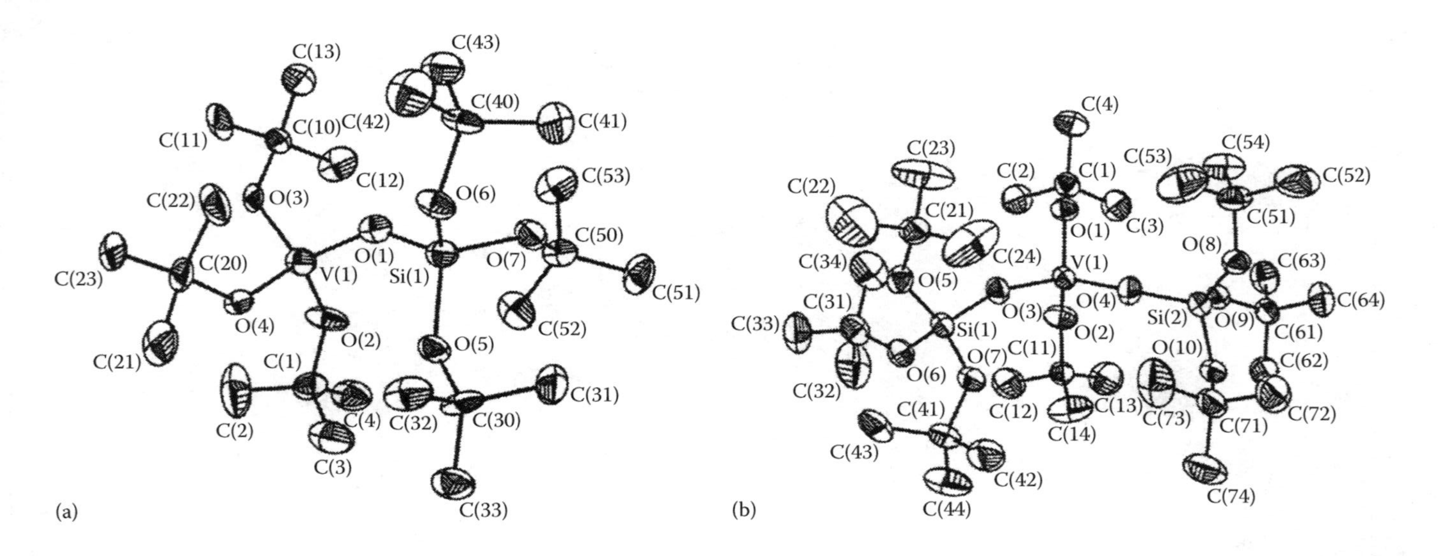

FIGURE 7.5 ORTEP diagrams of (a) **25** and (b) **26** at the 50% probability level. Methyl protons have been removed for clarity. (Reproduced from Fujdala, K.L. and Tilley, T.D., *Chem. Mater.*, 14, 1376, 2002. With permission.)

from the concentrated solution at −30 °C to obtain **30** and **31** in a 1:4 ratio. However, this reaction did not provide precise route to chromia–silica materials. The neat reactions of 1 and 2 equivalents of $HOSi(OtBu)_3$ with $Cr(OtBu)_4$ afforded the first ever Cr(IV) alkoxysiloxy complexes $(tBuO)_3CrOSi(OtBu)_3$ (**32**) and $(tBuO)_2Cr[OSi(OtBu)_3]_2$ (**33**), respectively.[26] Both **32** and **33** contain *pseudo* tetrahedral d^2 chromium centers (Figure 7.6). Low-temperature (≤180 °C) thermolysis of **32** and **33** results in chromia–silica materials that serve as very useful catalysts for the oxidative dehydrogenation of propane to propene.

Recently, Copéret has isolated Cr(II) and Cr(III) siloxides $[Cr[OSi(OtBu)_3]_2]_2$,[27] (**34**) and $[Cr[OSi(OtBu)_3]_2]_3 \cdot 2THF$,[27b] (**35**) and grafted them onto silica gel to obtain $(\equiv SiO)Cr_2(OSi(OtBu)_3)_3$ (**34a**) and $(\equiv SiO)Cr(OSi(OtBu)_3)_2(THF)$ (**35a**), respectively. Subsequent thermal treatment of the grafted materials produces $(\equiv SiO)_4Cr_2$ (**34b**) and $(\equiv SiO)_3Cr$ (**35b**) (Scheme 7.5).[27] The catalytic activity of **34b** for the ethylene polymerization is poor due to the presence of trace amounts of Cr(III). Treatment of **34b** with N_2O leads to the formation of the Cr(III) dimer $(\equiv SiO)_6Cr_2$ (**34c**), whose catalytic activity is similar to that of the Phillips catalyst. Nocera has reported first chromium(IV) siloxide complex $Cr(OSitBu_2Me)_4$ (**36**) from the reaction of $CrCl_3$ and excess $NaOSitBu_2Me$ that is stable in air and moisture.[28]

Molybdenum molecular precursors have been obtained via similar synthetic strategies. Tilley in 1997 synthesized first such complexes $Mo_2[O_2Si(OtBu)_2]_3$ (**37**) and $Mo_2[O_2Si(OtBu)_2]_3(Py)$ (**38**) by the reaction of $Mo_2(NMe_2)_6$ and 3 equivalents of $(HO)_2Si(OtBu)_2$ in pentane and pyridine, respectively.[29] Both these complexes have been isolated as single crystals and structurally characterized. In solid state, the complex **37** stabilizes in eclipsed conformation around the axis of dimolybdenum (Mo≡Mo) bond, with three di(*tert*-butoxy)silanediolato $(O_2Si(OtBu)_2)$ ligands bridging the center in a symmetrical fashion (Figure 7.7a). Pyrolysis of **37** at 550 °C for 1 h under argon resulted in a ceramic material with the approximate formula $Mo_2Si_3O_{10}$, which upon further heating at 1200 °C for 1 h yielded another ceramic material with the composition $Mo_{0.5}Si_3O_{6.5}$. Mononuclear complex $MoO[OSi(OtBu)_3]_4$ (**39**), featuring metal–oxo fragments and tri(alkoxy)siloxy ligands, has been synthesized by the reaction of $MoOCl_4$ and $LiOSi(OtBu)_3$.[30] Similarly, a thermally unstable complex $MoO_2[OSi(OtBu)_3]_2$ has been stabilized as ether adduct $MoO_2[OSi(OtBu)_3]_2(THF)$ (**40**) and isolated as single crystals.[30] The covalently grafted **39** onto SBA-15 catalyzes expoxidation of cyclohexene using *tert*-butyl hydroperoxide (TBHP) or aqueous H_2O_2 as the oxidant. The xerogels obtained from the thermolysis of **40** are a mesoporous material with surface area of 106 m^2g^{-1} and contain MoO_3 domains. A number of other molybdenum siloxides have also been reported, such as $Mo_2(NMe_2)_4[OSi(OtBu)_3]_2$ (**41**), $Mo_2(OtBu)_4[OSi(OtBu)_3]_2$ (**42**), and $Mo_2(NMe)_4\{OB[OSi(OtBu)_3]_2\}_2$ (**43**).[31] Complex **41** is an effective molecular precursor to Mo/Si/O materials, which upon thermolysis gives $2MoO_3 \cdot 2SiO_2$ (for sample heated at 600 °C) and $2MoO_{1.5} \cdot 2SiO_2$ (for sample heated at 700 °C).

Reaction of $W_2(NMe_2)_6$ with $(OH)_2Si(OtBu)_2$ yields air-stable tungsten precursor $W_2(NHMe_2)_2[O_2Si(OtBu)_2]_2[OSi(OH)(OtBu)_2]_2$ (**44**) as single crystals.[29] The structure of **44** contains two $-O_2Si(OtBu)_2$ ligands that bridge the W≡W bond in an η^1, η^1 fashion, and one $-OSi(OH)(OtBu)_2$ ligand coordinated to each W center (Figure 7.7b). As in the case of Mo complex **37**, complex **44** also adopts eclipsed

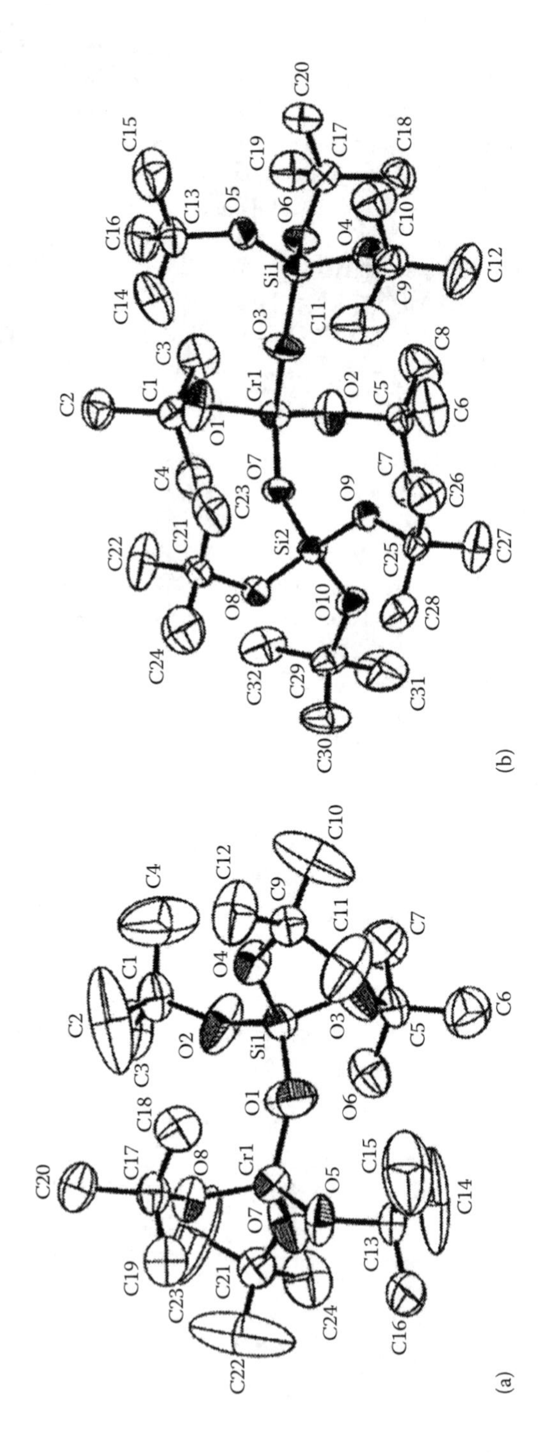

FIGURE 7.6 ORTEP diagrams of (a) **32** and (b) **33** at the 50% probability level. Methyl protons have been removed for clarity. (Reproduced from Fujdala, K.L. and Tilley, T.D., *Chem. Mater.*, 13, 1817, 2001. With permission.)

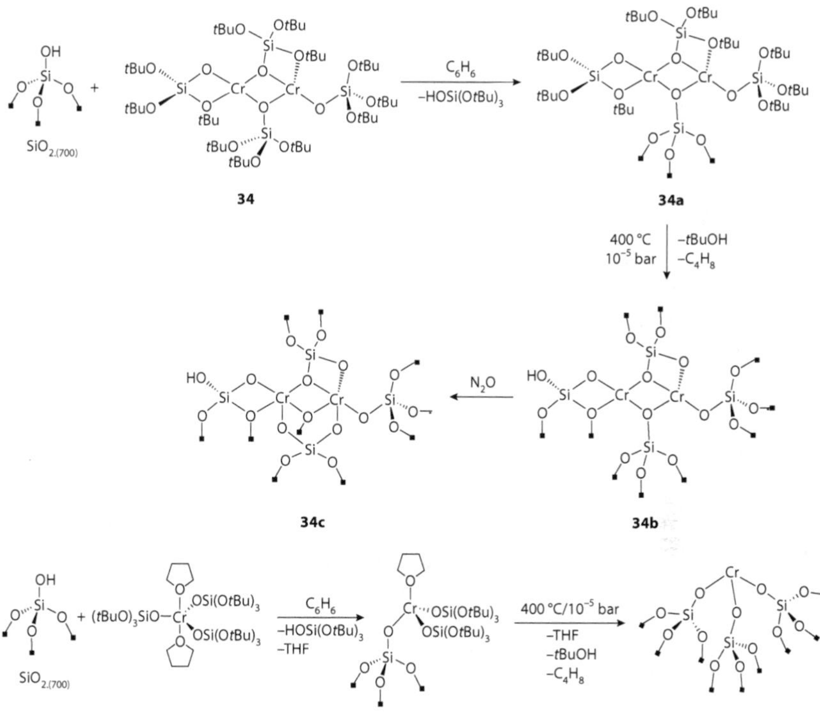

SCHEME 7.5 Schematic presentation of grafting of **34** and **35** on $SiO_{2(700)}$ and their subsequent thermal decomposition.

conformation in the solid state and possesses a twofold axis of symmetry perpendicular to W≡W bond. Each W(III) ion is further coordinated with one $NHMe_2$ ligand. Pyrolysis of **44** under argon produced ceramic materials $W_{2.2}Si_4O_{17.4}$ (amorphous) and $W_{2.2}Si_4O_{13}$ (crystalline) at 550 °C and 1200 °C, respectively.

7.3.2.4 Group 8 Metal Silicates

Iron-containing zeolite FeZSM-5 has attracted considerable attention due to its high activity as a catalyst for the reduction of nitrogen oxides (NO_x) and for the selective oxidation of hydrocarbons with nitrous oxide as the oxidant. Given the potential utility of iron-supported catalysts, grafting molecular iron complexes onto silica surface has attracted much attention. Tilley reported the synthesis and isolation of $Fe[OSi(OtBu)_3]_3(THF)$ (**45**) by the reaction of $NaOSi(OtBu)_3$ and $FeCl_3$ in THF.[32] In solid state, **45** exists as a distorted tetrahedral complex (toward a trigonal pyramid). The precursor **45**-grafted SBA-15 materials (≡SiO–$Fe[OSi(OtBu)_3]_2(THF)$ (**45a**), after calcination at 300 °C, exhibit high selectivity for oxidations of alkanes, alkenes, and arenes with hydrogen peroxide as the oxidant. The grafted materials, upon calcination at 1000 °C, yield $^1/_2Fe_2O_3 \cdot 3SiO_2$.

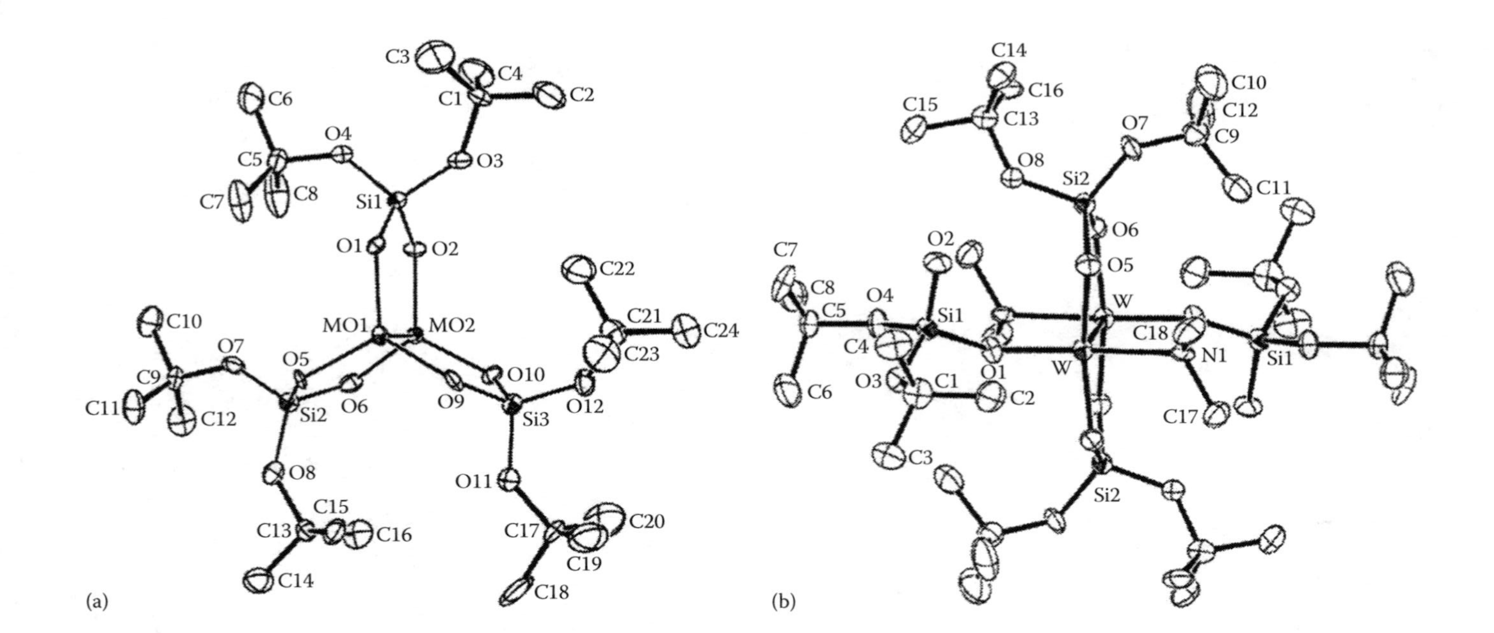

FIGURE 7.7 ORTEP diagrams of (a) **37** and (b) **44**. Methyl protons have been removed for clarity. (Reproduced from Su, K. and Tilley, T.D., *Chem. Mater.*, 9, 588, 1997. With permission.)

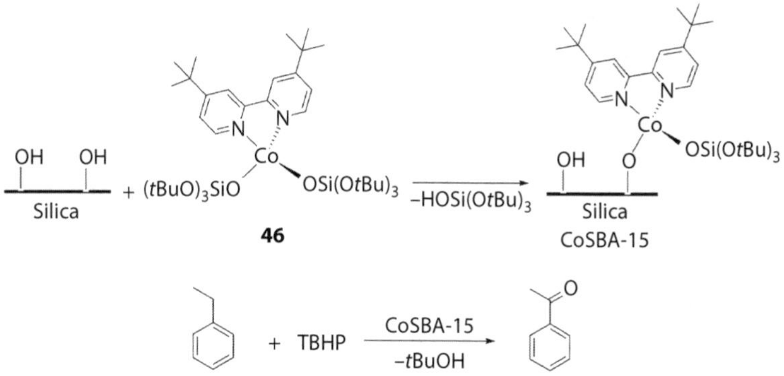

SCHEME 7.6 Preparation of site-isolated CoSBA-15 and oxidation of ethylbenzene.

7.3.2.5 Group 9 Metal Silicates

In search of new cobalt-based heterogeneous catalysts for the oxidation reaction, Tilley et al. have reported a *pseudo* tetrahedral Co(II) complex (4,4′-*t*Bu-bipy)$_2$Co[OSi(O*t*Bu)$_3$]$_2$ (**46**) obtained from the reaction of Co[N(SiMe$_3$)$_2$]$_2$, 4,4′-di-*tert*-butyl-bipyridine (4,4′-*t*Bu-bipy), and 2 equivalents of HOSi(O*t*Bu)$_3$ in toluene.[33] Immobilization of **46** on SBA-15 resulted in site-isolated CoSBA-15 silicate, which has been successfully utilized for the catalytic oxidization of alkylaromatic substrates in the presence of *tert*-butyl hydroperoxide (TBHP) (Scheme 7.6). Tilley et al. also reported rhodium siloxides [(COD)Rh(μ-OSi(O*t*Bu)$_3$)]$_2$ (**47**) and [(NBD)Rh(μ-OSi(O*t*Bu)$_3$)]$_2$ (**48**) (COD = 1,5-cyclooctadiene; NBD = norborna-2,5-diene) by the reaction of [(diene)RhCl]$_2$ with 2 equivalents of KOSi(O*t*Bu)$_3$.[34] Both these precursors have been isolated at −30 °C from pentane as yellow-orange crystals in high yield (82% for **47** and 76% for **48**); however, the x-ray quality crystals have not been isolated. Thermolysis of these precursors leads to the formation of Rh/Si/O materials. The former complex forms a black precipitate in toluene at 180 °C, which contains rhodium nanoparticles with an average diameter of *ca.* 22 nm. Marciniec and coworkers have reported a similar complex [(COD)Rh(μ-OSi(Me)$_3$)]$_2$ (**49**) and immobilized it onto silica, which is a highly active and stable catalyst for hydrosilylation reactions.[35]

7.3.2.6 Group 10 Metal Silicates

The chemistry of nickel siloxides is far less developed. The tris-(*tert*-butoxy)siloxy complexes of nickel Na$_3$(μ$_3$-I){Ni[μ$_3$-OSi(O*t*Bu)$_3$]$_3$I}·0.5THF·0.5C$_5$H$_{12}$ (**50**) and [(η^3-allyl)NiOSi(O*t*Bu)$_3$]$_2$ (**51**) have been obtained by the reaction of NiI$_2$(THF)$_2$ with NaOSi(O*t*Bu)$_3$ in THF and bis(η^3-allyl)nickel with HOSi(O*t*Bu)$_3$ in pentane, respectively (Figure 7.8).[36] The complex **50** forms a cube-like structure, with a Ni(II) ion, three sodium ions, three silanolate oxygen atoms, and an iodide ion at the vertices. The Ni$_2$O$_2$ core of the complex **51** is bent along the O–O vector, resulting in the fold angle of *ca.* 35°.

Pd(II), on the other hand, yields catalytically attractive silicate materials. For instance, (4,4′-*t*Bu-bipy)$_2$Pd[OSi(O*t*Bu)$_3$]$_2$ (**52**) has been synthesized by the reaction of

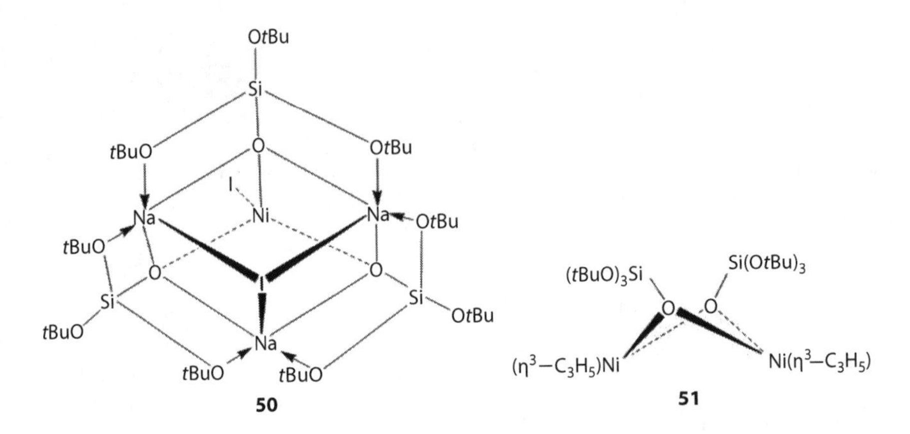

FIGURE 7.8 Nickel siloxides.

$KOSi(OtBu)_3$ with (4,4'-tBu_2-bipy)$PdCl_2$ in dichloromethane.[37] Similarly, Pt(IV) complex $Me_3Pt(tmeda)[OSi(OtBu)_3]$ (**53**) (tmeda=*N,N,N′,N′*-tetramethylethylenediamine) has been isolated. Covalently grafted Pd(II)SBA-15 and Pt(IV)SBA-15 under mild and nonaqueous conditions generate supported Pd(II) and Pt(IV) centers, which are catalytically active and selective for the semihydrogenation of 1-phenyl-1-propyne.

7.3.2.7 Group 11 Metal Silicates

Generally, group 11 metal ions form tetrameric complexes with alkoxysiloxy ligands. Relatively few trialkoxysiloxy complexes of copper are known in the literature. Schmidbaur et al. reported a tetrameric Cu(I) complex $[CuOSiMe_3]_4$ (**54**)

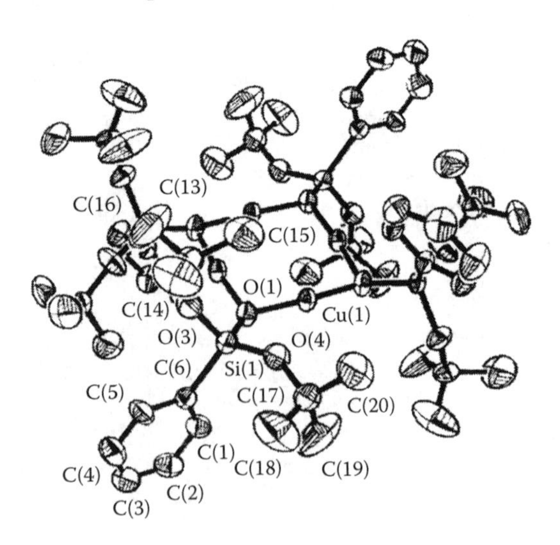

FIGURE 7.9 ORTEP diagram of molecular structure of **56**. Methyl protons have been removed for clarity. (Reproduced from Terry, K.W. et al., *Chem. Mater.*, 8, 274, 1996. With permission.)

by the reaction of CuCl with $NaOSiMe_3$, which was thought to possess a cube-like structure.[38] Later, Tilley et al. reported similar complex $[CuOSi(OtBu)_3]_4$ (**55**) by the silanolysis of $[CuOtBu]_4$ with $HOSi(OtBu)_3$ and decomposed it to Cu/SiO_2 and Cu_nO/SiO_2 (n = 1, 2) materials.[39a] The covalent grafting of **55** onto SBA and subsequent loss of the organic groups from $-OSi(OtBu)_3$ ligands results in the formation of $Si_{(surface)}$–O–Si linkages which significantly stabilized Cu(I) sites (100% of Cu(I) detected).[39b] Similarly, a related tetranuclear complex $[Cu[\mu\text{-}OSiPh(OtBu)_2]_4$ (**56**) has been isolated as single crystals and characterized by x-ray crystallography. This complex features a planar Cu_4O_4 core that resembles **54** with oxygen-bridged Cu atoms (Figure 7.9).[39a] A sublimable copper(I) complex $[NaCu(OSi(OtBu)_3)_2]_6$ (**57**) has been obtained by the reaction of $NaOSi(OtBu)_3$ and **54** or from the direct reaction of CuCl and 2 equivalents of $NaOSi(OtBu)_3$. Under argon atmosphere, **54** gives Cu(0) and Cu_2O nanoparticles dispersed in silica, whereas an oxygen atmosphere leads to the formation of CuO nanoparticles. Pyrolytic conversion of **55** leads to the clean formation of CuO/SiO_2 materials.

7.3.2.8 Group 12 Metal Silicates

Zinc orthosilicates exhibit high luminescence efficiency, such as manganese-doped Zn_2SiO_4 (willemite) is a green-emitting phosphor that serve as display and electroluminescent devices. Zinc tris(*tert*-butoxy)siloxide complex $[MeZnOSi(OtBu)_3]_2$ (**58**) and $[Zn(OSi(OtBu)_3)_2]_2$ (**59**) have been prepared by the reaction of $ZnMe_2$ and $HOSi(OtBu)_3$ in 1:1 and 1:2 molar ratios, respectively (Scheme 7.7).[40] Complex **59** is an asymmetric dimer with four $-OSi(OtBu)_3$ ligands, where each $-OSi(OtBu)_3$ ligand exhibits a unique coordination mode, which could be described as η^1; η^2; μ–η^1; η^1; μ–η^1; η^2. The reaction of $ZnMe_2$ with $(HO)_2Si(OtBu)_2$ in a 1:1 molar ratio results in the formation of one-dimensional polymer $[ZnOSi(OtBu)_2O]_n$ (**60**).[40] The structure of **60** has 4-coordinate zinc atoms bridged by-O(O*t*Bu)Si(O*t*Bu)O-linkages. These zinc precursors readily undergo thermolysis to zinc orthosilicate Zn_2SiO_4:SiO_2 particles of ~50–100 nm size.

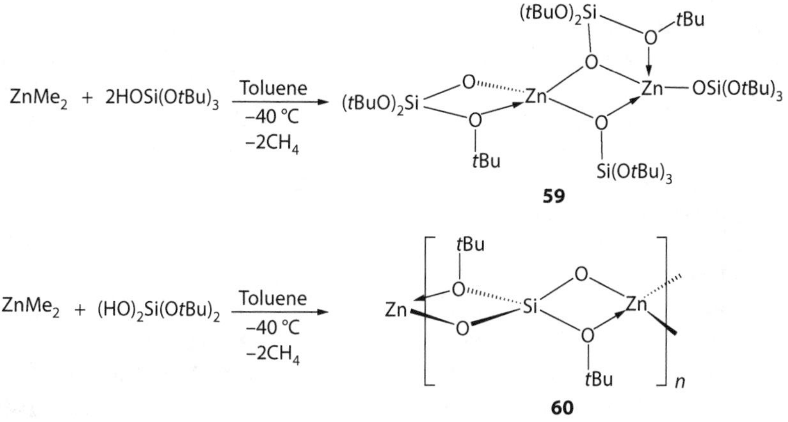

$ZnMe_2$ + $2HOSi(OtBu)_3$ —Toluene, −40 °C, $-2CH_4$→ **59**

$ZnMe_2$ + $(HO)_2Si(OtBu)_2$ —Toluene, −40 °C, $-2CH_4$→ **60**

SCHEME 7.7 Synthesis of zinc silicate precursors.

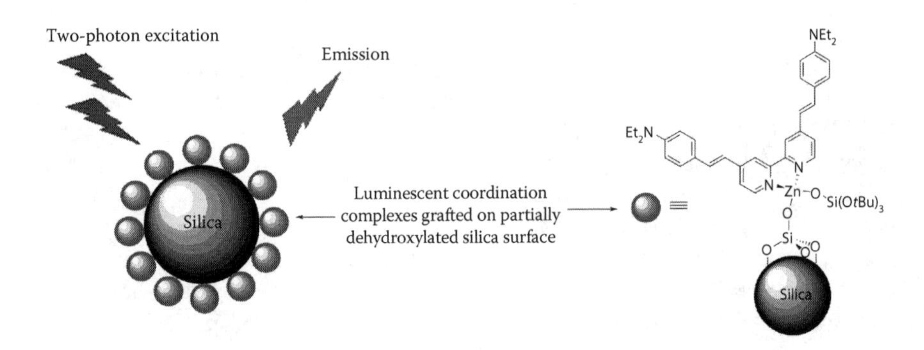

FIGURE 7.10 Schematic representation of surface functionalized luminescent silica nanoparticles (**61**) with two-photon excited luminescent Zn(II) complex.

Interestingly, grafting **59** on partially dehydroxylated silica (at 700 °C) followed by the coordination of 4,4′-(diethylaminostyryl)-2,2′-bipyridine (DEAS-bipy) acts as two-photon absorbing luminescent nanoparticles ($\equiv$SiO–Zn(OSi(O*t*Bu)$_3$ (DEAS-bipy) (**61**) (Figure 7.10).[41] The pyrolysis of Mn-doped zinc complex $[ZnOSi(OtBu)_2O]_n/[Mn(CH_2SiMe_3)_2]_m$ (**62**) results in the formation of $Zn_2SiO_4{\cdot}SiO_2$:Mn nanocomposite phosphors. $Zn_2SiO_4{\cdot}SiO_2$:Mn exhibits two photoluminescence emission bands centered at 535 (major) and 605 nm (minor).

7.3.2.9 Group 13 Metal Silicates

Group 13 silicates have been extensively studied as catalysts, catalyst supports, and structural material. Boron readily forms trivalent siloxides with HOSi(O*t*Bu)$_3$. For example, the reaction of B(O*t*Bu)$_3$ with 2 and 3 equivalents of HOSi(O*t*Bu)$_3$ leads to the formation of (*t*BuO)B[OSi(O*t*Bu)$_3$]$_2$ (**63**) and B[OSi(O*t*Bu)$_3$]$_3$ (**64**), respectively.[18] Both **63** and **64** exist as monomer in solid state, with a trigonal planar geometry around the boron centers. These complexes **63** and **64** are efficient single-source molecular precursors for $BO_{1.5}{\cdot}2SiO_2$ and $BO_{1.5}{\cdot}3SiO_2$ materials, respectively. In the presence of 1 equivalent of water, **63** gives boronous acid HOB[OSi(O*t*Bu)$_3$]$_2$, which indicates that B–O*t*Bu linkages are more susceptible to hydrolysis than that of the B–OSi(O*t*Bu)$_3$ fragments. Thus, the order of hydrolytic stability is B–O*t*Bu $<$ B–OSi $\ll$ Si–O*t*Bu.

Molecular aluminosilicate precursors [(*i*PrO)$_2$AlOSi(O*t*Bu)$_3$]$_2$ (**65**) and (acac) Al[OSi(O*t*Bu)$_3$]$_2$ (**66**) (acac = acetylacetonato) were initially synthesized by Abe et al.[42] The demand for new molecular materials has increased considerably due to the application of such complexes in the clean synthesis of aluminosilicate nanomaterials. More recently, several aluminum tris(*tert*-butoxy)siloxides have been discovered by Tilley and coworkers, including those based on Al–C linkages such as [Me$_2$AlOSi(O*t*Bu)$_3$]$_2$ (**67**) and [Me(*t*BuO)AlOSi(O*t*Bu)$_3$]$_3$ (**68**).[43] Further investigations led to the synthesis of new tris(*tert*-butoxy)siloxy complexes of aluminum that possess only Al–O and Si–O linkages (no Al–C bonds).[44] Reaction of [Al(O*i*Pr)$_3$]$_4$ with 12 equivalents of HOSi(O*t*Bu)$_3$ in THF resulted in the formation of Al[OSi(O*t*Bu)$_3$]$_3$(THF) (**69**), which is unstable at room temperature and decomposes to insoluble Al/Si/O materials even in inert conditions. Attempts have been made to achieve stable materials with higher Al/Si ratio (1:1) through the reaction of [Al(O*i*Pr)$_3$]$_4$ with 4 equivalents of

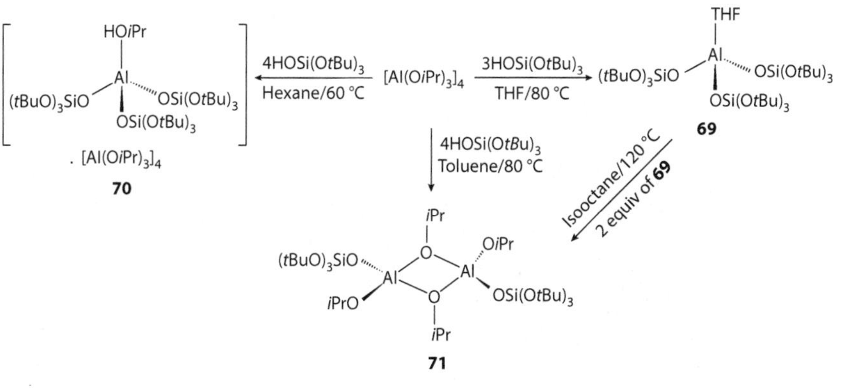

SCHEME 7.8 Synthesis of aluminum siloxide precursors.

$HOSi(OtBu)_3$ in hexane to form structurally characterized $Al[OSi(OtBu)_3]_3(HOiPr)\cdot 0.5[Al(OiPr)_3]_4$ (**70**). A related dimeric complex $[(iPrO)_2AlOSi(OtBu)_3]_2$ (**71**) has been isolated from the reaction of 2 equivalents of $[Al(OiPr)_3]_4$ with **69** in isooctane at 120 °C or $[Al(OiPr)_3]_4$ with 4 equivalents of $HOSi(OtBu)_3$ in toluene at 80 °C (Scheme 7.8). Thermolytic decomposition of **70** and **71** leads to the formation of aluminosilicate materials $\frac{1}{2}Al_2O_6\cdot 3SiO_2$ and $\frac{3}{2}Al_2O_6\cdot 3SiO_2$, respectively.

7.4 MOLECULAR METALLOPHOSPHATES

Metallophosphates represent another class of interesting materials that are widely found in naturally occurring phosphate minerals. Over the years, metallophosphates have found potential applications in various fields such as cation exchange, ion conduction, proton conduction, catalysis, magnetic materials, and optical materials.[45] Murugavel et al. and Tilley et al. have unveiled that di-*tert*-butyl phosphate, $HOP(O)(OtBu)_2$ (abbreviated hereafter as dtbp-H), preferably forms discrete molecules (or clusters) that serve as excellent SSPs for the clean synthesis of phosphate nanomaterials.[4] The dtbp-H ligand was originally reported in 1957 by Goldwhite and Saunders, which could best be stored below –20 °C.[46] At room temperature (<24 h), dtbp-H decomposes into an insoluble phosphate material. This ligand could be stored at room temperature for an extended period in the form of potassium salt, $KOP(O)(OtBu)_2$.[47] A cold (0 °C) aqueous solution of $KOP(O)(OtBu)_2$ with concentrated HCl gives a white precipitate, which, upon rapid filtration and subsequent crystallization, affords pure dtbp-H in good yields (>80%).

7.4.1 Alkali and Alkaline Earth Metal Phosphates

7.4.1.1 Alkali Metal Phosphates

Sodium and potassium salts of alkyl and aryl phosphate ligands have been mainly used as the starting reagents for the preparation of other metal phosphates.[48] They are often generated *in situ* and therefore no detailed studies have been carried out to establish their molecular structures. However, the solid-state structure of $KOP(O)(OtBu)_2$

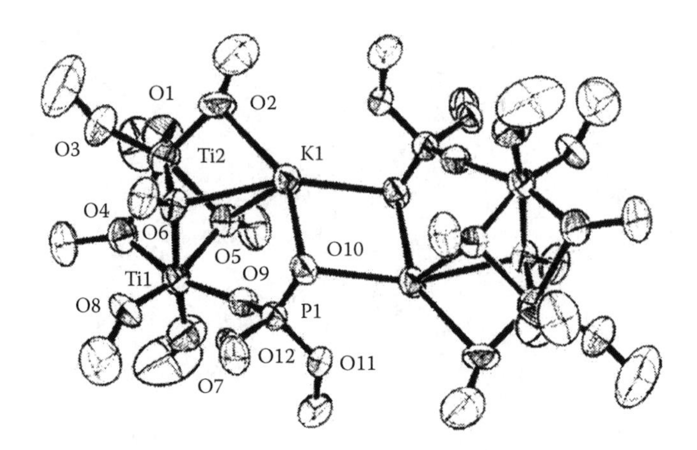

FIGURE 7.11 ORTEP view of **72** with thermal ellipsoids drawn at 50% probability. Methyl groups have been removed for clarity. (Reproduced from Lugmair, C.G. and Tilley, T.D., *Inorg. Chem.*, 37, 1821, 1998. With permission.)

has been determined, which crystallizes as a tetramer $[(KO)(O)P(OtBu)_2 \cdot 3HOiPr]_4$ (**72**), featuring a cube-like core structure (Figure 7.11).[49] This is a low melting compound (−30 °C) that can be crystallized from concentrated 2-propanol solution at −80 °C. The metal phosphates incorporating alkali metal ions along with other metals from other groups are discussed later under the groups of the second metal ion.

7.4.1.2 Alkaline Earth Metal Phosphates

Although dtbp complexes of alkaline earth metals are not reported so far, similar phosphate complexes of Mg(II), Sr(II), and Ba(II) have been reported. Ezra and Collin in 1973[50a] reported structurally characterized magnesium diethylphosphate complex $Mg(O_2P(OEt)_2)_2$ (**73**), which was earlier reported by the same group.[50b] Each Mg(II) center in **73** is coordinated to four phosphoryl oxygen atoms in a nearly tetrahedral geometry. Burns et al. reported structurally characterized strontium and barium di-*n*-butylphosphate complexes $M[(O_2P(OnBu)_2)_2(H_2O)(18\text{-crown-}6)]$, where M = Sr (**74**) or Ba (**75**).[51] Both the complexes are structurally identical where the metal ion is buried inside the 18-crown-6 cavity and is coordinated on either side by the oxygen atom of a di-*n*-butylphosphate ligand. Additionally, a water molecule is coordinated to the metal ion from one of the sides of the macrocyclic ring. The first barium phosphate, $[Ba(O_2P(OEt)_2)_2]_n$ (**76**), was reported much earlier by Kyogoku et al.,[52] which was synthesized from triethylphosphate and barium hydroxide in the presence of hydrochloric acid. The central barium ion of phosphate polymer **76** is coordinated to eight oxygen atoms from six different diethylphosphate ligands.

7.4.2 Transition Metal Phosphates

7.4.2.1 Titanium Phosphates

Titanium phosphates have been studied for a variety of applications. For example, potassium titanyl phosphate $[(KTi_2(PO_4)_3]$ is an important nonlinear optics.[53a]

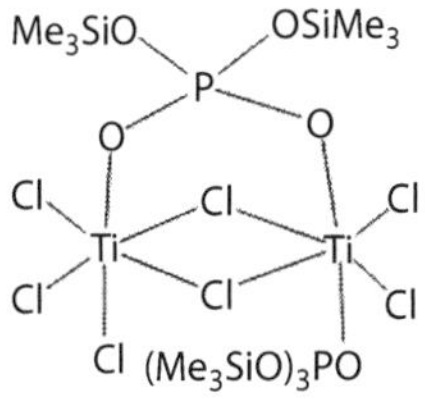

FIGURE 7.12 Titanium phosphates.

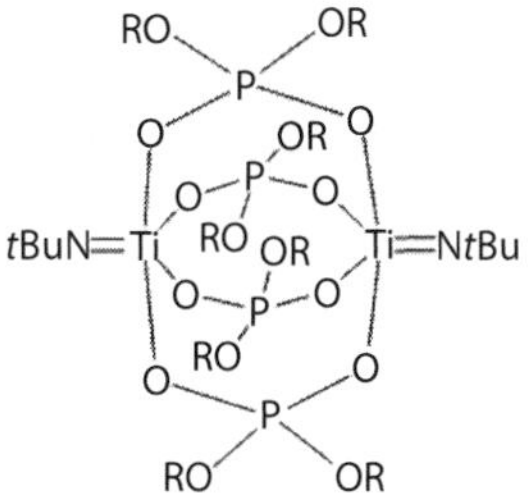

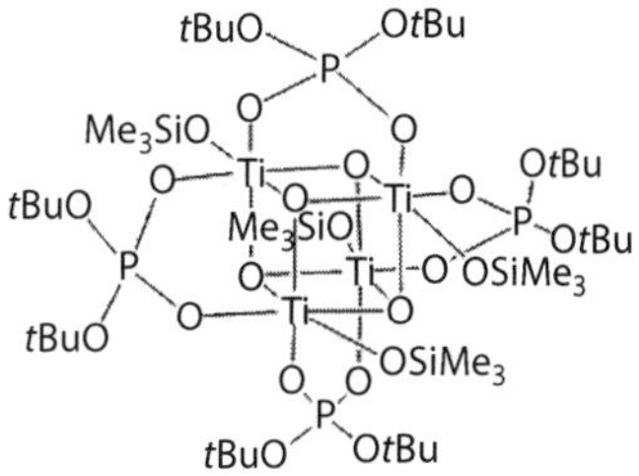

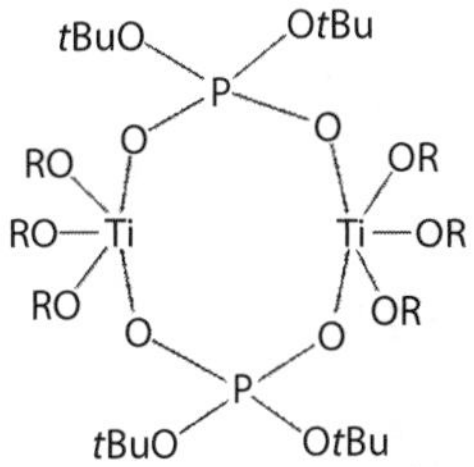

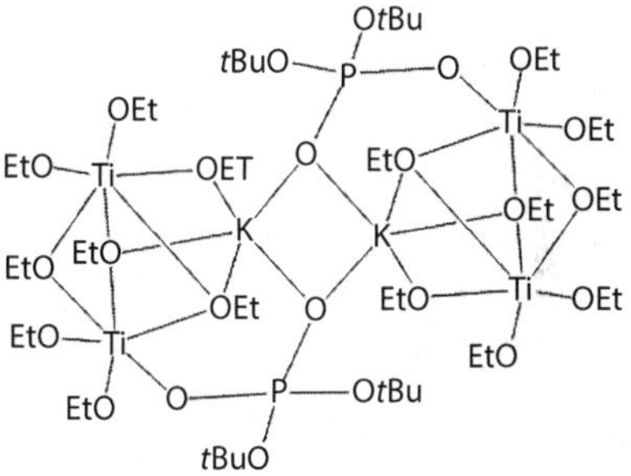

Phosphato-titanium esters have been used as flame retardants and hardening agents, ion exchanger, and fast ion conductors.[53] With the objective to synthesize structurally well-characterized titanium phosphates, Thorn and Harlow reported three different phosphato-titanium esters, namely, chlorotitanium, imidotitanium, and oxo-titanium phosphates, using trimethylsilyl and *tert*-butyl phosphates as the ligands (Figure 7.12).[53a] Chlorotitanium phosphate, $Ti_2Cl_7(O_2P(OSiMe_3)_2)(OP(OSiMe_3)_3)$ (**77**), has been synthesized by the reaction of $TiCl_4$ and $(Me_3SiO)_3P{=}O$. The first example of a dimeric titanium compound $[t\text{BuN}{=}Ti(O_2P(OSiMe_3)_2)_2]_2$ (**78**) with a terminal imido group has been obtained by the reaction between $(Me_3SiO)_3P{=}O$ and $[(Me_2N)_2Ti(\mu\text{-N}t\text{Bu})]_2$. Oxo-titanium complex $[TiO(OSiMe_3)(dtbp)_2]_4$ (**79**) has been obtained from the reaction of $Ti(OSiMe_3)_4$ with dtbp-H. The molecular structure of **79** is very similar to the well-known phosphinato-bridged titanium oxo-alkoxides cubane cluster $[Ti(\mu_3\text{-O})(O i\text{Pr})(Ph_2PO_2)]_4\cdot 0.5DMSO$, reported by Mutin et al.[54]

Tilley et al. synthesized dinuclear titanium phosphates $[Ti(OR)_3(dtbp)]_2$ (R = Et (**80**) or *i*Pr (**81**) from the reaction of dtbp-H and the corresponding titanium alkoxide precursors.[49] In the solid state, **81** exists as a centrosymmetric dimer containing 5-coordinate metals bridged by the dtbp ligands. An attempt was made to prepare potassium titanyl phosphate (KTP) precursor $[Ti_2K(OEt)_8dtbp]_2$ (**82**) via the reaction of **80** with 2 equivalents of KOEt. In the solid state, compound **82** exists as a centrosymmetric dimer containing two Ti-centered, face-sharing *pseudo* octahedra in the unique half of the dimer.

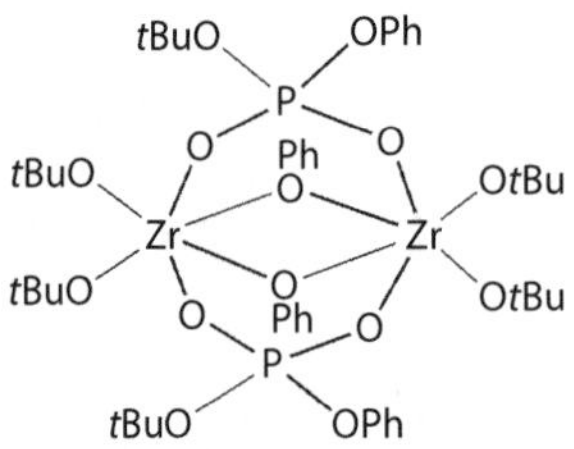

FIGURE 7.13 Zirconium phosphates.

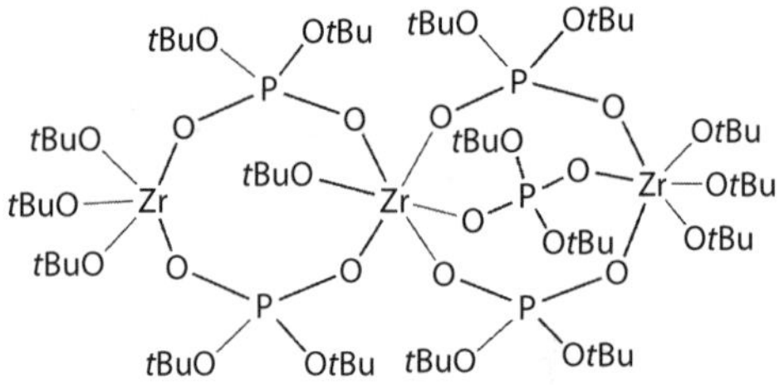

7.4.2.2 Zirconium Phosphates

The chemistry of molecular zirconium phosphates is limited to a very few complexes. Kumara Swamy et al. have synthesized dinuclear and trinuclear zirconium phosphates $[Zr\{\mu,\mu'-O_2P(OtBu)(OPh)\}(\mu-OPh)(OtBu)_2]_2$ (**83**) and $[Zr_3(dtbp)_5(OtBu)_7]\cdot{}^1/_2C_6H_5CH_3$ (**84**) from the reactions of $Zr(OtBu)_4$ with diphenyl phosphate and dtbp-H, respectively (Figure 7.13).[55] These complexes are unstable in solution such as benzene, suggesting that there is a dynamic equilibrium as earlier observed by Tilley in titanium phosphates.[49] The molecular structure of **83** features two hexacoordinated zirconium ions with two bridging phenoxide ions, two *tert*-butyl phenyl phosphate (tbpp), and four terminal *tert*-butoxide ions. The trimeric zirconium compound **84** consists of three different zirconium ions where one of the terminal zirconium ions is octahedrally coordinated by three bridging dtbp and three *tert*-butoxide ions, while the other terminal zirconium ion is pentacoordinated through two bridging dtbp and three *tert*-butoxide ions. The middle zirconium ion is hexacoordinated by five bridging dtbp and one *tert*-butoxide ligands.

7.4.2.3 Vanadium and Niobium Phosphates

Vanadium phosphates have been known as catalysts in various organic transformations in the form of vanadyl pyrophosphates $(VO)_2P_2O_7$.[56] Soluble and processable molecular precursors to vanadium phosphates can be readily synthesized and decomposed by pyrolysis to yield metal phosphates. Thorn et al. have synthesized dinuclear and trinuclear vanadium diethyl phosphates $[(dipic)V(O)(O_2P(OEt)_2)]_2$ (**85**) (dipic-H_2 = pyridine-2,6-dicarboxylic acid) and $[(VO)_3(O_2P(OEt)_2)_6]\cdot CH_3CN$ (**86**) via the reactions of diethylphosphate with (dipic)V(O)(O*i*Pr) or vanadyl tris-isopropoxide, respectively (Scheme 7.9).[57] The trinuclear precursor **86** has been converted to $VO(PO_3)_2$ through calcination at 500 °C, which is catalytically active for oxidation of butane to maleic anhydride.

Also there exists one report in the literature on a well-characterized niobium phosphate complex $[Nb(OiPr)_4(O_2P(OtBu)_2]_2$ (**87**), which has been synthesized by the reaction of $Nb(OiPr)_5$ with dtbp-H in pentane.[58] The thermal decomposition of **87** at 1100 °C results in the formation of *t*-$NbPO_5$ (tetragonal phase) with small amount of *m*-$NbPO_5$ (monoclinic phase) of an average particle size of 26 nm.

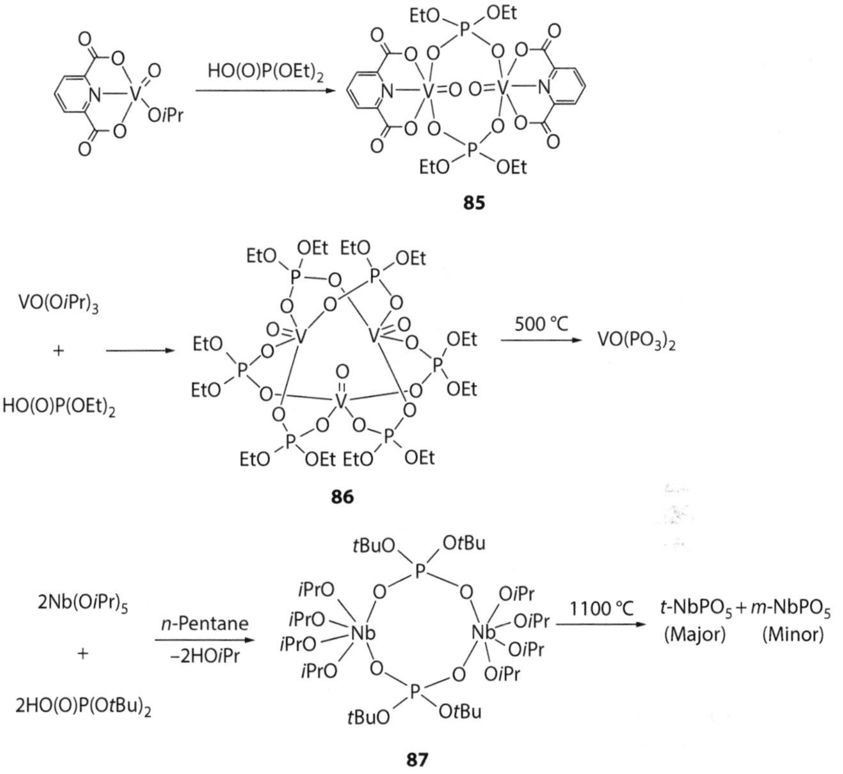

SCHEME 7.9 Synthesis of vanadium and niobium phosphate precursors.

7.4.2.4 Molybdenum Phosphates

Since the first isolation of a triply bonded Mo_2L_6 complex by Wilkinson et al. in 1971,[59] various metal–metal bonded molybdenum dtbp complexes have been isolated with an objective to synthesize efficient catalysts for oxidative transformations. In this connection, Tilley and coworkers reported a series of structurally characterized dimolybdenum(III) phosphates $Mo_2(NMe_2)_2[\mu\text{-}O_2P(O\textit{t}Bu)_2]_2[O_2P(O\textit{t}Bu)_2]_2$ (**88**), $Mo_2(NMe_2)_2[OSi(O\textit{t}Bu)_3]_2[\mu\text{-}O_2P(O\textit{t}Bu)_2]_2$ (**89**), and $Mo_2(NMe_2)_2[\mu\text{-}O_2P(O\textit{t}Bu)_2]_2\{OB[OSi(O\textit{t}Bu)_3]_2\}_2$ (**90**) starting from dtbp-H and tri-*tert*-butoxy silanol (Scheme 7.10).[31] Both *cis*- and *trans*-isomers of **88** (**88a** and **88b**) and **89** (**89a** and **89b**) have been isolated and structurally characterized. Solution thermolysis of these precursors in toluene leads to the formation of mixed oxide (Mo/Si/P/O) materials. The *cis*-isomer **88a** results in the formation of $2MoO_{1.5}{\cdot}2P_2O_5$ ($MoOP_{xg}$), whereas both **89a** and **89b** form $2MoO_{1.5}{\cdot}2P_2O_5{\cdot}2SiO_2$ (*c*-$MoOPOSi_{xg}$ and *t*-$MoOPOSi_{xg}$) xerogels. The xerogels obtained from **88a** exhibit low activity and poor selectivity for the oxidative dehydrogenation of propane to propylene. Cothermolysis of **88a** and $Bi[OSi(O\textit{t}Bu)_3]_3$ resulted in Bi/Mo/P/Si/O xerogels, exhibiting improved selectivity for the oxidation reactions.[59,60]

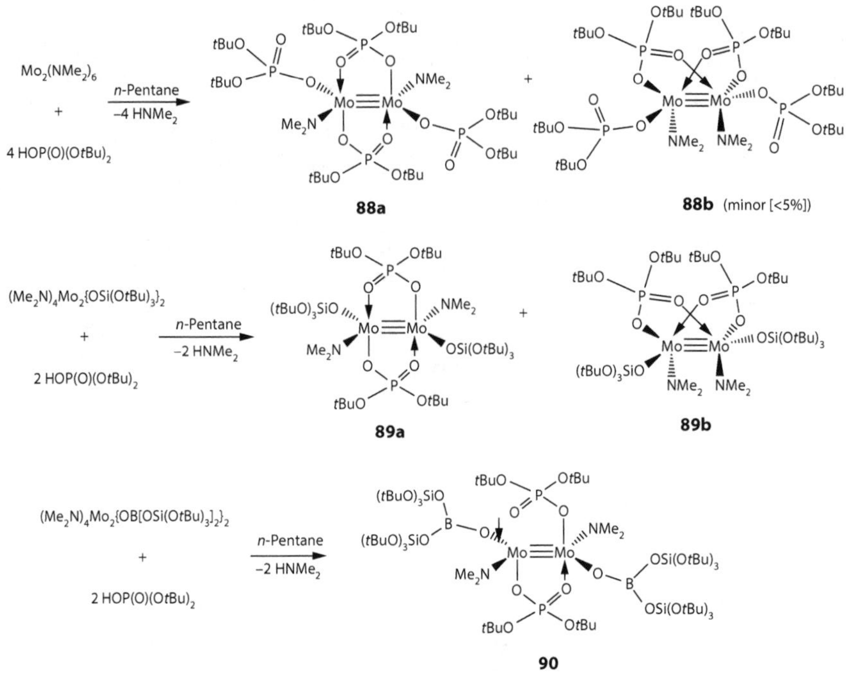

$Mo_2(NMe_2)_6$ + 4 $HOP(O)(OtBu)_2$ —(*n*-Pentane, −4 $HNMe_2$)→ **88a** + **88b** (minor [<5%])

$(Me_2N)_4Mo_2\{OSi(OtBu)_3\}_2$ + 2 $HOP(O)(OtBu)_2$ —(*n*-Pentane, −2 $HNMe_2$)→ **89a** + **89b**

$(Me_2N)_4Mo_2\{OB[OSi(OtBu)_3]_2\}_2$ + 2 $HOP(O)(OtBu)_2$ —(*n*-Pentane, −2 $HNMe_2$)→ **90**

SCHEME 7.10 Synthesis of molybdenum dtbp precursors.

7.4.2.5 Manganese Phosphates

The discovery of manganese carboxylate, $Mn_{12}O_{12}(O_2CPh)_{16}(H_2O)_4$,[61] one of the most studied complexes as single-molecule magnets (SMMs), has led to an exponential growth in manganese-based complexes. Carboxylate ligands allow isolation of metal complexes with relatively short distance between metal centers, and therefore such materials exhibit magnetic order at relatively low temperatures.[62] The similarity in coordination behavior of carboxylate and phosphate diesters has sparked interest in the phosphate-based magnetic materials. With the objective of developing SSPs to manganese phosphate materials, Murugavel et al. have reported one-dimensional manganese dtbp-phosphate $[Mn(dtbp)_2]_n$ (**91**) via the reaction of $Mn(OAc)_2{\cdot}4H_2O$ with dtbp-H in a 1:2 molar ratio in MeOH followed by crystallization in MeOH/THF medium.[63] The formation of **91** takes place via a tetrameric phosphate cluster $Mn_4(O)(dtbp)_6$ (**92**), which has also been isolated in pure form. The structure of **91** features alternating triple and single dtbp bridges between the adjacent Mn(II) ions along the polymer axis (Figures 7.14 and 7.15). This polymer undergoes a facile transformation to a non-interpenetrated two-dimensional grid, $[Mn(dtbp)_2(4,4'\text{-bpy})_2{\cdot}2H_2O]_n$ (**93**), upon treating with 4,4′-bipyridine.[64] Mn(II) ions in this extended solid have octahedral geometry with four 4,4′-bipyridine moieties and two dtbp ligands in a trans arrangement. The neighboring two-dimensional layers are stacked in a staggered fashion in order to minimize the repulsion between terminally coordinated dtbp ligands. The use of chelating N-donor diamines such as 1,10-phenanthroline

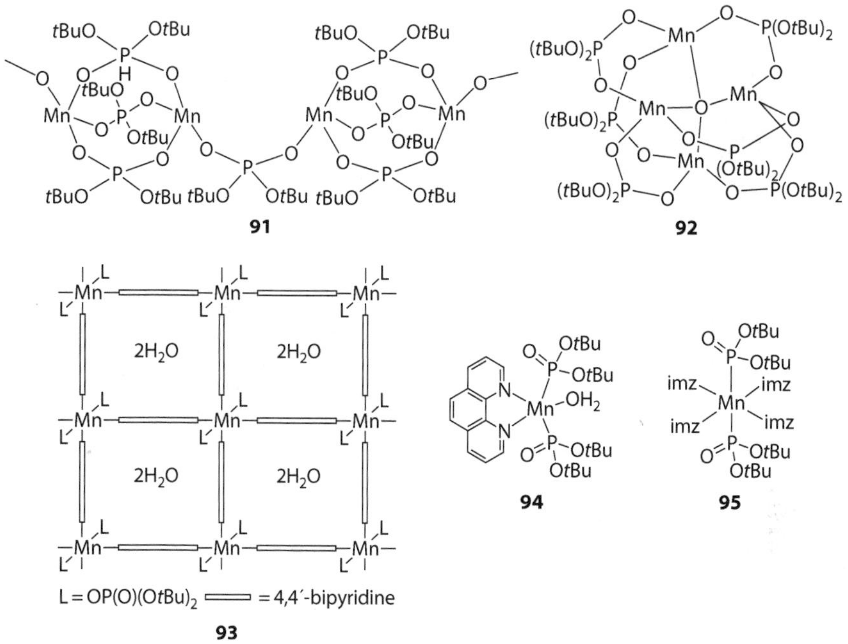

FIGURE 7.14 Manganese phosphates.

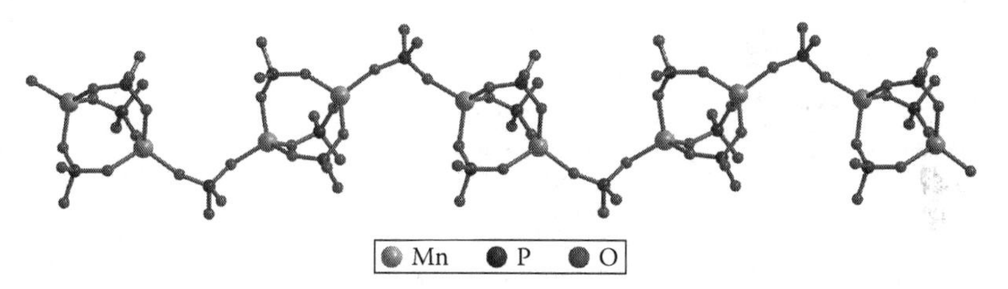

FIGURE 7.15 A section of the x-ray single-crystal structure of **91**. Methyl groups have been removed for clarity. (Reproduced from Sathiyendiran, M. and Murugavel, R., *Inorg. Chem.*, 41, 6404, 2002.)

(phen) leads to the isolation of monomeric phosphate $Mn(dtbp)_2(phen)(H_2O)$ (**94**), which features intramolecular P=O·H–O hydrogen bonding between the free P=O groups of dtbp and the coordinated aqua ligand.[65] The use of imidazole (imz) in place of 1,10-phenanthroline results in the formation of mononuclear complex $Mn(dtbp)_2(imz)_4$ (**95**).[66] Thermal decomposition of **91** and **93** results in metaphosphate material $Mn(PO_3)_2$ below 500 °C. Similarly, calcination of **92** yields a mixture of metaphosphates $Mn(PO_3)_2$ and pyrophosphates $Mn_2P_2O_7$.

Recently, Tilley et al. have isolated Mn_4O_4 dtbp cubane complex $Mn_4O_4[(O_2P(OtBu)_2]_6$ (**96**) from the reaction of $[(bpy)_2Mn(\mu_2\text{-}O)_2Mn(bpy)_2](ClO_4)_3$ with NBu_4[dtbp] in acetonitrile (Scheme 7.11).[67] The molecular structure of **96** features Mn_4O_4 core with each Mn_2O_2 face of the cube bridged diagonally by a dtbp

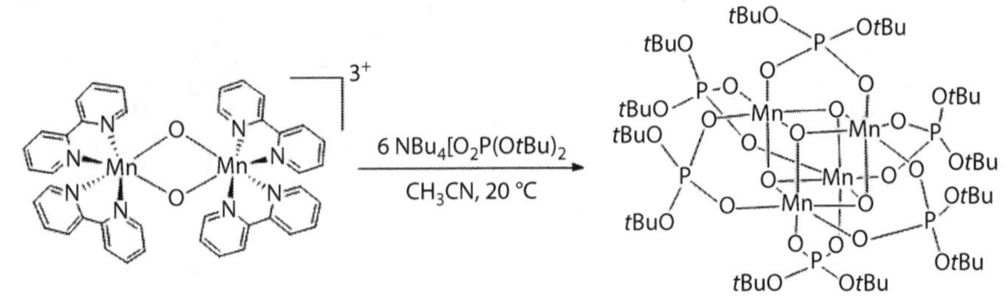

SCHEME 7.11 Synthesis of $Mn_4O_4[(O_2P(OtBu)_2]_6$ (**96**).

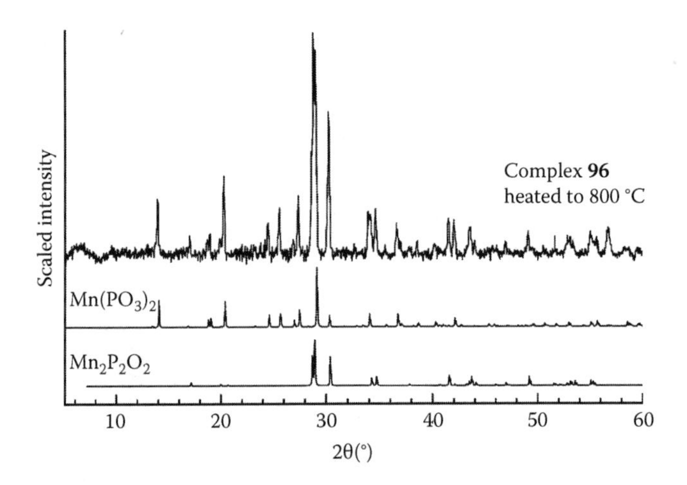

FIGURE 7.16 PXRD patterns of **96**, heated to 800 °C during TGA. (Reproduced from Van Allsburg, K.M. et al., *Chem. Eur. J.*, 21, 4646, 2015. With permission.)

ligand. Unequal Mn–O_{oxo} bond distances reflect the presence of two Mn(III) and two Mn(IV) ions. This structure resembles the oxygen-evolving complex (OEC) of photosystem-II. Variable temperature magnetic susceptibility data indicate extensive magnetic coupling within the cubane with μ_{eff} of 7.3 BM at 300 K. The cubane complex **96** transfers an oxo atom to triethylphosphine to form a butterfly-like compound, $Mn_4O_2\{O_2P(OtBu)_2\}_6(OPEt_3)_2$ (**97**). This event suggests that **96** could actively serve as a possible source of oxygen atoms in oxidation reactions such as water oxidation. Solid-state thermolysis of **96** at 800 °C leads to a crystalline mixture of metaphosphate/pyrophosphate [$2Mn(PO_3)_2$–$Mn_2P_2O_7$ ($Mn_4P_6O_{19}$)], as determined from PXRD (Figure 7.16).

7.4.2.6 Cobalt Phosphates

Cobalt phosphates are useful materials for energy storage and catalysis.[68] Polymeric and dinuclear cobalt phosphates of di-*n*-butylphosphate $[Co(O_2P(OnBu)_2)_2]_n$ (**98**),[69] and

$[Co(O_2P(OnBu)_2)_2(py)_3]_2 \cdot CHCl_3$ (**99**) were reported in the 1980s.[70] The structures of these phosphates were predicted to be polymeric and cyclic with the geometry around cobalt being tetrahedral and octahedral, respectively. Recent studies on the exploitation of dtbp complexes of transition metals for the preparation of fine-particle phosphate materials have been extended to cobalt complexes by Murugavel et al. The reaction of $Co(OAc)_2 \cdot 4H_2O$ with dtbp-H in a 4:6 molar ratio in MeOH or THF yielded an isolable tetranuclear metal phosphate cluster $Co_4(\mu_4\text{-}O)(dtbp)_6$ (**100**).[71] The structure of **100** is similar to that of the manganese tetramer **92**. In the presence of auxiliary ligands imidazole (imz), ethylenediamine (en), or 3,5-dimethylpyrazole (3,5-dmp) $Co(OAc)_2 \cdot 4H_2O$ results mononuclear complexes $Co(dtbp)_2(imz)_4$ (**101**), $Co(dtbp)_2(en)_2$ (**102**) or $Co(dtbp)_2(3,5\text{-dmp})_2$ (**103**), respectively. One-dimensional cobalt-dtbp molecular wire $[Co(dtbp)_2]_n$ (**104**) has also been isolated from the reaction of cobalt acetate with dtbp-H in the presence of a very weak Lewis base 1,3-bis(3,5-dimethylpyrazol-1-yl)propan-2-ol.[64b] Upon stirring with 2 equivalents of 4,4′-bpy, **104** transforms into two-dimensional grid $[Co(dtbp)_2(4,4'\text{-bpy})_2.(H_2O)_2]_n$ (**105**) (Figure 7.17).[64a,71] The thermal decomposition of **99** above 500 °C forms a 2:1 mixture of crystalline metaphosphate $Co(PO_3)_2$ and amorphous pyrophosphate $Co_2P_2O_7$. The thermal decomposition of **100–104** results in exclusive formation of metaphosphate $Co(PO_3)_2$.

Recently, with the objective of electrocatalysis for water oxidation, Tilley et al. have employed polymer **104** to prepare nanostructured $Co(PO_3)_2$ materials.[72]

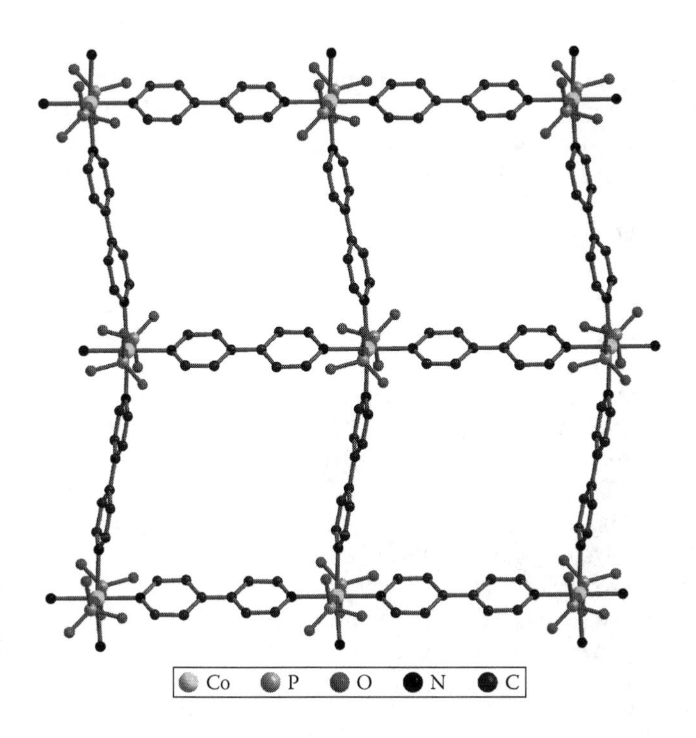

FIGURE 7.17 A section of the single-crystal x-ray structure of **105**. *Tert*-butyl groups are omitted for clarity. (Reproduced from Pothiraja, R. et al., *Inorg. Chem.*, 44, 7585, 2004.)

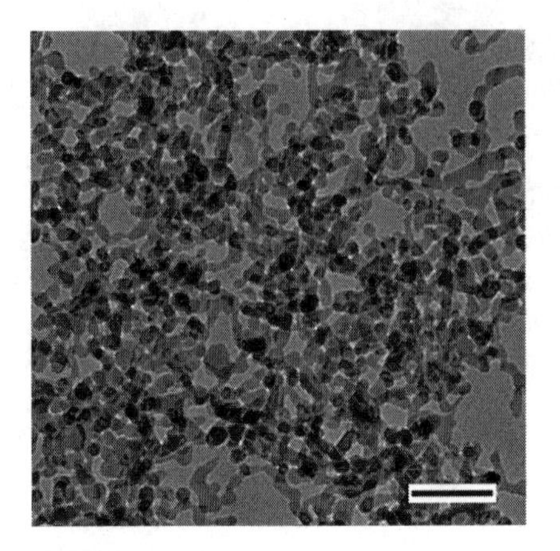
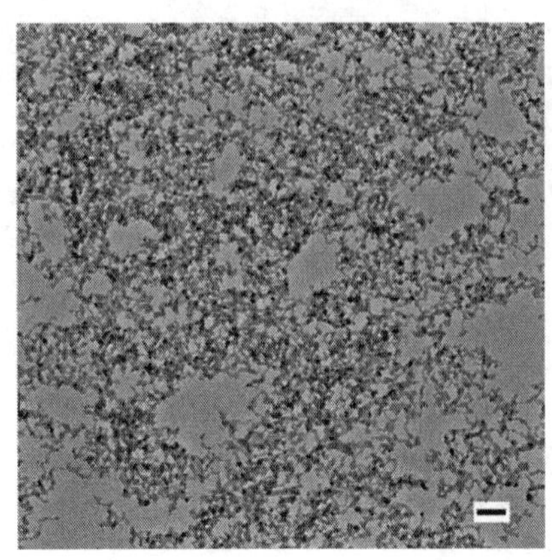

FIGURE 7.18 Transmission electron microscopy image of web-like structure of $Co(PO_3)_2$ (both scale bars are 100 nm). (Reproduced from Ahn, H.S. and Tilley, T.D., *Adv. Funct. Mater.*, 23, 227, 2013.With permission.)

The xerogels prepared via solution-based heating of polymer **104** resulted in high surface area metaphosphate nanoparticles (Figure 7.18), which acts as an efficient electrocatalyst for water oxidation at a relatively low overpotential of about 310 mV.

7.4.2.7 Copper Phosphates

The use of dtbp-H as the phosphate ligand in copper chemistry has led to the isolation of structurally interesting molecules. Murugavel and coworkers have synthesized one-dimensional polymer $[Cu(dtbp)_2]_n$ (**106**) from the reaction of copper acetate with dtbp-H in a 1:2 molar ratio (Scheme 7.12).[63] Unlike the manganese and cobalt phosphate polymers (**91** and **104**), **106** features uniform double dtbp bridges between the adjacent Cu(II) centers across the polymeric chain. The structure consists of eight-membered nearly planar $Cu_2O_4P_2$ rings as repeating units with Cu···Cu separation of 5.045 Å along the chain axis. The addition of 4 equivalents of imidazole to **106** yields mononuclear complex $Cu(dtbp)_2(imz)_4$ (**107**), which can also be prepared via a direct reaction of copper acetate, dtbp-H, and imidazole in a 1:2:4 molar ratio.[66] The monomeric distorted square planar and trigonal bipyramidal complexes $Cu(phen)(dtbp)_2(OH_2)$ (**108**)[65] and $Cu(dtbp)_2(collidine)_2$ (**109**) have been isolated from the reactions of dtbp-H, copper acetate, and 1,10-phenanthroline or collidine.[73] On the other hand, pyridine yields the one-dimensional polymeric complex $[Cu(dtbp)_2(py)_2(\mu\text{-}OH_2)]_n$ (**110**) as blue hollow crystalline tube in DMSO/THF/CH_3OH solvent system.[73] The copper atoms in **110** possess octahedral geometry and are surrounded by two terminal dtbp ligands, two pyridine molecules, and two bridging water molecules. The complexes **108** and **110** feature intramolecular P=O·H–O hydrogen bonding between the dtbp and coordinated aqua ligands, which provides snapshots of various steps of metal-catalyzed phosphate ester hydrolysis. The recrystallization of **109** from a DMSO/THF/CH_3OH solvent mixture leads to the formation of more stable copper cluster $Cu_4(\mu_3\text{-}OH)_2(dtbp)_6(py)_2$ (**111**), consisting of a tetranuclear core $Cu_4(\mu_3\text{-}OH)_2$, which is coordinated with six bidentate

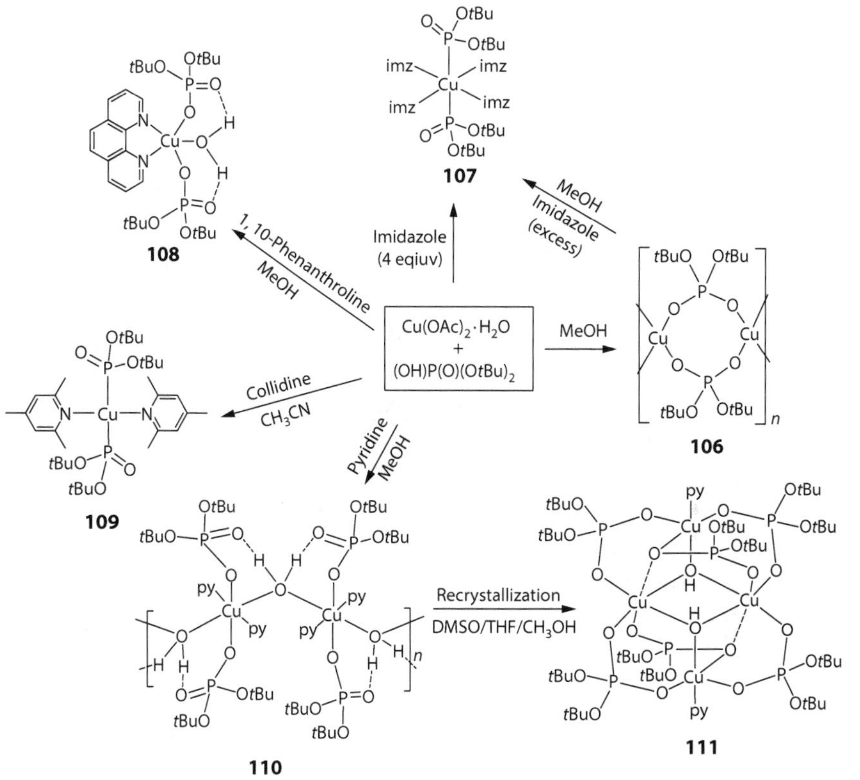

SCHEME 7.12 Synthesis of copper dtbp precursors.

bridging dtbp ligands. While two of the copper(II) ions are pentacoordinated with trigonal bipyramidal geometry, the other two copper ions feature a *pseudo* octahedral geometry with five normal Cu–O bonds and an elongated Cu–O linkage. Each pentacoordinated copper center bears an axial pyridine ligand. The short Cu·Cu nonbonding distances in the tetranuclear core of **111** facilitates an antiferromagnetic coupling at ~20 K.

The thermolysis of **106** up to 500 °C yields crystalline metaphosphate material $Cu(PO_3)_2$ through the loss of four *iso*-butene molecules at low temperature (150 °C–200 °C). Solid-state thermolysis of other mononuclear copper complexes in bulk at 500 °C–510 °C produces copper pyrophosphate $Cu_2P_2O_7$ along with small quantities of $Cu(PO_3)_2$ as dictated in Chart 7.2.

7.4.2.8 Zinc Phosphates

Zinc phosphates are important for biological studies due to the occurrence of zinc(II) ions in several nuclease and phosphatase enzymes. There have been a number of reports on the synthesis of zinc-based compounds and catalysts for the hydrolysis of model phosphate esters. Stucky et al. reported zinc diethyl and dimethyl phosphates,

$[Cu(dtbp)_2]_n \xrightarrow[-H_2C{=}CMe_2]{} [Cu(O_2P(OH)_2)]_n \xrightarrow[-H_2O]{} Cu(PO_3)_2$

$[Cu(dtbp)_2(imz)_4] \xrightarrow[-\text{Volatiles}]{} Cu(PO_3)_2$

$[Cu(dtbp)_2(py)_2(\mu\text{-}OH_2)]_n \xrightarrow[-\text{Volatiles}]{} Cu(PO_3)_2 \xrightarrow[-\delta P_2O_5]{} Cu_2P_2O_7 \text{ (Major)} + Cu(PO_3)_2 \text{ (Minor)}$

$[Cu_4(\mu_3\text{-}OH)_2(dtbp)_6(py)_2] \xrightarrow[-\text{Volatiles}]{} Cu(PO_3)_2 + Cu_2P_2O_7 \xrightarrow[-\delta P_2O_5]{} Cu_2P_2O_7 \text{ (Major)} + Cu(PO_3)_2 \text{ (Minor)}$

$[Cu(dtbp)_2(collidine)_2] \xrightarrow[-\text{Volatiles}]{} Cu(PO_3)_2 \xrightarrow[-\delta P_2O_5]{} Cu_2P_2O_7 \text{ (Major)} + Cu(PO_3)_2 \text{ (Minor)}$

CHART 7.2 Suggested decomposition patterns of copper dtbp precursors.

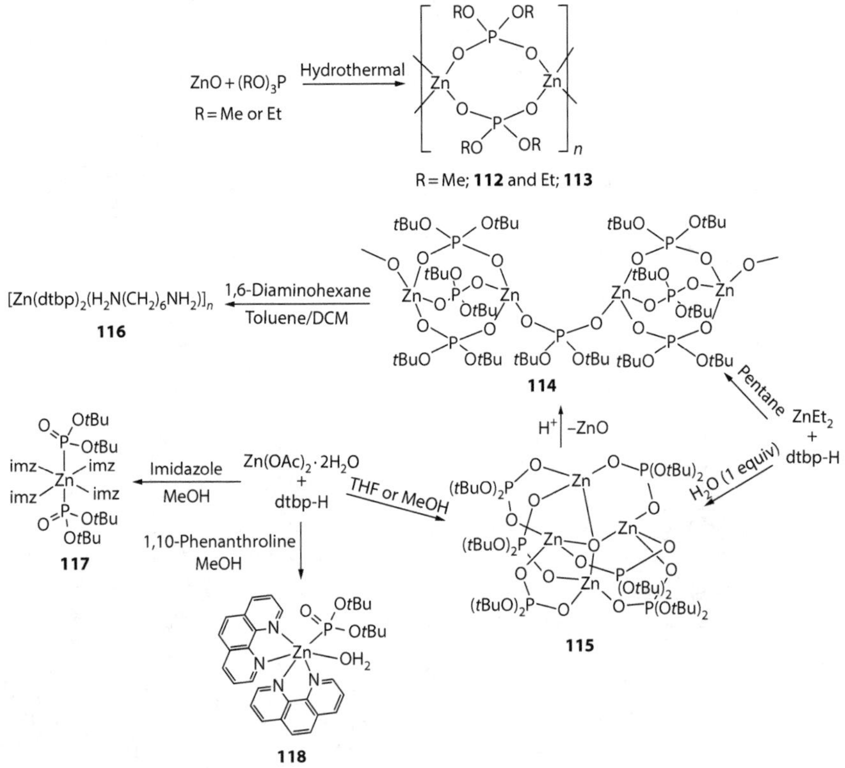

SCHEME 7.13 Synthesis of zinc dtbp precursors.

$[Zn(O_2P(OMe)_2)_2]_n$ (**112**) and $[Zn(O_2P(OEt)_2]_n$ (**113**) via hydrothermal reaction of ZnO with $(RO)_3P$ (R = Me or Et) (Scheme 7.13).[74] These isostructural complexes consist of infinite one-dimensional chains of corner-sharing tetraderal ZnO_4 and PO_4 units. Tilley et al. prepared more thermally labile zinc phosphates using dtbp-H as the ligand.[75] The reaction of 2 equivalents of dtbp-H with $ZnEt_2$ in pentane resulted in an insoluble polymer material $[Zn(dtbp)_2]_n$ (**114**) as white precipitate and a soluble oxo-centered tetranuclear cluster $Zn_4(\mu_4\text{-}O)(dtbp)_6$ (**115**), which was crystallized from the supernatant. The tetranuclear cluster **115** has been independently synthesized by the reaction of dtbp-H with $ZnEt_2$ in 2:3 molar ratio in a slightly aqueous condition (1 equivalent water). In the solid state, **114** adopts a *zig-zag* structure with zinc atoms coordinated alternately by one and then three bridging dtbp groups, as in the case of zirconium (**84**) and cobalt (**91**) phosphate polymers. Slow diffusion of a toluene solution of $Zn_4(\mu_4\text{-}O)(dtbp)_6$ into a dichloromethane solution of 1,6-hexanediamine leads to the isolation of "pleated sheet" polypeptide types of polymer $[Zn(dtbp)_2(H_2N(CH_2)_6NH_2)]_n$ (**116**) with the elimination of ZnO. The polymeric strands of **116** are interconnected by hydrogen bonds between N–H and P=O functionalities. Mononuclear zinc–dtbp complexes $Zn(dtbp)_2(imidazole)_4$ (**117**) and $Zn(phen)_2(dtbp)(OH_2)][dtbp](MeOH)(MeCOOH)(H_2O)_3$ (**118**) have also been isolated by Murugavel et al.[71,76]

The thermal decomposition of **113** shows melting followed by decomposition (loss of diethyl ether), leading to the formation of crystalline metaphosphate $Zn(PO_3)_2$. Similarly, the thermolysis of **114** at 800 °C yields quantitative pyrolytic conversion to crystalline $\beta\text{-}Zn(PO_3)_2$. The tetranuclear cluster **115** is thermally more labile and undergoes elimination of isobutene and water over the temperature range 130 °C–220 °C and yields pure $Zn_4P_6O_{19}$ material ($\alpha\text{-}Zn_2P_2O_7$ and $\beta\text{-}Zn(PO_3)_2$) at 900 °C. When heated in ethanol at 85 °C for 30 h, **115** converts to **114** and ZnO. This transformation is facilitated at room temperature by organic acids (terephthalic or benzoic acid). Upon heating an ethanol suspension of **115** in pyrex tube at 60 °C (16 h), 85 °C (56 h), and 130 °C (56 h), yielded $Zn_3(PO_4)_2{\cdot}4H_2O$ (hopeite), $Zn_3(PO_4)_2{\cdot}H_2O$, and $\alpha\text{-}Zn_3(PO_4)_2$, respectively. The mononuclear complex **117** completely converts into the corresponding metaphosphate $Zn(PO_3)_2$ before 800 °C by the stepwise elimination of imidazole, isobutene, and water, whereas **116** and **118** result in the formation of a mixture of both $Zn(PO_3)_2$ and ZnP_2O_7.

7.4.2.9 Cadmium Phosphates

Faller et al. in 1983 isolated one-dimensional cadmium ethylphosphate, $[Cd(O_2P(OEt)_2]_n$ (**119**) from an ethanolic mixture of potassium diethyl phosphate and cadmium nitrate.[77] The compound is a highly interlocked polymer with each adjacent pair of Cd(II) ions bridged by two diethylphosphate ligands on either side. Additionally, one of the oxygen atoms of the phosphate ligand bridges a third cadmium ion. Murugavel et al. have reported one-dimensional linear polymer $[Cd(dtbp)_2(H_2O)]_n$ (**120**) via the reaction of cadmium acetate with dtbp-H in a 1:2 molar ratio in methanol (Scheme 7.14).[63] The Cd(II) ions in this polymer are pentacoordinated. The polymer structure closely resembles the structure of copper polymer **106** in the sense that adjacent cadmium ions are bridged by two dtbp ligands. A water molecule additionally coordinates each cadmium ion. Also, the polymer readily transforms into non-interpenetrating rectangular grid structure $[Cd(dtbp)_2(4,4'\text{-}bpy)_2.(H_2O)_2]_n$ (**121**), by addition of 4,4′-bpy spacer in methanol (Figure 7.19).[64a] The solid-state thermal decomposition of both **120** and **121** leads to the formation of $Cd(PO_3)_2$ below 500 °C and 700 °C, respectively.

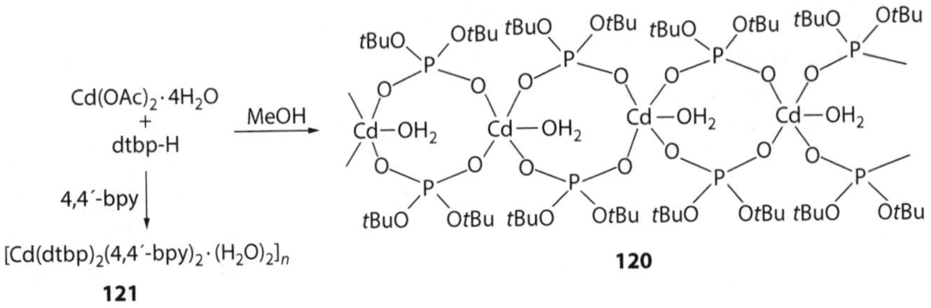

$Cd(OAc)_2 \cdot 4H_2O$ + dtbp-H $\xrightarrow{MeOH}$ **120**

4,4′-bpy ↓

$[Cd(dtbp)_2(4,4'\text{-}bpy)_2 \cdot (H_2O)_2]_n$

121

SCHEME 7.14 Synthesis of cadmium dtbp precursors.

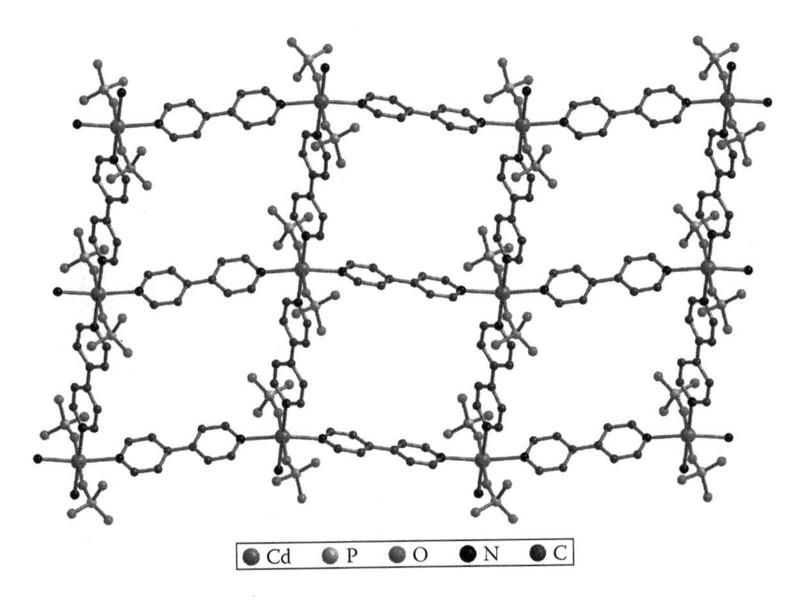

FIGURE 7.19 Ball and stick model of the two-dimensional grid structure formed in the crystal structure of **121** (view down the crystallographic *a*-axis). *Tert*-butyl groups are omitted for clarity. (Reproduced from Pothiraja, R. et al., *Inorg. Chem.*, 43, 6314, 2005.)

7.4.2.10 Group 13 Phosphates

Group 13 phosphates are useful as catalysts and catalyst supports and therefore making soluble aluminum phosphate precursors for ceramic phosphates and exploring their properties is of significant importance. Tilley et al. reported two aluminum phosphates $[Me_2AlO_2P(OtBu)_2]_2$ (**122**) and $[Al(OiPr)_2O_2P(OtBu)_2]_4$ (**123**) by the reaction of dtbp-H with Al_2Me_6 and $[Al(OiPr)_3]_4$, respectively (Scheme 7.15).[78] In the solid state, **122** is a centrosymmetric dimer consisting of two 4-coordinate aluminum atoms bridged by two dtbp ligands, while **123** exists as a centrosymmetric tetramer in which the unique half of the tetramer consists of two aluminum atoms bridged by two dtbp ligands. Solution-phase thermolysis of **122** and **123** in organic solvents followed by air-drying leads to the formation of transparent and opaque xerogels, respectively.

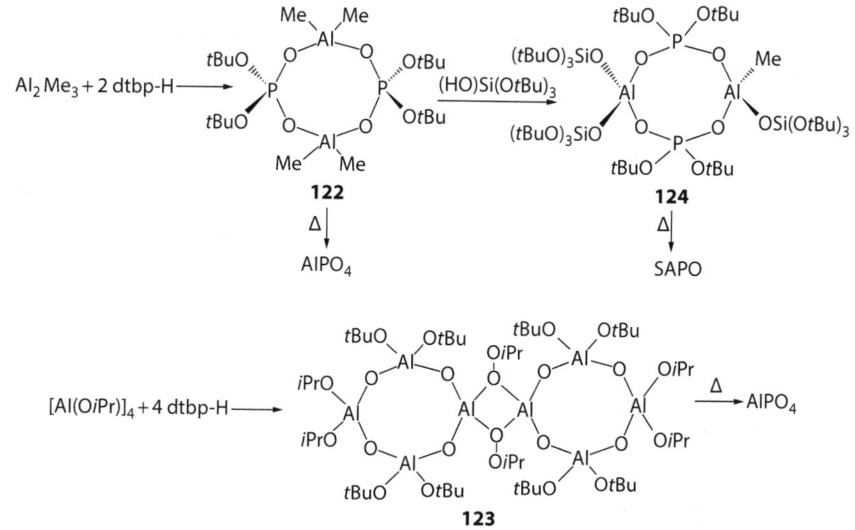

SCHEME 7.15 Synthesis of aluminum dtbp precursors.

The reaction of **122** with excess $HOSi(OtBu)_3$ leads to the isolation of silicoaluminophosphates (SAPO) $(tBuO)_3SiO)_2Al(O_2P(OtBu)_2)_2AlCH_3$-$(OSi(OtBu)_3$ (**124**). The thermal decomposition of **124** yields mesoporous SAPO with large surface area of >500 m^2g^{-1} with large pores.[79]

7.5 CONCLUSION AND OUTLOOK

SSPs method offers advantages over more established routes for the generation of nanostructured silicate and phosphate materials. These materials are useful for heterogeneous catalysts, catalyst supports, and electronics. The most important step in this chemistry involves the synthesis of organically soluble molecular precursors, $L_nM(O_xE(OR)_y]_z$, where E = Si, P, B, and so on, and R = *t*Bu or $Si(OtBu)_3$. Calcination or solution thermolysis of the molecular precursors with preexisting covalent M–O–E bonds enables control over stoichiometry, elemental dispersion (avoiding phase separation, which is a common nuisance in solid-state synthetic protocols), surface structure, site-isolated catalytic sites, and particle size of the materials. Apart from being useful as SSPs, metal phosphate complexes containing aqua ligands have proven to be useful models for phosphate ester hydrolysis reactions.

Owing to the outstanding inherent properties such as large surface area, excellent mechanical and thermal stability, and nanometer size, the applications of silicate and phosphate nanomaterials will continue to attract scientific interest in the years to come. In recent years, much effort has been directed toward the development of various types of catalysts and energy storage materials using silicates and phosphates. These nanocatalysts serve as powerful tools for the conversion of raw materials into useful chemicals of both industrial and pharmaceutical significance, such as oxidation, dehydrogenation, and dehydration. Recently, cobalt phosphates have emerged as the efficient elecrocatalysts for hydrogen generation from water splitting and lithium ion batteries.

ACKNOWLEDGMENT

The authors thank DST-SERB (Nanomission and J. C. Bose Fellowship projects) and DAE (DAE-SRC Outstanding Research Investigator Award).

REFERENCES

1. Milton, R. M. Commercial development of molecular sieve technology. *Molecular Sieves*, Papers read at the conference held at the School of Pharmacy, University of London, April 4–6, 1967. Society of Chemical Industry, London, U.K., 1968, pp. 199–203.
2. Wilson, S. T.; Lok, B. M.; Messina, C. A.; Cannan, T. R.; Flanigen, E. M. *J. Am. Chem. Soc.* 1982, *104*, 1146.
3. Fujdala, K.; Brutchey, R.; Tilley, T. D. In *Surface and Interfacial Organometallic Chemistry and Catalysis*, Copéret, C., Chaudret, B., eds. Springer-Verlag Berlin Heidelberg, 2005; Vol. *16*, p. 69.
4. (a) Murugavel, R.; Walawalkar, M. G.; Dan, M.; Roesky, H. W.; Rao, C. N. R. *Acc. Chem. Res.* 2004, *37*, 763; (b) Murugavel, R.; Choudhury, A.; Walawalkar, M. G.; Pothiraja, R.; Rao, C. N. R. *Chem. Rev.* 2008, *108*, 3549; (c) Gupta, S. K.; Kuppuswamy, S.; Walsh, J. P. S.; McInnes, E. J. L.; Murugavel, R. *Dalton Trans.* 2015, *44*, 5587; (d) Gupta, S. K.; Dar, A. A.; Rajeshkumar, T.; Kuppuswamy, S.; Langley, S. K.; Murray, K. S.; Rajaraman, G.; Murugavel, R. *Dalton Trans.* 2015, *44*, 5961.
5. McMullen, A. K.; Tilley, T. D.; Rheingold, A. L.; Geib, S. J. *Inorg.Chem.* 1989, *28*, 3772.
6. Kriesel, J. W.; Tilley, T. D. *J. Mater. Chem.* 2001, *11*, 1081.
7. Michel, O.; König, S.; Törnroos, K. W.; Maichle-Mössmer, C.; Anwander, R. *Chem. Eur. J.* 2011, *17*, 11857.
8. Marciniec, B.; Maciejewski, H. *Coord. Chem. Rev.* 2001, *223*, 301.
9. Terry, K. W.; Tilley, T. D. *Chem. Mater.* 1991, *3*, 1001.
10. Dhayal, V.; Chaudhary, A.; Choudhary, B. L.; Nagar, M.; Bohra, R.; Mobin, S. M.; Mathur, P. *Dalton Trans.* 2012, *41*, 9439.
11. Coles, M. P.; Lugmair, C. G.; Terry, K. W.; Tilley, T. D. *Chem. Mater.* 1999, *12*, 122.
12. Narula, C. K.; Varshney, A.; Riaz, U. *Chem. Vap. Depos.* 1996, *2*, 13.
13. Gunji, T.; Kashara, T.; Abe, Y. *J. Sol-Gel Sci. Technol.* 1998, *13*, 975.
14. Brutchey, R. L.; Mork, B. V.; Sirbuly, D. J.; Yang, P.; Tilley, T. D. *J. Mol. Catal. A* 2005, *238*, 1.
15. Jarupatrakorn, J.; Tilley, T. D. *J. Am. Chem. Soc.* 2002, *124*, 8380.
16. Terry, K. W.; Lugmair, C. G.; Tilley, T. D. *J. Am. Chem. Soc.* 1997, *119*, 9745.
17. Lugmair, C. G.; Tilley, T. D. *Inorg. Chem.* 1998, *37*, 764.
18. Fujdala, K. L.; Oliver, A. G.; Hollander, F. J.; Tilley, T. D. *Inorg. Chem.* 2003, *42*, 1140.
19. (a) Rulkens, R.; Male, J. L.; Terry, K. W.; Olthof, B.; Khodakov, A.; Bell, A. T.; Iglesia, E.; Tilley, T. D. *Chem. Mater.* 1999, *11*, 2966; (b) Rulkens, R.; Tilley, T. D. *J. Am. Chem. Soc.* 1998, *120*, 9959.
20. Fujdala, K. L.; Tilley, T. D. *Chem. Mater.* 2002, *14*, 1376.
21. Brutchey, R. L.; Lugmair, C. G.; Schebaum, L. O.; Tilley, T. D. *J. Catal.* 2005, *229*, 72.

22. (a) Ruddy, D. A.; Tilley, T. D. *Chem. Commun.* 2007, 3350; (b) Cordeiro, P. J.; Tilley, T. D. *Langmuir* 2011, *27*, 6295.
23. Wolczanski, P. T. *Polyhedron* 1995, *14*, 3335.
24. (a) McDaniel, M. P. In *Advances in Catalysis*; Bruce, C. G. and Helmut, K., eds. Academic Press, 2010; Vol. 53, p. 123; (b) Groppo, E.; Damin, A.; Otero Arean, C.; Zecchina, A. *Chem. Eur. J.* 2011, *17*, 11110.
25. Terry, K. W.; Gantzel, P. K.; Tilley, T. D. *Inorg. Chem.* 1993, *32*, 5402.
26. Fujdala, K. L.; Tilley, T. D. *Chem. Mater.* 2001, *13*, 1817.
27. (a) Conley, M. P.; Delley, M. F.; Núñez-Zarur, F.; Comas-Vives, A.; Copéret, C. *Inorg. Chem.* 2015, *54*, 5065; (b) Delley, M. F.; Núñez-Zarur, F.; Conley, M. P.; Comas-Vives, A.; Siddiqi, G.; Norsic, S.; Monteil, V.; Safonova, O. V.; Copéret, C. *Proc. Natl. Acad. Sci.* 2014, *111*, 11624; (c) Conley, M. P.; Delley, M. F.; Siddiqi, G.; Lapadula, G.; Norsic, S.; Monteil, V.; Safonova, O. V.; Copéret, C. *Angew. Chem., Int. Ed.* 2014, *53*, 1872; (d) Conley, M.; Copéret, C. *Top. Catal.* 2014, *57*, 843.
28. Marshak, M. P.; Nocera, D. G. *Inorg. Chem.* 2013, *52*, 1173.
29. Su, K.; Tilley, T. D. *Chem. Mater.* 1997, *9*, 588.
30. Jarupatrakorn, J.; Coles, M. P.; Tilley, T. D. *Chem. Mater.* 2005, *17*, 1818.
31. Fujdala, K. L.; Tilley, T. D. *Chem. Mater.* 2004, *16*, 1035.
32. Nozaki, C.; Lugmair, C. G.; Bell, A. T.; Tilley, T. D. *J. Am. Chem. Soc.* 2002, *124*, 13194.
33. Brutchey, R. L.; Drake, I. J.; Bell, A. T.; Tilley, T. D. *Chem. Commun.* 2005, 3736.
34. Jarupatrakorn, J.; Tilley, T. D. *Dalton Trans.* 2004, 2808.
35. Marciniec, B.; Szubert, K.; Potrzebowski, M. J.; Kownacki, I.; Łęszczak, K. *Angew. Chem. Int. Ed.* 2008, *47*, 541.
36. McMullen, A. K.; Tilley, T. D.; Rheingold, A. L.; Geib, S. J. *Inorg. Chem.* 1990, *29*, 2228.
37. Choi, Y. S.; Moschetta, E. G.; Miller, J. T.; Fasulo, M.; McMurdo, M. J.; Rioux, R. M.; Tilley, T. D. *ACS Catal.* 2011, *1*, 1166.
38. Schmidbaur, H.; Adlkofer, J.; Shiotani, A. *Chem. Ber.* 1972, *105*, 3389.
39. (a) Terry, K. W.; Lugmair, C. G.; Gantzel, P. K.; Tilley, T. D. *Chem. Mater.* 1996, *8*, 274; (b) Fujdala, K. L.; Drake, I. J.; Bell, A. T.; Tilley, T. D. *J. Am. Chem. Soc.* 2004, *126*, 10864.
40. Su, K.; Tilley, T. D.; Sailor, M. J. *J. Am. Chem. Soc.* 1996, *118*, 3459.
41. Rendón, N.; Bourdolle, A.; Baldeck, P. L.; Le Bozec, H.; Andraud, C.; Brasselet, S.; Coperet, C.; Maury, O. *Chem. Mater.* 2011, *23*, 3228.
42. Kijima, I.; Yamamoto, T.; Abe, Y. *Bull. Chem. Soc. Jpn.* 1971, *44*, 3193.
43. Terry, K. W.; Ganzel, P. K.; Tilley, T. D. *Chem. Mater.* 1992, *4*, 1290.
44. Lugmair, C. G.; Fujdala, K. L.; Tilley, T. D. *Chem. Mater.* 2002, *14*, 888.
45. (a) Clearfield, A. *Chem. Rev.* 1988, *88*, 125; (b) Martin, S. W. *J. Am. Ceram. Soc.* 1991, *74*, 1767; (c) Umeyama, D.; Horike, S.; Inukai, M.; Itakura T.; Kitagawa, S. *J. Am. Chem. Soc.* 2012, *134*, 12780; (d) Monceaux, L.; Courtine, P. *Eur. J. Solid State Inorg. Chem.* 1991, *28*, 233; (e) Dean, N. S.; Mokry, L. M.; Bond, M. R.; O'Connor, C. J.; Carrano, C. J. *Inorg. Chem.* 1996, *35*, 3541; (f) Ungashe, S. B.; Wilson, W. L.; Katz, H. E.; Scheller, G. R.; Putvinski, T. M. *J. Am. Chem. Soc.* 1992, *114*, 8717.
46. Goldwhite, H.; Saunders, B. C. *J. Chem. Soc.* 1957, 2409.
47. Zwierzak, A.; Kluba, M. *Tetrahedron* 1971, *27*, 3163.
48. Murugavel, R.; Choudhury, A.; Walawalkar, M. G.; Pothiraja, R.; Rao, C. N. R. *Chem. Rev.* 2008, *108*, 3549.
49. Lugmair, C. G.; Tilley, T. D. *Inorg. Chem.* 1998, *37*, 1821.
50. (a) Ezra, F. S.; Collin, R. L. *Acta Crystallogr.* 1973, *B29*, 1398; (b) Scanlon, J.; Collin, R. L. *Acta Crystallogr.* 1954, *7*, 781.
51. Burns, J. H.; Kessler, R. M. *Inorg. Chem.* 1987, *26*, 1370.

52. Kyogoku, Y.; Iitaka, Y. *Acta Crystallogr.* 1966, *21*, 49.
53. (a) Thorn, D. L.; Harlow, R. L. *Inorg. Chem.* 1992, *31*, 3917; (b) Munowitz, M.; Jarman, R. H.; Harrison, J. F. *Chem. Mater.* 1992, *4*, 1296; (c) Wang, B.; Greenblatt, M.; Wang, S.; Hwu, S. J. *Chem. Mater.* 1993, *5*, 23.
54. Guerrero, G.; Mehring, M.; Hubert Mutin, P.; Dahan, F.; Vioux, A. *J. Chem. Soc. Dalton Trans.* 1999, 1537.
55. Kumara Swamy, K. C.; Veith, M.; Huch, V.; Mathur, S. *Inorg. Chem.* 2003, *42*, 5837.
56. (a) Johnston, D. C.; Johnson, J. W. *J. Chem. Soc. Chem. Commun.* 1985, 1720; (b) Bond, M. R.; Mokry, L. M.; Otieno, T.; Thompson, J.; Carrano, C. J. *Inorg. Chem.* 1995, *34*, 1894.
57. (a) Thorn, D. L.; Harlow, R. L.; Herron, N. *Inorg. Chem.* 1996, *35*, 547; (b) Herron, N.; Thorn, D. L.; Harlow, R. L.; Coulston, G. W. *J. Am. Chem. Soc.* 1997, *119*, 7149.
58. Lugmair, G. C.; Tilley, T. D. *Monatsh. Chem.* 2006, *137*, 557.
59. Huq, F.; Mowat, W.; Shortland, A.; Skapski, A. C.; Wilkinson, G. *J. Chem. Soc. Chem. Commun.* 1971, 1079.
60. Terry, K. W.; Su, K.; Tilley, T. D.; Rheingold, A. L. *Polyhedron* 1998, *17*, 891–897.
61. Mirebeau, I.; Hennion, M.; Casalta, H.; Andres, H.; Gudel, H. U.; Irodova, A. V.; Caneschi, A. *Phys. Rev. Lett.* 1999, *83*, 628.
62. Bhat, G. B.; Vishnoi, P.; Gupta, S. K.; Murugavel, R. *Inorg. Chem. Commun.* 2015, 59, 84.
63. Sathiyendiran, M.; Murugavel, R. *Inorg. Chem.* 2002, *41*, 6404.
64. (a) Pothiraja, R.; Sathiyendiran, M.; Butcher, R. J.; Murugavel, R. *Inorg. Chem.* 2005, *43*, 6314; (b) Pothiraja, R.; Sathiyendiran, M.; Butcher, R. J.; Murugavel, R. *Inorg. Chem.* 2004, *44*, 7585.
65. Murugavel, R.; Sathiyendiran, M.; Pothiraja, R.; Butcher, R. J. *Chem. Commun.* 2003, 2546.
66. Murugavel, R.; Sathiyendiran, M. *Chem. Lett.* 2001, *30*, 84.
67. Van Allsburg, K. M.; Anzenberg, E.; Drisdell, W. S.; Yano, J.; Tilley, T. D. *Chem. Eur. J.* 2015, *21*, 4646.
68. (a) Xi, Y.; Dong, B.; Dong, Y.; Mao, N.; Ding, L.; Shi, L.; Gao, R.; Liu, W.; Su, G.; Cao, L. *Chem. Mater.* 2016, *28*, 1355; (b) Liu, T.; Duan, Y.; Zhang, G.; Li, M.; Feng, Y.; Hu, J.; Zheng, J.; Chen, J.; Pan, F. *J. Mater. Chem. A* 2016, *4*, 4479.
69. Kantacha, A.; Wongnawa, S. *Inorg. Chim. Acta* 1987, *134*, 135.
70. Huang, C.; Zhang, D.; Xu, G.; Zhang, S. *Sci. Sin. Ser. B (Engl. Ed.)* 1988, *31*, 1153.
71. Murugavel, R.; Sathiyendiran, M.; Walawalkar, M. G. *Inorg. Chem.* 2001, *40*, 427.
72. Ahn, H. S.; Tilley, T. D. *Adv. Funct. Mater.* 2013, *23*, 227.
73. Murugavel, R.; Sathiyendiran, M.; Pothiraja, R.; Walawalkar, M. G.; Mallah, T.; Riviére, E. *Inorg. Chem.* 2004, *43*, 945.
74. (a) Harrison, W. T. A.; Nenoff, T. M.; Gier, T. E.; Stucky, G. D. *Inorg. Chem.* 1992, *31*, 5395; (b) Harrison, W. T. A.; Nenoff, T. M.; Gier, T. E.; Stucky, G. D. *J. Mater. Chem.* 1994, *4*, 1111.
75. Lugmair, C. G.; Tilley, T. D.; Rheingold, A. L. *Chem. Mater.* 1997, *9*, 339.
76. Pothiraja, R.; Shanmugan, S.; Walawalkar, M. G.; Nethaji, M.; Butcher, R. J.; Murugavel, R. *Eur. J. Inorg. Chem.* 2008, *2008*, 1834.
77. Miner, V. W.; Prestegard, J. H.; Faller, J. W. *Inorg. Chem.* 1983, *22*, 1862.
78. Lugmair, C. G.; Tilley, T. D.; Rheingold, A. L. *Chem. Mater.* 1999, *11*, 1615.
79. Fujdala, K. L.; Tilley, T. D. *J. Am. Chem. Soc.* 2001, *123*, 10133.

8 Molecular Tools for Controlling Nanoparticle Size/Morphologies

Puspanjali Sahu, Jayesh Shimpi, and B. L. V. Prasad

CONTENTS

8.1 INTRODUCTION

The preparation of nanoparticles (especially those of noble metals) predates the coinage of the term *nano* itself by several centuries. The earliest example of the preparation of nanoparticles has been linked to Roman artisans who created colored glass by probably mixing gold chloride and other salts of noble metals while making glass. Many artistic marvels such as the Lycurgus cup and the glass panels in many European cathedrals were created by this simple procedure. Furthermore, many believe that medicinal preparation such as "aurum potabile" and formulations injected for treating rheumatic arthritis were made of colloidal gold. Several people such as the fourteenth-century Bolognese painters, the alchemists Paracelsus, Andreus Cassius, and Johannes Kunckel are all believed to have made the colloidal form of gold, which is the most studied form among the metallic nanoparticles. However, none of these methods of preparations or people were really interested to see in what form the metals existed in composite/solution or whether these existed as small ("nano") particles. Actually, it was Michael Faraday who first conjectured that the ruby red color imparted to the glass/ceramics could be due to the reduction of gold salts to metallic state.[1] He in fact prepared a "sol" of gold by reacting a gold chloride solution with red phosphorous, thus causing the reduction of gold (III) ions to metallic gold. He was also the first person who inferred that the color imparted to the glass/ceramics could be due to the existence of gold in a "finely divided metallic state." Several systematic studies by Richard Adolf Zsigmondy, Theodor Svedberg, and Gustav Mie[2–5] on gold nanoparticles followed. However, the preparations in all these studies have been mainly curiosity driven and practical applications, if any, have been rather limited. One of the causes for this limitation could be related to the difficult preparative procedures. This major hurdle has been crossed thanks to the pioneering works first by Turkevich et al.[6] and later by Frens[7,8] that provided the first major breakthrough as far as the simplification of gold colloid preparation was concerned. Procedures developed by these people were also the first ones where an organic molecule, trisodium citrate, was used for nanoparticle synthesis (Figure 8.1).

The preparation of gold/silver colloids mediated by trisodium citrate also allowed systematic studies to be carried out on these gold sols. These studies revealed that apart from causing reduction, the unoxidized citrate ions and their oxidized products form a charged layer (called discrete double layer) on the surface of the formed nanoparticles. The repulsive interactions of these charged particles oppose the van der Waals (vdW)

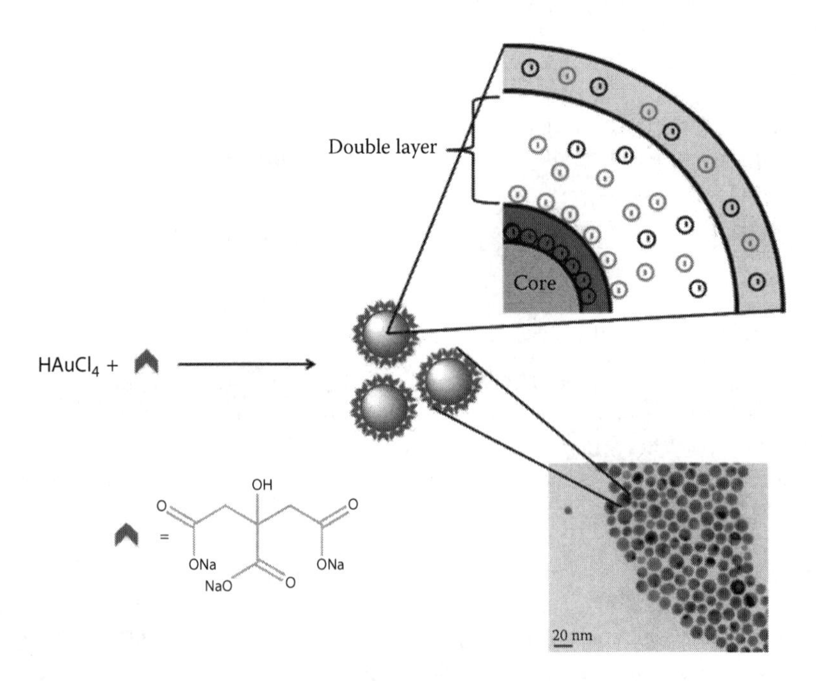

FIGURE 8.1 A typical schematic representation of charge stabilized gold nanoparticle preparation using trisodium citrate as reducing and capping agent. Bottom right inset: A representative TEM image of the nanoparticles. Top right inset: The pictorial representation of the charge double layer. The molecular structure of trisodium citrate is also shown.

attractive forces between the metallic cores, providing a dispersional stability to the system. The nature of these interactions and the factors that affect this stability have been worked out in great detail by Derjaguin and Landau and by Verwey and Overbeek and is known as the DLVO theory.[9] This theory gives the quantitative value of total potential energy as a function of the interparticle distance between two approaching charged particles considering the effect of both the vdW attraction between the cores and the electrostatic repulsion due to the so-called double layer of counterions. But the stability of nanoparticles arising from the electrostatic repulsion of the charged surface is highly sensitive and can be switched from a stabilized state to a destabilized state with minor external disturbances, such as small variations in electrolyte concentrations or pH.

Contrary to this, it has been realized that the stability of nanoparticle/colloidal dispersions can be enhanced enormously by the adsorption of thick ligand/polymeric shells on their surfaces. This enhanced stability is due to the repulsive steric force that arises mainly from the unfavorable entropy associated with the compression of adsorbed polymer/molecular chains between the surfaces at separations lower than the distance where the outer segments of the polymers/molecules on adjacent particles begin to overlap.[10] Such stabilization is not normally perturbed by minor external conditions such as pH variations or electrolyte additions as mentioned. Thus began the utilization of long alkyl chains containing molecules/polymers to improve the chemical stability of nanomaterials/particles. However, the interaction between the molecules, which have been used to impart stabilization, and the solid nanoparticle surface remained to a large extent a nonspecific one. This scenario changed rapidly following the demonstration that small organic molecules possessing certain functional groups can interact specifically with inorganic surfaces and form the so-called self-assembled monolayers.[11,12] In the rest of the chapter, we refer to such molecules that can interact with the surface of the inorganic materials/nanoparticles specifically through their functional groups as ligands. Initially, ligands were used during nanoparticle synthesis to improve their dispersional stability in solvents with certain characteristics (e.g., polar vs. nonpolar). From this humble beginning of their usage to just increase the dispersional stability, the utility of molecular tools in the area of nanoparticle synthesis took rapid strides and these are now being used for many complex tasks. The most notable among them are targeted drug delivery, theranostics, and the preparation of multitasking "chimeric" nanoparticles.[13]

8.2 MOLECULAR TOOLS FOR SIZE CONTROL OF NANOPARTICLES

As mentioned, ligands when attached on nanoparticle surface impart strong repulsive forces against the attractive vdW forces existing between the metal nanoparticles, imparting stability to the nanoparticle dispersion. One of the first preparations where molecules/ligands that are known to form self-assembled monolayers have been used as capping/passivating agents of nanoparticle surface in organic solvents is popularly referred to as the Brust–Schiffrin protocol.[14] The Brust–Schiffrin method is a two-phase synthetic method where the gold ions are first transferred into an organic solvent using a phase transfer agent/surfactant. Subsequently, these gold ions are reduced in a water-in-oil reverse micelle/microemulsion and the small gold nanoparticles that are formed are immediately capped by the adsorption of thiol molecules

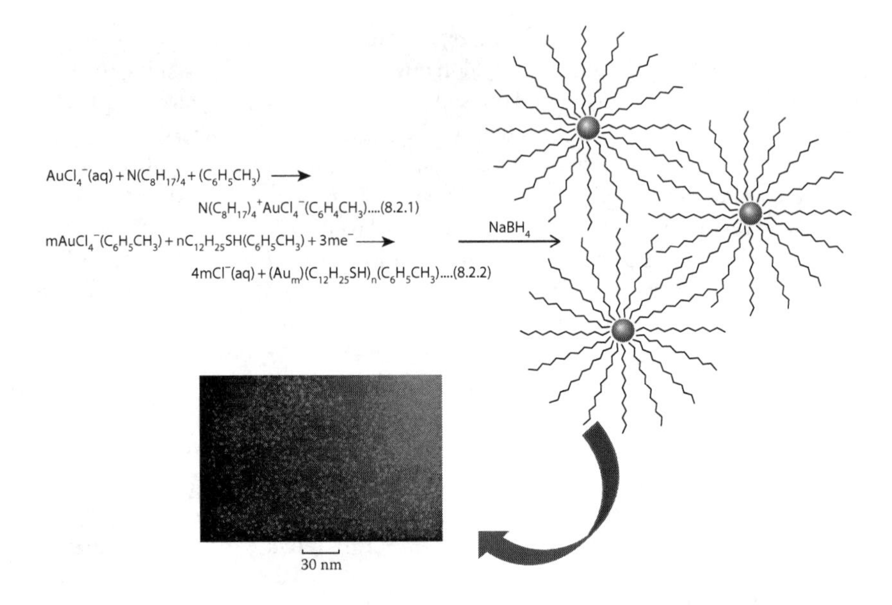

FIGURE 8.2 Proposed reaction scheme for the preparation of alkanethiol-protected gold nanoparticle dispersions in organic solvent media as reported by Brust–Schiffrin. The original TEM picture of these gold nanoparticles is included. (Brust, M., Walker, M., Bethell, D., Schiffrin, D.J., and Whyman, R.J., *J. Chem. Soc., Chem. Commun.*, 1994, 801–802. Reproduced by permission of The Royal Society of Chemistry.)

that were previously added in the organic solvent. The resulting particles could be precipitated as powder and stored. Furthermore, this powder could be redispersed in nonpolar organic solvents whenever desired. The simplicity of the procedure and the convenience with which the resulting particles could be handled (similar to the handling of organic/inorganic molecules/complexes) made it one of the most practiced methods (Figure 8.2).

Thus, this is probably the first method where ligands that can get anchored to the surface of nanoparticles were used during their synthesis. Very soon it was realized that these ligands, apart from being able to improve the stability of nanoparticle dispersion, can play a very important role in controlling many aspects of nanoparticles, such as their size, shape, and dispersional stability.[15,16] In the rest of the chapter, we will describe how these "molecular tools" have been used, very effectively, to control the size, shape, and surface chemistry of the nanoparticles/materials prepared. To make reading simpler, we will restrict the examples mostly to gold and silver nanoparticles, adding a note that the chemistry described is more general in nature and that it can be easily extended to many other systems such as transition metal, metal oxide, and metal chalcogenide nanoparticles.

8.2.1 Steric Effect

Since many of the ligands utilized contain carbon chains, one of the first variations researchers tried was to change their bulkiness using ligands of different lengths and multiple substitution. As the ligand that covers nanoparticle surface becomes bulkier, it will need more space for its free movement, thus preferring a more curved surface. Hence, smaller-sized nanoparticles are expected to form when bulkier ligands are used as capping agents. Second, once bound on nanoparticle surface, bulky ligands prevent transportation of atoms/clusters during the growth stage and hence can restrict nanoparticle growth. Based on these facts, researchers in the area follow a simple thumb rule that "bulkier ligands make smaller particles." The bulkiness of a ligand can be increased by using either a molecule with longer alkyl chain or a bulkier motif (Figure 8.3).

The work by Schadt et al.[17] is one of the earliest that demonstrated that the size of nanoparticles can be tuned by manipulating the chain length of ligands. By heating gold nanoparticles capped by thiols of different chain lengths, they observed that nanoparticle cores capped by shorter chain length thiols led to the formation of larger particles compared to those capped by longer chain length thiols. For example, the size of the nanoparticles obtained using pentanethiol (5 carbon atoms) is 7.6 nm ± 0.43, whereas upon increasing the chain length to 15 carbon atoms, the particle diameter is reduced to 5.8 nm ± 0.31.[18]

Similar results have been observed by several other groups with a library of ligands where the chain length, functional group, etc., have been varied.[19–22] For example, usages of bulky ligands like trioctylphosphine are found to result in the formation of smaller particles. Similarly, several bulky thiols such as thiophenol, glutathione, and tiopronin have been used during the synthesis of nanoparticles,[23–27] and in almost all cases, these bulky ligands led to certain discrete-sized nanoclusters (NCs). Initially, the formation of such NCs was ascribed to increased geometrical stability of the metal core or electronic effects.[28–32] But it was soon realized that the stability cannot be ascribed to the electronic/geometric configuration of Au core alone.[33–37] Zeng et al.[38] reported new magic-sized gold NCs of atomic precision using secondary cyclohexanethiolate with a composition of $Au_{64}(S\text{-}c\text{-}C_6H_{11})_{32}$. The same synthetic protocol with a less bulky ligand, that is, primary phenylethylthiolate (SC_2H_4Ph), resulted in the formation of Au_{144} $(SC_2H_4Ph)_{60}$ NCs.[39] This clearly demonstrates that varying the ligand structure has a huge influence on the nanoparticle size and that bulky ligands favor smaller size. Krommenhoek et al.[40] synthesized Au nanoparticles with 1-adamantanethiol (AdSH), cyclohexanethiol (c-HexSH), and n-hexanethiol (n-HexSH). Here, the thiol group is attached to a tertiary, secondary, and primary carbon atom in AdSH, c-HexSH, and n-HexSH, respectively. Here, again the size of the nanoparticles varied as 2.18 nm (with n-HexSH), compared with 1.53 nm (with c-HexSH) and 1.47 nm (with AdSH). Please refer to Figure 8.3b, e, and f for details.

8.2.2 Metal–Ligand Binding Strength

Several ligands are known to attach on the nanoparticle surface by forming covalent bonds (Au–thiol being the best example). In such a scenario, the stabilization of a particular size of nanoparticles when a specific ligand is used is mainly governed by thermodynamics rather than kinetic pathway of the reaction. It is hypothesized that the size obtained for a particular metal nanoparticle–ligand system is the consequence of a competition between curvature-dependent surface energy and

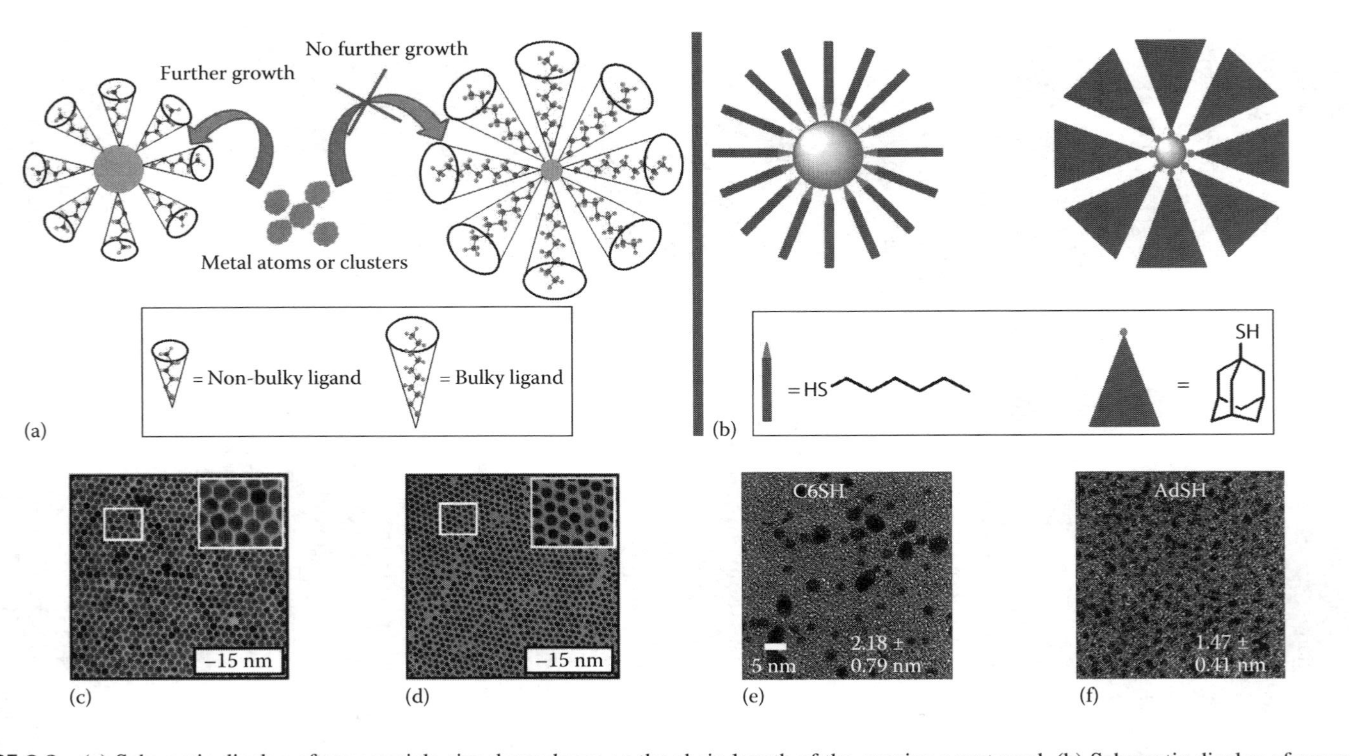

FIGURE 8.3 (a) Schematic display of nanoparticle size dependence on the chain length of the capping agent used. (b) Schematic display of nanoparticle size dependence on the bulkiness of the capping agent used. TEM images of metal nanoparticles prepared with (c) short-chain and (d) long-chain ligands. TEM images of metal nanoparticles prepared with ligands that have (e) less steric hindrance and (f) high steric hindrance. (d: Reprinted with permission from Schadt, M.J., Cheung, W., Luo, J., and Zhong, C.-J., *Chem. Mater.*, 18, 5147. Copyright 2006 American Chemical Society; f: Reprinted with permission from Krommenhoek, P.J., Wang, J., Hentz, N., Johnston-Peck, A.C., Kozek, K.A., Kalyuzhny, G., and Tracy, J.B., *ACS Nano*, 6, 4903. Copyright 2012 American Chemical Society.)

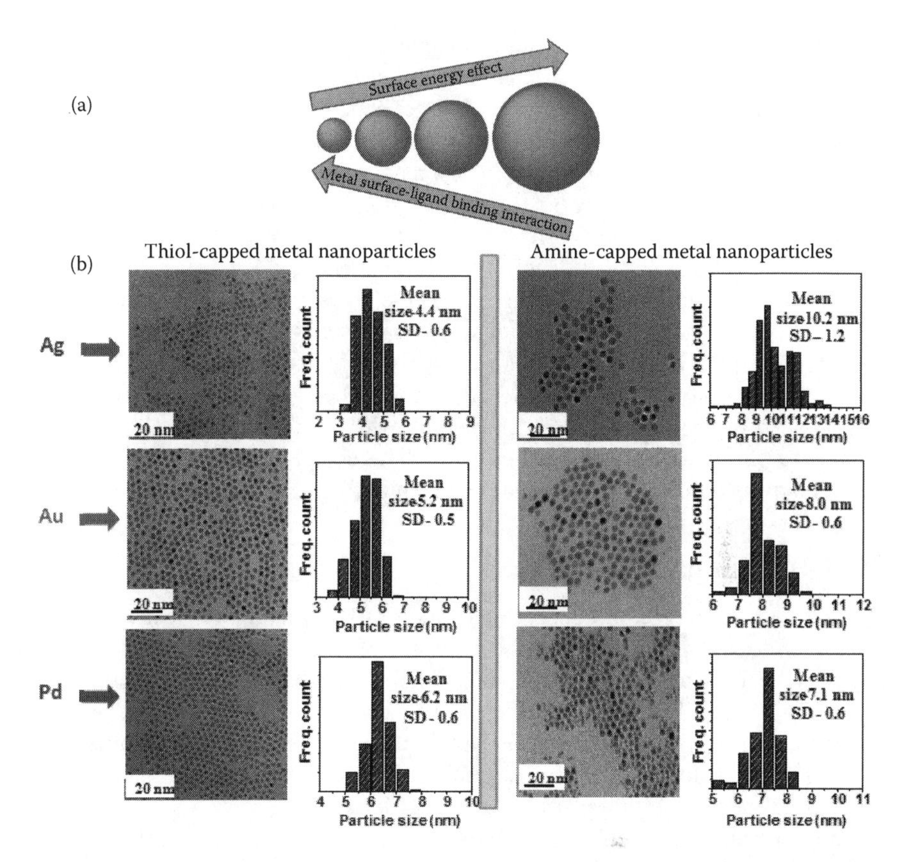

FIGURE 8.4 (a) The relation between nanoparticle size and the nature of ligand used in the digestive ripening process. (b) Particle size distributions obtained with different metals when dodecanethiol and dodecylamine are used as digestive ripening agents.

metal–ligand binding energy effects.[41] Obviously, smaller particles are featured with larger curvature and hence higher surface energy. Thus, based on surface energy arguments alone the formation of larger particles would be favored. On the other hand, the total metal–ligand binding energy increases linearly with the number of total ligand–nanoparticle surface interactions. This in turn requires more surface area and hence favors the formation of smaller-sized particles (Figure 8.4).

This assertion is supported by the work of Prasad et al.[41] They have carried out the synthesis of Au nanoparticles with two ligand systems, namely, dodecanethiol and dodecylamine. Both the ligands have 12 carbon atoms and only differ in their head group, that is, –SH in dodecanethiol and $-NH_2$ in dodecylamine. Their results convincingly prove that with dodecanethiol smaller-sized particles (~4 nm) were formed while with dodecylamine bigger particles of size ~8 nm were obtained under the same preparative conditions. As these two ligands have the same chain length, particle size variation can be solely attributed to the metal–ligand head group interaction, omitting any influence of steric factors. It is well known that Au binds with thiol more preferentially than amine. This binding energy difference can be attributed to hard–soft acid–base interactions. Every metal in its zero oxidation state behaves like a soft acid and prefers to bind with a soft base. Among dodecanethiol and dodecylamine, thiol being a soft base possesses strong affinity for Au surface, and to accommodate a large number of thiol molecules, smaller particles (having large surface area) are obtained. For weakly binding amine groups, surface curvature effect dominates, leading to bigger-sized particles.

This result is further affirmed by another study by Sahu and Prasad where nanoparticles of Ag, Au, and Pd were synthesized via the digestive ripening procedure using dodecanethiol and dodecylamine as digestive ripening agents.[42] Here again the particles formed with dodecanethiol were smaller with all the three metallic systems as compared to those obtained with dodecylamine. These conclusions were recently corroborated by the work of de la Llave et al. where using Uv photoelectron spectroscopic (UPS) investigations authors have convincingly proved that amines bind weakly to the surface of gold as compared to thiol.[43]

8.2.3 Concentration of Ligand

The Royce Murray group pioneered the study of the ligand place-exchange reactions on gold nanoparticles/clusters.[26] By many systematic studies their group and other groups have shown that the ligands that are attached to the nanoparticle surface exist in equilibrium with the ligands present in solution when the nanoparticles are dispersed in a solvent. Very soon it was realized that this concept can be used to control the size of nanoparticles as well. This is based on the premise that during the preparation of nanoparticles if ligands are present in excess, more and more ligands will prefer to attach on nanoparticle surface. This demands more surface and hence small-sized particles are expected to form. This idea has been proven to be correct by many studies.[44–49] For example, Lin et al.[48] preformed a systematic study to understand the effects of ligand concentration on the size of nanoparticles. Their results indicated that without the addition of dodecanethiol, the average size of the gold nanoparticles was determined to be ~20 nm. Adding extra amounts of dodecanethiol of 0.22 and 0.37 g reduced the maximum particle size to about 14 nm, and further increase in dodecanethiol amount, that is, 0.74 g reduced the maximum particle size to about ~5 nm. Increase in concentration of ligand above this did not bring any change in particle size. This inverse relation of particle size with concentration of ligand is effective until a certain concentration of ligand is reached, beyond which the nanoparticle size does not change. Leff et al. proposed a simple model to explain these observations.[50] According to their model, as the ligand concentration increases, the number of ligand molecules preferring to attach to Au nanoparticle surface increases. This necessitates increase in the total surface area, that is, small-sized particles. However, below a certain size, the surface energy increases much more rapidly, which does not get compensated by the ligand binding energy. So once a limiting size of nanoparticle is reached (determined by Au–ligand binding energy), extra ligands prefer to present as monomers (Figure 8.5a, c, and d).

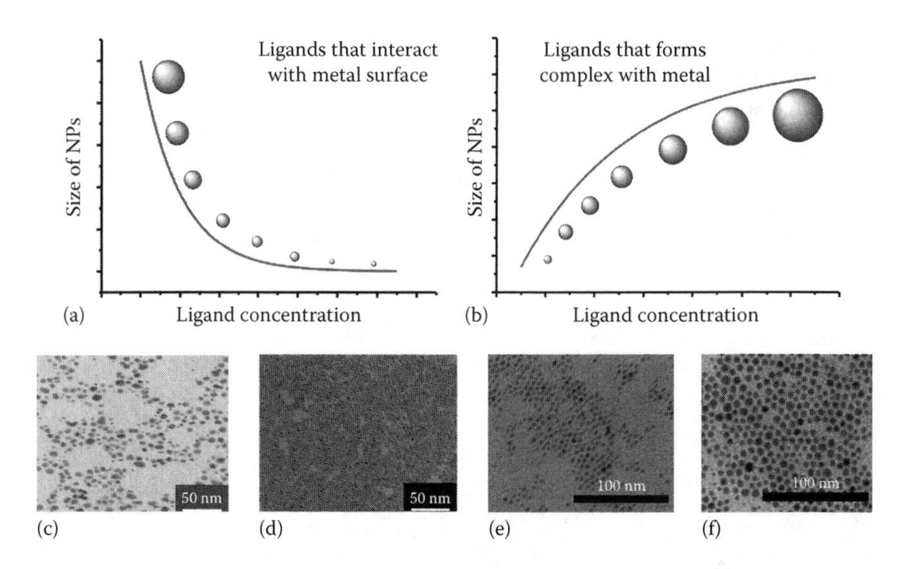

FIGURE 8.5 Summary of the particle size distribution trends observed when nanoparticles synthesis is carried in presence of (a) ligands that interact with metal surface only and (b) ligands that form complexes with metal. TEM images of metal nanoparticles prepared with ligands interacting with the metal surface at (c) low and (d) high concentration and with ligands that form complex with metals at (e) low and (f) high ligand concentrations. (d: Reprinted from *Colloids Surf. A Physicochem. Eng. Asp.*, 448, Lin, M.-L., Yang, F., and Lee, S., 16, Copyright 2014, with permission from Elsevier; f: Reprinted with permission from Sardar, R. and Shumaker-Parry, J.S., *J. Am. Chem. Soc.*, 133, 8179. Copyright 2011 American Chemical Society.)

An opposite effect, that is, an increase in particle size with increase in concentration of ligand was reported by few groups[51,52] Au nanoparticles within an average diameter of 2.6–4.3 nm were obtained by varying the ligand concentration (octadecanethiol) from 0.12 to 2.50 mmol. The opposite trend of nanoparticle size with ligand concentration was attributed to complex formation between the ligand and reducing agent used. Complexation of ligands with the reducing agent or metal ion not only controls the nanoparticle size by controlling nucleation rate but also influences the growth process. Panaček et al. introduced a simple one-step method for silver nanoparticle (nanoparticles) preparation in a wide range of sizes, that is, from 28 to 77 nm. The resulting size of the prepared particles was controlled by the addition of long-chain poly acrylic acid (PAA) of different concentrations during the synthesis. Increase in PAA concentration caused an increase in particle size from 28 nm (without PAA) to 77 nm (highest PAA conc.) The presence of poly (acrylic acid) has been argued to influence both the nucleation process and the subsequent stage of nanoparticle growth. According to them, PAA slows down the nucleation rate by forming a strong complex with silver ions and this is proportional to PAA concentration. The adsorption layer of PAA on the surface of the emerging silver nuclei prevents direct contact between the silver ions and nuclei, thus preventing further growth. Because of this restriction, the silver nuclei grow up to their final size by the aggregation mechanism. This study proves that ligand concentration controls particle size not only by affecting nucleation time but also by regulating growth mechanism. The same trend was also observed for α-Cyclodextrin-protected Au nanoparticles.[53] In this study, Au nanoparticles were prepared by borohydride reduction of $HAuCl_4{\cdot}3H_2O$ in the presence of α-CD-SH/Au. When α-CD-SH/Au ratio was greater than or equal to 1, smaller-sized particles <2.0 nm were obtained and their size was found to increase with increase in α-CD-SH/Au ratio. This increase in particle size is attributed to the inter-hydrogen bond between the α-CD-SH molecules at higher concentrations. This reduces the availability of α-CD-SH for the stabilization of nanoparticles, helping in particle growth, and therefore favoring bigger-sized particles.

So, it can be concluded that when ligands do not interfere in nucleation process, increase in concentration of ligand causes reduction in nanoparticle size. This decrease happens till an equilibrium size is reached, decided by the metal–ligand interaction strength. Once the equilibrium size is reached, the ligand addition remains ineffective toward further decrease in nanoparticle size. But if the ligand influences the nucleation process, an opposite trend between the particle size and ligand amount can be observed. Either way, ligands can be used to manipulate nanoparticle size by playing with concentration.

8.2.4 Multivalency of Ligands

In the previous section (8.2.2), it has been explained that by modulating the interaction energy between the ligands and nanoparticle surface, the size of the nanoparticles could be controlled. It can be summarized that, in general, a stronger interaction between the nanoparticle surface and the ligand leads to smaller nanoparticle formation. This obviously leads to a question—what happens when multivalent ligands are used as opposed to monovalent ligands that have been discussed so far? Multivalent ligands offer stronger interactions when compared to monovalent ones and this generated a great curiosity in researchers to think about such ligands as a size control tool during nanoparticle synthesis.[54,55] It is expected that because of the strong binding affinity, multivalent ligands will bind to nanoparticles soon after nucleation and will not allow further growth of particles. Hence, smaller-sized particles are expected with multivalent ligands. This expectation was supported by many studies.[54,56] For example, carrying out the preparation of Au nanoparticles in the presence of porphyrin-based ligands having four sulfur atoms yielded much smaller-sized particles compared to those obtained with a monovalent variant of the same structural unit and also with dodecanethiols. Different binding affinity of monovalent and divalent ligands is also known to affect nanoparticle size and size distribution. The study by Oh et al. demonstrated that a molecule containing two thiol motifs leads to faster and stronger binding with Au ions.[57] This faster rate of metal–ligand precursor formation was shown to result in smaller-sized particles.[58,59] So, multiple coordination sites provide a strong binding to nanoparticles and can significantly affect particle size.

8.3 MOLECULAR TOOLS FOR SHAPE CONTROL OF NANOPARTICLES

One of the prominent research areas that emanated from the usage of ligands in nanoparticle synthesis is shape control. While some of the initial discoveries in this area can be termed as serendipitous, these were followed by thorough and systematic studies that enriched the mechanistic understanding of anisotropic nanoparticle growth and the role of ligands in the same to a large extent. This enhanced understanding of shape control mechanism has in turn enabled the synthesis of metal NCs of various shapes. In the following sections (8.3.1, 8.3.2, and 8.3.3), we will try to briefly describe the ligand effects on nanoparticle shape. As enormous studies were done in order to understand the mechanisms involved, we will confine the description to anisotropic nanostructures of plasmonic material such as Au and Ag and also limit to a few shapes such as nanorods (NRs) and nanowires (NW), with little emphasis on other nanostructures.

8.3.1 NANORODS

Initial syntheses of gold NRs involved electrochemical reduction of Au salts within the pores of prefabricated hard silica and alumina templates in the presence of a surfactant solution (often cetyltrimethylammonium bromide [CTAB]).[60,61] CTAB was originally chosen because they were effective electrolytes. Later, as CTAB was known to form cylindrical micelles above its second critical micelle concentration, its role as a soft template was suspected even though the synthesis was being carried out inside a hard template. While the procedure mentioned earlier led to the formation of NRs of micrometer scale dimensions (100 nm by 100–200 μm),[61] the synthetic procedures that were devoid of the hard alumina template (which in the literature are referred to as soft template methods) led to the formation of NRs of much smaller dimensions.

The most celebrated of these soft template–based methods is the seed-mediated synthesis of AuNRs. In this method, a preformed, small, gold nanoparticle seed is added to a series of growth solutions containing Au(I) and CTAB. In this approach, a precursor compound is either decomposed or reduced to generate zero-valent atoms in the presence of a seed. The earliest formulation of the seeded growth approach involved the addition of 3.5 nm citrate-stabilized Au nanoparticles to a growth solution containing $HAuCl_4$, CTAB, and ascorbic acid. After a short period of time, an aliquot of this growth solution is added to a new growth solution and then to another. It has also been observed that the yield and the quality of NRs could be dramatically improved by the inclusion of small amounts of $AgNO_3$ in the growth solution. These findings became the topic of one of the most thoroughly researched areas and proved that in NR formation ligands play a significant role, which is described in detail here (Figure 8.6).

The first step in the successful synthesis of NRs of good quality and higher yield seems to be the preparation of 3.5 nm spherical, single crystalline, "seed" particles. In the presence of a growth solution, which contains Au(I) ions and CTAB, these seeds initially grow (in all directions) into isometric twinned particles with a decahedral morphology possessing 10 well-defined {111} faces and 5 {100} side edges

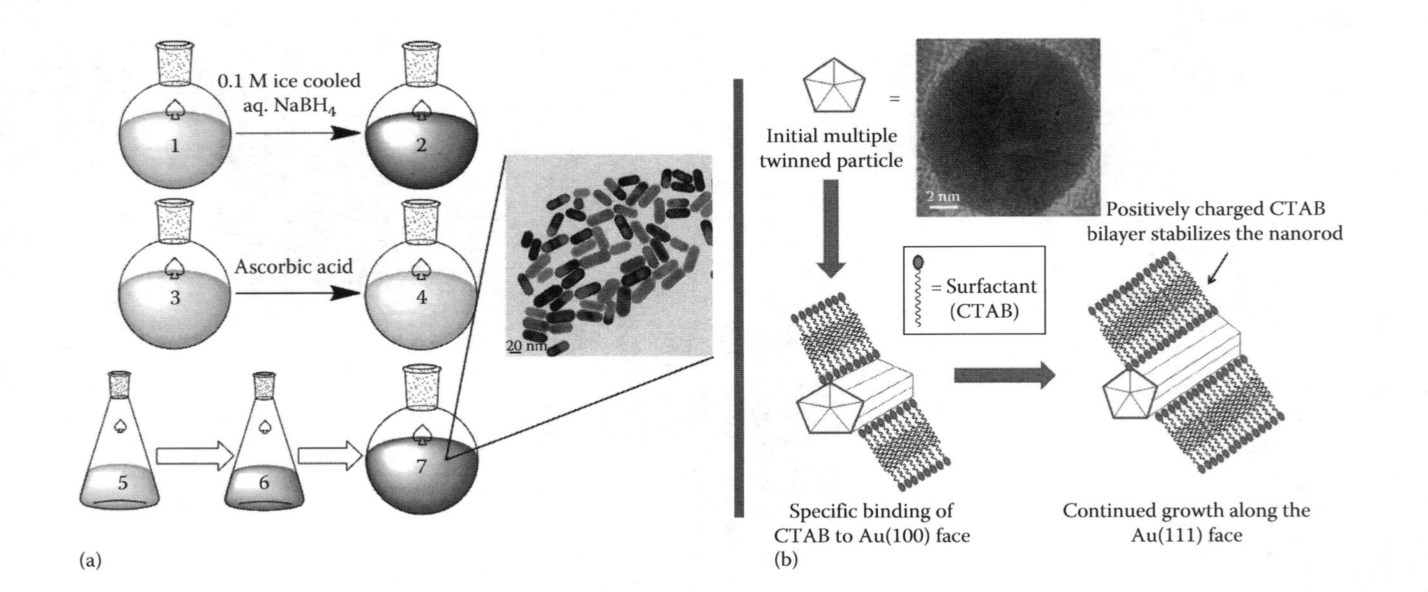

FIGURE 8.6 (a) Schematic of the Au nanorod synthetic protocol (1, 0.25 mM $HAuCl_4$ + 0.25 mM Na-citrate; 2, AuNPs seeds (~4 nm size); 3, 0.25 mM $HAuCl_4$ + 0.1M CTAB; 4, Colorless Au+1 ions; 5, 1 mL of solution 2 + 9 mL of solution 4; 6, 1 mL of solution 5 + 9 mL of solution 4; 7, 10 mL of solution 6 + 9 mL of solution 4 which leads to formation of Au nanorods). (b) The suggested mechanistic growth of Au nanorods. (From Jana, N.R. et al., *Adv. Mater.*, 13, 1389, 2001; Jana, N.R. et al., *J. Phys. Chem. B*, 105, 4065, 2001; Elechiguerra, J.L., Jose, R., and Yacaman, M.J., *J. Mater. Chem.*, 2006, 16, 3906–3919. Reproduced by permission of The Royal Society of Chemistry.)

around a common [110] central axis.[62,63] Based on many detailed studies, it has been suggested that during the subsequent growth phase the {100} side edges are effectively blocked from further growth compared with the {111} end faces. This is attributed to the preferential binding of the CTA^+ groups to the {100} edges than the {111} end faces. Several reasons have been proposed for this preferential binding. First, the surface energy of the {100} edges is more than the {111} end faces. Therefore to reduce this surface energy the CTA^+ groups bind strongly to them. Second, the spacing of Au atoms on the {100} edges is more comparable to the size of the CTA^+ head group than that found on the {111} face of gold where Au atoms are more closely packed. Consequently, the CTA^+ molecules preferentially bind to the {100} edges. Third, the CTA^+ molecules form a bilayer on the {100} gold surfaces in a dynamic way via the interdigitation of hydrocarbon tails, providing a stabilization during gold nanorod growth, in a "zipper" like fashion. Finally, to explain the role of Ag^+ ions, it has been suggested that under experimental conditions AgBr is more likely to form, which also has some preferential binding to the {100} surfaces as surfactants, thus preventing growth in that direction. Any combination of these factors or all these factors could be working in tandem leading to the formation of Au NRs. Thus, Au NR formation remains one of the most eloquent methods that highlight the complex role ligands play in nanoparticle shape control.[64–68]

8.3.2 Nanowires

Like NRs, NWs can also be synthesized with the help of templates formed by ligands. For example, arrays of Ag NWs with high aspect ratio were synthesized by electrodeposition in a micellar phase composed of sodium bis(2-ethylhexyl) sulfosuccinate (AOT), p-xylene, and water.[69] In this case, it has been concluded that the initially formed rod shaped micelles further assemble into a hexagonal liquid crystalline phase as the concentration of surfactant is increased and the NW formation is assisted by them.

Block copolymers were also exploited as template for the synthesis of NWs. Like ligands, mixture of polymers can be arranged into different phases under appropriate conditions. The templates were decorated with metal precursors either by physical adsorption or by chemical coordination, with the subsequent reduction of the metal precursor leading to NW formation. A range of block copolymers were used as templates. For example, carbosilane dendrimers and polyisocyanopeptides were used as templates for the formation of Ag NWs. Later on, this strategy was extended to biomolecules such as DNA and naturally occurring fibers.[70–73]

One of the most elegant methods reported for the preparation of Au NWs employs oriented attachment of small Au nanoparticles.[74] In this protocol, chloroauric acid is added to oleylamine (OA) in toluene medium. This produces a colorless solution in which Au is in the +1 oxidation state. The intermediate solution is stable at room temperature for a few hours and gradually turns pale pink indicating the formation of gold nanoparticles. The addition of ascorbic acid and aging the dispersion for longer time lead to the formation of ultrathin NWs. It is proposed that the formation of NWs proceeds by an oriented attachment of the nanoparticles in the solution phase on the basis that the diameter of the NWs is virtually identical to the diameter of the nanoparticles.

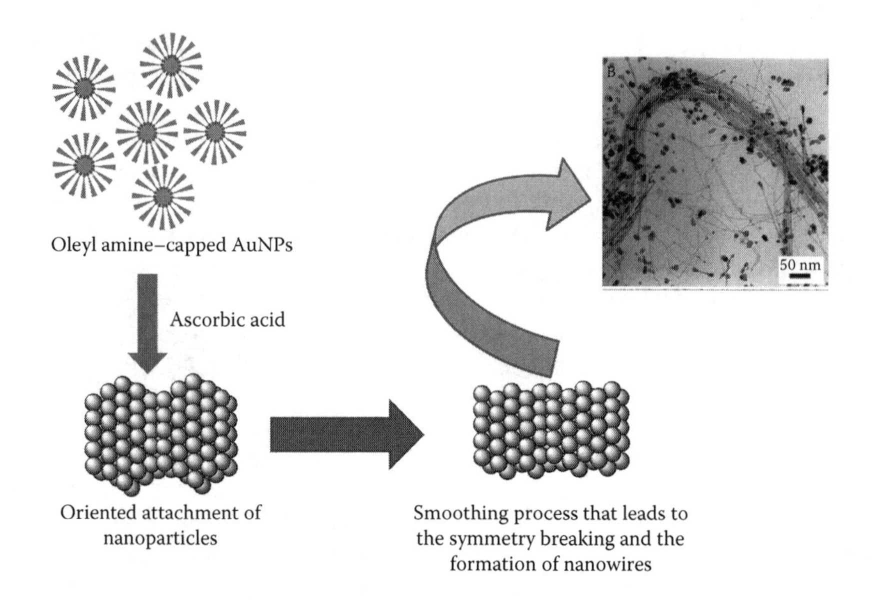

FIGURE 8.7 The suggested mechanistic path for the Au nanowires by the oriented attachment of oleylamine capped of Au nanoparticles. (Halder, A. and Ravishankar, N.: *Adv. Mater.* 2007. 19. 1854–1858. Copyright Wiley-VCH Verlag GmbH & Co. KGaA. Reproduced with permission.)

Also the defects seen in the NWs are characteristic of the oriented-attachment process. Difference in the binding affinity of amines toward different faces of Au is explained as the cause for NW growth. Density functional theory (DFT) calculations indicated that the binding of amine with high-index ridge surfaces is much higher than that with the {100} and the {111} surfaces and that the fraction of bound sites on {111} is much lower than that on {100} surfaces. The difference in amine/gold binding energies on the different facets of gold enables preferential removal of the amines from {111} facet. These surface facets then attach with each other followed by Au atom migration promoting the formation of NWs (Figure 8.7).

Few other reports oppose oriented attachment of small nanoparticles as the cause for NW formation. For example, Loubat et al. synthesized ultrathin Au NWs by reducing $HAuCl_4$ in a solution of OA in heaxane.[75] Here, they claim that small nanoparticles act as seed and NW growth is driven by the phase change of the ligands used. These contradictory mechanisms clearly indicate that this is a work in progress and needs more inputs based on systematic and thorough studies.

8.3.3 Other Nanostructures

Apart from NRs and NWs, ligands also influence the growth of other anisotropic structures such as cubes, plates, and prisms. In most of the cases, preferential attachment of ligands toward specific planes is shown as the cause of formation of these anisotropic particles. For example, difference in binding affinities of different ligands can be used to construct different Ag nanostructure. Sun and Xia showed that citrate group can be used to construct triangular nanoplates of Ag.[76] Citric acid was found to bind Ag (111) surface more strongly compared to Ag (100) surface. A computational study showed the binding energy of citric acid on Ag(111) surface is 13.8 kcal/mol compared to 3.7 kcal/mol for (100) surface.[77] This huge difference in binding energy is attributed to symmetry matching of citric acid and Ag (111) surfaces. The approximate threefold symmetry of citric acid matches that of the Ag (111) surface leading to stronger binding. It was reported by Sun and Xia.[76] that when $AgNO_3$ was heated in the presence of the polymer poly (N-vinyl-2-pyrrolidone) (PVP) and citrate, citrate could bind to (111) facets of Ag preferentially leading to the formation of Ag nanoplates.

Zeng et al. demonstrated that Ag nanoplates could grow in a lateral or vertical mode depending on the ligands used.[78] In presence of sodium citrate, lateral growth was more favorable leading to the formation of Ag thin plates with large lateral dimensions. In contrast, thicker Ag plates were obtained because of vertical growth when PVP was used instead of sodium citrate. This sharp difference observed in the course of Ag deposition was attributed to the different affinities of sodium citrate and PVP toward {111} and {100} facets of Ag. It is known that both the top and bottom of Ag nanoplate are terminated by {111} plane. So when PVP is added to Ag nanoplate, it binds more strongly to {100} than {111} facets of Ag and can thereby reduce the growth rate along the {100} direction and the plate grows vertically.

Conformational changes in the intrinsic chemical structure of ligands also can direct the anisotropic growth of nanostructures. For example, Kedia and Kumar[79] showed that conformational changes of PVP can direct different anisotropic size/shaped gold nanoparticles synthesis in the seedless synthetic procedure. It is known that PVP can interact with Au ions in various ways and this interaction between gold metal ions and PVP at the given specific monomer to metal ratio leads to ligand exchange and the structure directing property of PVP then influences the resultant shape. Second, PVP attaches to the metal nanoparticles via different coordination sites in different solvents, which can affect their reduction potential and hence affect the resultant nanoparticle size.

Ligands also can drive the anisotropic growth of nanoparticles by affecting the nucleation process, that is, reduction rate. The reduction potential of metal is strongly influenced by its local ligand environment. Thus, selection of metal precursor and reducing agent became important for shape control of nanoparticles. Wang et al. have demonstrated that Ag nanocubes with controlled edge lengths below 30 nm can be obtained by replacing ethylene glycol (EG) with diethylene glycol (DEG) as both a solvent and a reductant.[80] Compared with EG, DEG, owing to the increase in hydrocarbon chain length, is more viscous and has lower reducing power. This slow reduction produces a large number of single-crystal Ag seeds, followed by a relatively slow growth, and hence produces small Ag nanocubes with uniform, tightly controlled sizes.

Another report by Ravi Kumar et al. showed that reducing agents alone can direct the shape of nanoparticles.[81] They synthesized gold nanostructures using different dicarboxylic (viz., oxalic, malonic, succinic, glutaric, and adipic) acids as reducing agents in the absence of any other additives or surfactants. Various anisotropic structures such as kites, tadpoles, triangular/hexagonal plates, and twinned particles were seen to evolve depending on the molar ratio of dicarboxylic acid to $HAuCl_4$ used. It was also demonstrated that gold nanostructure formation is hampered as the chain length/distance between two carboxylic acid groups increases.

This discussion clearly exemplifies the fact that a slight change in ligand nature affects the shape of nanoparticles drastically and that the involved mechanism needs deep investigation.

8.4 LIGANDS AS A TOOL TO DISPERSE NANOPARTICLES IN DIFFERENT SOLVENT MEDIA

The usage of ligands in nanoparticle synthesis was initiated with an intention of increasing their dispersional stability. The steric crowding associated with ligands not only acts against the strong vdW force of attraction between the metal cores but also helps nanoparticles to remain dispersed in nonpolar solvent media.[82,83] Furthermore, such ligand-stabilized nanoparticles could be easily separated from the solvent in which they are dispersed by precipitation by simply adding a solvent in which they are not dispersible. These characteristics are not difficult to explain. In case of a ligand with single functional group, the functional group would bind to the nanoparticle surface and the tail $-CH_2-(CH_2)_n-CH_3$ is exposed to the solvent. For example, in case of alkanethiols like dodecanethiol and Au nanoparticles, the –SH group is attached to the gold surface and the hydrophobic tail is exposed to the solvent. Thus, such nanoparticles are easily dispersed in nonpolar solvents such as toluene, hexane, and chloroform. Since the interaction between the hydrophobic tail and polar/protic solvents is not a favorable one, these particles get precipitated when solvents like ethanol are added to the alkanethiol-capped Au nanoparticle dispersion in solvents like chloroform or toluene.

Once this picture of how a ligand molecule gets attached to the nanoparticle surface became clear, researchers started devising methods of preparing monolayer-protected metal nanoparticles that could be dispersed in polar solvents and water. For this, a simple strategy of using bifunctional ligands is generally employed. For example, changing the ligand from dodecanethiol as mentioned earlier to PEG-dodecanethiol has been seen to dramatically enhance the dispersibility of Au nanoparticles in water.[84] Lévy et al.[85] using combinatorial methods designed a pentapeptide ligand, CALNN, which converts citrate-stabilized gold nanoparticles into extremely stable, water-dispersible gold nanoparticles.

Though the initial choice of CALNN came through combinatorial methods, systematic studies carried out subsequently unraveled the reasons for the enhanced stability of CALNN-capped Au NPs in water. Here, the thiol group of the N-terminal cysteine (C) forms a covalent bond with the gold nanoparticle surface. Moreover, it has been shown by Gordillo and coworkers that the presence of a positively charged ammonium group in the vicinity of the thiol significantly accelerates the adsorption kinetics of thiols onto citrate-stabilized gold nanoparticles.[86] Alanine (A) and leucine (L) in positions 2 and 3 possess hydrophobic side chains and are chosen to promote the self-assembly of the peptide on the nanoparticle surface and also prevent water from attacking the nanoparticle surface. Formation of a peptide layer on the surface provides stability to Au nanoparticles. Many other ligands, for example, calixarene, were also developed, which gives stable water-soluble nanoparticles.[87] Zwitterionic molecules were also shown to provide stability to nanoparticles and make it water dispersible. For example, 4-dimethylaminopyridine (DMAP) molecules physisorbed onto the Au nanoparticle surface by forming a labile donor–acceptor complex with the surface atoms of the nanoparticles through the endocyclic nitrogen atoms.[88] The surface charge, which arose from the partial protonation of the exocyclic nitrogen atom, prevented the metallic nanoparticles from aggregation and agglomeration. Later on, many zwitterion-appended ligands were designed, which provide stability to nanoparticles.[89] In all the cases, one end of the ligand possesses an anchoring group that binds strongly to the nanoparticle surface, while the charge present on the other end of the ligand make nanoparticle water dispersible and stable (Figure 8.8).

8.5 MOLECULES AS BOTH REDUCING AND STABILIZING AGENTS

Formation of metal nanoparticle, especially with a monolayer molecular coating, often involves three reagents, namely, (1) a surfactant or a phase transfer agent to make the metal ions dispersed in the nonpolar organic solvents, (2) a reducing agent, and (3) a passivating agent (the ligand). But this conventional approach often results in undesirable side products and obviously involves multiple steps. This stimulated a desire to develop reagents that play the dual role of effective reducing agents and of stabilizers in one step (Figure 8.9). A review by Dumur et al.[90] describes role of these (DFAs) in the formation of Au nanoparticles extensively. Among all those discussed, acids and amines are listed as the most commonly used DFAs. One of the celebrated reagents that perform these dual functions is trisodium citrate, as discussed in the beginning of this chapter (Figure 8.1).

Like acids, amines also act efficiently as dual functioning ligands. Many aliphatic and aromatic amines have been reported to act as the DFAs.[91] OA is one of the common reagents used as DFA in nanoparticle synthesis.[92] These amines act as electron transfer agents at elevated temperature to reduce ions to nanoparticles, and once produced these nanoparticles are prevented from aggregation by the steric repulsion between the long chains of these amines, which get attached to the nanoparticle surface. Chen et al.[93] used Fourier transform infrared (FTIR) spectroscopy to study the formation of Ag nanoparticles from Ag(I) salts in an OA–paraffin solution. On the basis of the presence of nitriles and imines in the final product, they proposed that the Ag nanoparticles formation starts with the formation of a stable complex of Ag(I) salt with amines, followed by a one-electron transfer process from OA to the salt at a higher temperature, leading to the formation of amino radicals. These species then get deprotonated to form imines, and subsequently nitriles, which is followed by passivating and stabilizing silver nanoparticles. Contrary to this, Lu et al.[94] showed initial complex formation of C=C of OA with Au (I). They studied Au nanoparticles formation with ODA and OA. Nearly monodisperse Au nanoparticles of average size ~12 nm were obtained when OA was used, while ODA resulted in much larger particles of around 100 nm. This is attributed to slow decomposition of stable complex of AuCl with OA compared to ODA as coordination between AuCl and C=C bond is

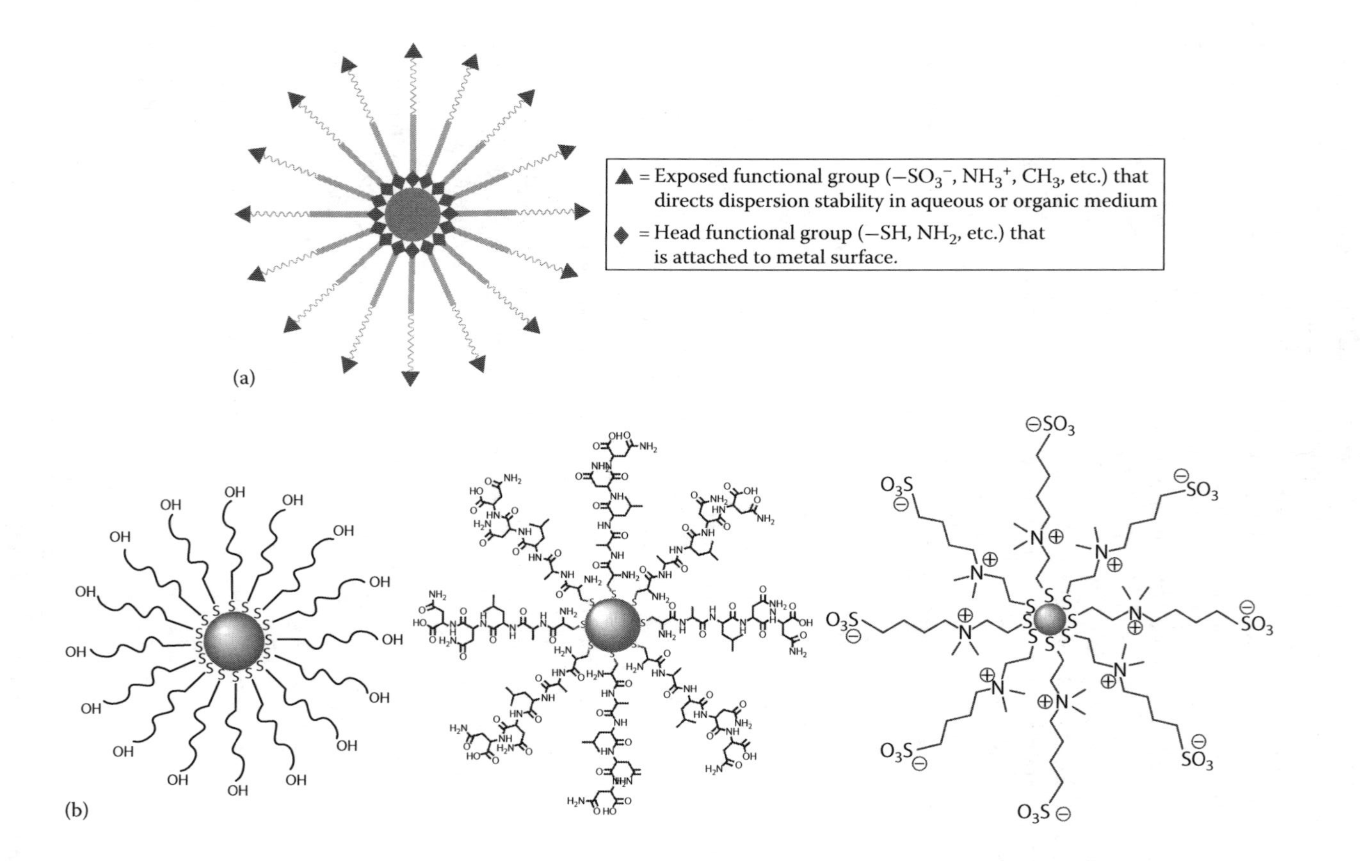

FIGURE 8.8 (a) General ligand design that can help in the preparation of water/organic solvent dispersible monolayer-protected nanoparticles. (b) Three different examples of ligand-protected water dispersible Au nanoparticles as reported in the literature.

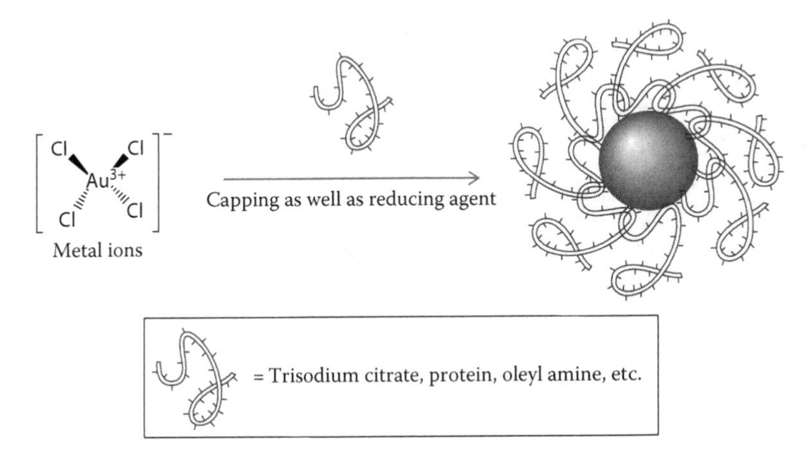

FIGURE 8.9 General synthetic strategy being used to prepare ligand-protected nanoparticles in one step.

stronger than that between AuCl and NH_2. The Au^+-OA complex formation was further confirmed by NW formation.[95–97] It is well demonstrated that the Au^+-OA complex serves as growth templates that govern one-dimensional growth in the nanoscale. Due to the remarkable reducing and stabilizing capabilities of amines, amino acids and their derivatives were also studied for the synthesis of metallic nanoparticles. It has been noticed that amino acids such as tyrosine, aspartic acid, tryptophan, and glutamic acid also act as DFA in nanoparticle formation, providing a wide control over nanoparticle size and shape.[90,98–100] Apart from these, reducing sugars are also very popular as DFAs. For example, glucose has been effectively used as a reducing and capping agent for the synthesis of Ag and Au nanoparticles.[101,102]

Extending this concept of sugars as DFAs, Prasad's group has proposed the use of glycolipids as DFAs. For example, sophorolipids (SLs), a class of glycolipids, have been successfully used as DFA for the synthesis of Au and Ag nanoparticles.[103,104] The use of SLs for Ag nanoparticles synthesis not only eliminates the necessity for exogenous reducing agents but provides better stability to the nanoparticles. Several other microorganisms and plant extracts are also known to serve as simultaneous reducing and stabilizing agents for nanoparticles.[105–107]

Though a lot more is needed to be done to achieve fine control over nanoparticle size and shape using DFAs, the use of these molecules assures the elegant formation of nanoparticles, free of byproducts, and hence it is one of the major attractions of current research. Another advantage of using DFA is the possibility of making nanoparticles via continuous flow methods.[108]

8.6 CONCLUSION

The examples given in the chapter illustrate the importance of ligands in size- and shape-controlled synthesis of metal nanoparticle dispersions. Uncovering the roles that ligands played in fine-tuning of nanoparticle, size and shape can help design new synthetic methods to obtain various nanostructures for a desired application and synergistically open up pathways for further studies.

REFERENCES

1. Faraday, M. *Philosophical Transactions of the Royal Society of London* 1857, *147*, 145–181.
2. Zsigmondy, R. *Journal of the American Chemical Society* 1909, *31*, 951–952.
3. Hergert, W.; Wriedt, T. *The Mie Theory: Basics and Applications*. Springer, Berlin Heidelberg, 2012.
4. Sharma, V.; Park, K.; Srinivasarao, M. *Materials Science and Engineering: R: Reports* 2009, *65*, 1–38.
5. Zeng, S.; Yu, X.; Law, W.-C.; Zhang, Y.; Hu, R.; Dinh, X.-Q.; Ho, H.-P.; Yong, K.-T. *Sensors and Actuators B: Chemical* 2013, *176*, 1128–1133.
6. Turkevich, J.; Hillier, J.; Stevenson, P. C. *Discussions of the Faraday Society* 1951, *11*, 55.
7. Frens, G. *Nature (London), PhysicalScience* 1973, *241*, 20.
8. Frens, G. *Colloid and Polymer Science* 1972, *250*, 736–741.
9. Elimelech, M.; Gregory, J.; Jia, X.; Williams, R. A. In *Particle Deposition and Aggregation: Measurement, Modelling and Simulation*. Butterworth-Heinemann, Woburn, MA, 1995, pp. 402–425.
10. Israelachvili, J. N. *Intermolecular and Surface Forces: Revised Third Edition*. Academic Press, Waltham, MA, 2011.
11. Laibinis, P. E.; Whitesides, G. M.; Allara, D. L.; Tao, Y. T.; Parikh, A. N.; Nuzzo, R. G. *Journal of the American Chemical Society* 1991, *113*, 7152–7167.
12. Nuzzo, R. G.; Allara, D. L. *Journal of the American Chemical Society* 1983, *105*, 4481–4483.
13. Slowing, I. I.; Vivero-Escoto, J. L.; Wu, C.-W.; Lin, V. S. Y. *Advanced Drug Delivery Reviews* 2008, *60*, 1278–1288.
14. Brust, M.; Walker, M.; Bethell, D.; Schiffrin, D. J.; Whyman, R. J. *Journal of the Chemical Society, Chemical Communications* 1994, *7*, 801–802.
15. Puntes, V. F.; Krishnan, K. M.; Alivisatos, P. *Applied Physics Letters* 2001, *78*, 2187–2189.
16. Wu, Z.; MacDonald, M. A.; Chen, J.; Zhang, P.; Jin, R. *Journal of the American Chemical Society* 2011, *133*, 9670–9673.
17. Schadt, M. J.; Cheung, W.; Luo, J.; Zhong, C.-J. *Chemistry of Materials* 2006, *18*, 5147–5149.
18. Ulman, A. *Chemical Reviews* 1996, *96*, 1533–1554.
19. Tolman, C. A. *Chemical Reviews* 1977, *77*, 313–348.
20. Mingos, D. M. P. *Inorganic Chemistry* 1982, *21*, 464–466.
21. Chen, J.; Zhang, Q.-F.; Bonaccorso, T. A.; Williard, P. G.; Wang, L.-S. *Journal of the American Chemical Society* 2014, *136*, 92–95.
22. Nishigaki, J.-i.; Tsunoyama, R.; Tsunoyama, H.; Ichikuni, N.; Yamazoe, S.; Negishi, Y.; Ito, M.; Matsuo, T.; Tamao, K.; Tsukuda, T. *Journal of the American Chemical Society* 2012, *134*, 14295–14297.
23. Stamplecoskie, K. G.; Kamat, P. V. *Journal of the American Chemical Society* 2014, *136*, 11093–11099.
24. Negishi, Y.; Nobusada, K.; Tsukuda, T. *Journal of the American Chemical Society* 2005, *127*, 5261–5270.
25. Negishi, Y.; Takasugi, Y.; Sato, S.; Yao, H.; Kimura, K.; Tsukuda, T. *The Journal of Physical Chemistry B* 2006, *110*, 12218–12221.

26. Hostetler, M. J.; Templeton, A. C.; Murray, R. W. *Langmuir* 1999, *15*, 3782–3789.
27. Price, R. C.; Whetten, R. L. *Journal of the American Chemical Society* 2005, *127*, 13750–13751.
28. Bartlett, P. A.; Bauer, B.; Singer, S. J. *Journal of the American Chemical Society* 1978, *100*, 5085–5089.
29. Mingos, D. M. P. *Chemical Society Reviews* 1986, *15*, 31–61.
30. Pyykkö, P. *Angewandte Chemie International Edition* 2004, *43*, 4412–4456.
31. Zhang, H.-F.; Stender, M.; Zhang, R.; Wang, C.; Li, J.; Wang, L.-S. *The Journal of Physical Chemistry B* 2004, *108*, 12259–12263.
32. Zheng, J.; Zhang, C.; Dickson, R. M. *Physical Review Letters* 2004, *93*, 077402.
33. Guo, R.; Murray, R. W. *Journal of the American Chemical Society* 2005, *127*, 12140–12143.
34. Schaaff, T. G.; Knight, G.; Shafigullin, M. N.; Borkman, R. F.; Whetten, R. L. *The Journal of Physical Chemistry B* 1998, *102*, 10643–10646.
35. Schaaff, T. G.; Whetten, R. L. *The Journal of Physical Chemistry B* 2000, *104*, 2630–2641.
36. Yang, Y.; Chen, S. *Nano Letters* 2002, *3*, 75–79.
37. Negishi, Y.; Tsukuda, T. *Chemical Physics Letters* 2004, *383*, 161–165.
38. Zeng, C.; Chen, Y.; Li, G.; Jin, R. *Chemistry of Materials* 2014, *26*, 2635–2641.
39. Qian, H.; Jin, R. *Nano Letters* 2009, *9*, 4083–4087.
40. Krommenhoek, P. J.; Wang, J.; Hentz, N.; Johnston-Peck, A. C.; Kozek, K. A.; Kalyuzhny, G.; Tracy, J. B. *ACS Nano* 2012, *6*, 4903–4911.
41. Prasad, B. L. V.; Stoeva, S. I.; Sorensen, C. M.; Klabunde, K. J. *Chemistry of Materials* 2003, *15*, 935–942.
42. Sahu, P.; Prasad, B. L. V. *Chemical Physics Letters*, *525*, 101–104, 2012.
43. de la Llave, E.; Clarenc, R.; Schiffrin, D. J.; Williams, F. J. *The Journal of Physical Chemistry C* 2013, *118*, 468–475.
44. Hostetler, M. J.; Wingate, J. E.; Zhong, C. J.; Harris, J. E.; Vachet, R. W.; Clark, M. R.; Londono, J. D. et al. *Langmuir* 1998, *14*, 17.
45. Wolfe, R. L.; Murray, R. W. *Analytical Chemistry* 2006, *78*, 1167–1173.
46. Lai, S.-F.; Chen, W.-C.; Wang, C.-L.; Chen, H.-H.; Chen, S.-T.; Chien, C.-C.; Chen, Y.-Y. et al. *Langmuir* 2011, *27*, 8424–8429.
47. Mafune, F.; Kohno, J.-y.; Takeda, Y.; Kondow, T.; Sawabe, H. *The Journal of Physical Chemistry B* 2000, *104*, 9111–9117.
48. Lin, M.-L.; Yang, F.; Lee, S. *Colloids and Surfaces A: Physicochemical and Engineering Aspects* 2014, *448*, 16–22.
49. Bhaskar, S. P.; Vijayan, M.; Jagirdar, B. R. *The Journal of Physical Chemistry C* 2014, *118*, 18214–18225.
50. Leff, D. V.; Ohara, P. C.; Heath, J. R.; Gelbart, W. M. *The Journal of Physical Chemistry* 1995, *99*, 7036–7041.
51. Sardar, R.; Shumaker-Parry, J. S. *Journal of the American Chemical Society* 2011, *133*, 8179–8190.
52. Wang, Y.-C.; Gunasekaran, S. *Journal of Nanoparticle Research C7—1200* 2012, *14*, 1–11.
53. Paau, M. C.; Lo, C. K.; Yang, X.; Choi, M. M. F. *The Journal of Physical Chemistry C* 2010, *114*, 15995–16003.
54. Ohyama, J.; Hitomi, Y.; Higuchi, Y.; Shinagawa, M.; Mukai, H.; Kodera, M.; Teramura, K.; Shishido, T.; Tanaka, T. *Chemical Communications* 2008, *47*, 6300–6302.
55. Ohyama, J.; Hitomi, Y.; Higuchi, Y.; Tanaka, T. *Topics in Catalysis* 2009, *52*, 852–859.
56. Negishi, Y.; Tsukuda, T. *Journal of the American Chemical Society* 2003, *125*, 4046–4047.
57. Oh, E.; Susumu, K.; Goswami, R.; Mattoussi, H. *Langmuir* 2010, *26*, 7604–7613.

58. Pettibone, J. M.; Reardon, N. R. *Nanoscale* 2012, *4*, 5593–5596.
59. Wang, Z.; Tan, B.; Hussain, I.; Schaeffer, N.; Wyatt, M. F.; Brust, M.; Cooper, A. I. *Langmuir* 2007, *23*, 885–895.
60. Jain, P. K.; Huang, X.; El-Sayed, I. H.; El-Sayed, M. A. *Accounts of Chemical Research* 2008, *41*, 1578–1586.
61. Eustis, S.; El-Sayed, M. A. *Chemical Society Reviews* 2006, *35*, 209–217.
62. Jana, N. R.; Gearheart, L.; Murphy, C. J. *Advanced Materials* 2001, *13*, 1389.
63. Jana, N. R.; Gearheart, L.; Murphy, C. J. *The Journal of Physical Chemistry B* 2001, *105*, 4065.
64. Ye, X.; Jin, L.; Caglayan, H.; Chen, J.; Xing, G.; Zheng, C.; Doan-Nguyen, V. et al. *ACS Nano* 2012, *6*, 2804–2817.
65. Straney, P. J.; Andolina, C. M.; Millstone, J. E. *Langmuir* 2013, *29*, 4396–4403.
66. Wadams, R. C.; Fabris, L.; Vaia, R. A.; Park, K. *Chemistry of Materials* 2013, *25*, 4772–4780.
67. Lohse, S. E.; Murphy, C. J. *Chemistry of Materials* 2013, *25*, 1250–1261.
68. Wang, Z. L.; Gao, R. P.; Nikoobakht, B.; El-Sayed, M. A. *The Journal of Physical Chemistry B* 2000, *104*, 5417–5420.
69. Huang, L. M.; Wang, H. T.; Wang, Z. B.; Mitra, A.; Bozhilov, K. N.; Yan, Y. S. *Advanced Materials* 2002, *14*, 61–64.
70. Gazit, E. *FEBS Journal* 2007, *274*, 317–322.
71. Seeman, N. C. *Annual Review of Biophysics and Biomolecular Structure* 1998, *27*, 225–248.
72. Braun, E.; Eichen, Y.; Sivan, U.; Ben-Yoseph, G. *Nature* 1998, *391*, 775–778.
73. Patolsky, F.; Weizmann, Y.; Lioubashevski, O.; Willner, I. *Angewandte Chemie International Edition* 2002, *41*, 2323–2327.
74. Halder, A.; Ravishankar, N. *Advanced Materials* 2007, *19*, 1854–1858.
75. Loubat, A.; Imperor-Clerc, M.; Pansu, B.; Meneau, F.; Raquet, B.; Viau, G.; Lacroix, L.-M. *Langmuir* 2014, *30*, 4005–4012.
76. Sun, Y.; Xia, Y. *Advanced Materials* 2003, *15*, 695–699.
77. Kilin, D. S.; Prezhdo, O. V.; Xia, Y. *Chemical Physics Letters* 2008, *458*, 113–116.
78. Zeng, J.; Xia, X.; Rycenga, M.; Henneghan, P.; Li, Q.; Xia, Y. *Angewandte Chemie International Edition* 2011, *50*, 244–249.
79. Kedia, A.; Kumar, P. S. *The Journal of Physical Chemistry C* 2012, *116*, 23721–23728.
80. Wang, Y.; Zheng, Y.; Huang, C. Z.; Xia, Y. *Journal of the American Chemical Society* 2013, *135*, 1941–1951.
81. Ravi Kumar, D. V.; Kumavat, S. R.; Chamundeswari, V. N.; Patra, P. P.; Kulkarni, A. A.; Prasad, B. L. V. *RSC Advances*, 2013, *3*, 21641–21647.
82. Sastry, M. *Current Science* 2003, *85*, 1735–1745.
83. Medintz, I. L.; Uyeda, H. T.; Goldman, E. R.; Mattoussi, H. *Nature Materials* 2005, *4*, 435–446.
84. Shimmin, R. G.; Schoch, A. B.; Braun, P. V. *Langmuir* 2004, *20*, 5613–5620.
85. Lévy, R.; Thanh, N. T.; Doty, R. C.; Hussain, I.; Nichols, R. J.; Schiffrin, D. J.; Brust, M.; Fernig, D. G. *Journal of the American Chemical Society* 2004, *126*, 10076–10084.
86. Bellino, M. G.; Calvo, E. J.; Gordillo, G. *Physical Chemistry Chemical Physics* 2004, *6*, 424–428.
87. Tauran, Y.; Brioude, A.; Kim, B.; Perret, F.; Coleman, A. *Molecules*, *18*, 5993, 2013.
88. Gandubert, V. r. J.; Lennox, R. B. *Langmuir* 2005, *21*, 6532–6539.
89. Aldeek, F.; Muhammed, M. H.; Palui, G.; Zhan, N.; Mattoussi, H. *ACS Nano* 2013, *7*, 2509–2521.
90. Dumur, F.; Guerlin, A.; Dumas, E.; Bertin, D.; Gigmes, D.; Mayer, C. R. *Gold Bulletin* 2011, *44*, 119–137.

91. Shen, C.; Hui, C.; Yang, T.; Xiao, C.; Tian, J.; Bao, L.; Chen, S.; Ding, H.; Gao, H. *Chemistry of Materials* 2008, *20*, 6939–6944.
92. Mourdikoudis, S.; Liz-Marzán, L. M. *Chemistry of Materials*, *25*, 1465–1476, 2013.
93. Chen, M.; Feng, Y.-G.; Wang, X.; Li, T.-C.; Zhang, J.-Y.; Qian, D.-J. *Langmuir* 2007, *23*, 5296–5304.
94. Lu, X.; Tuan, H.-Y.; Korgel, B. A.; Xia, Y. *Chemistry—A European Journal* 2008, *14*, 1584–1591.
95. Lu, X.; Yavuz, M. S.; Tuan, H.-Y.; Korgel, B. A.; Xia, Y. *Journal of the American Chemical Society* 2008, *130*, 8900–8901.
96. Pazos-Perez, N.; Baranov, D.; Irsen, S.; Hilgendorff, M.; Liz-Marzan, L. M.; Giersig, M. *Langmuir* 2008, *24*, 9855–9860.
97. Kura, H.; Ogawa, T. *Journal of Applied Physics* 2010, *107*, 074310.
98. Mandal, S.; Selvakannan, P. R.; Phadtare, S.; Pasricha, R.; Sastry, M. *Journal of Chemical Sciences* 2002, *114*, 513–520.
99. Selvakannan, P. R.; Mandal, S.; Phadtare, S.; Gole, A.; Pasricha, R.; Adyanthaya, S. D.; Sastry, M. *Journal of Colloid and Interface Science* 2004, *269*, 97–102.
100. Shao, Y.; Jin, Y.; Dong, S. *Chemical Communications* 2004, *9*,1104–1105.
101. Kemp, M. M.; Kumar, A.; Mousa, S.; Park, T.-J.; Ajayan, P.; Kubotera, N.; Mousa, S. A.; Linhardt, R. J. *Biomacromolecules* 2009, *10*, 589–595.
102. Raveendran, P.; Fu, J.; Wallen, S. L. *Green Chemistry* 2006, *8*, 34–38.
103. Singh, S.; D'Britto, V.; Prabhune, A. A.; Ramana, C. V.; Dhawan, A.; Prasad, B. L. V. *New Journal of Chemistry* 2010, *34*, 294–301.
104. Kasture, M. B.; Patel, P.; Prabhune, A. A.; Ramana, C. V.; Kulkarni, A. A.; Prasad, B. L. V. *Journal of Chemical Sciences* 2008, *120*, 515–520.
105. Nune, S. K.; Chanda, N.; Shukla, R.; Katti, K.; Kulkarni, R. R.; Thilakavathy, S.; Mekapothula, S.; Kannan, R.; Katti, K. V. *Journal of Materials Chemistry* 2009, *19*, 2912–2920.
106. Kasthuri, J.; Kathiravan, K.; Rajendiran, N. *Journal of Nanoparticle Research* 2009, *11*, 1075–1085.
107. Ankamwar, B.; Damle, C.; Ahmad, A.; Sastry, M. *Journal of Nanoscience and Nanotechnology* 2005, *5*, 1665–1671.
108. Kumar, D. V. R.; Kasture, M.; Prabhune, A. A.; Ramana, C. V.; Prasad, B. L. V.; Kulkarni, A. A. *Green Chemistry* 2010, *12*, 609–615.
109. Elechiguerra, J. L.; Jose, R.; Yacaman, M. J. *Journal of Materials Chemistry* 2006, *16*, 3906–3919.

9 Synthesis of Nano- and Microcrystallites of Metals Using Metal-Organic Precursors

Giridhar U. Kulkarni, Gangaiah Mettela, and S. Kiruthika

CONTENTS

In the past few decades, a number of metal-organic precursors/complexes (MOPs) have been synthesized and well characterized [1–3]. In MOPs, metal ions are surrounded by organic ligands via covalent or noncovalent interactions [4,5]. Commonly, MOPs are used as catalysts in various organic reactions [6,7]. With recent development in nanoscience, MOPs are being explored as metal ion sources for synthesizing well-defined metal nanocrystallites [8]. Unlike inorganic metal salts, MOPs offer interesting synthetic conditions due to their solubility in various polar [9] and nonpolar solvents [10], wide range of thermal decomposition temperatures [11], and sensitivity to radiation [12–14]. Such properties prove quite handy in tuning the size [11,15,16], shape [15], surface morphology [17], monodispersity [16], and sometimes even the lattice structure of the metal crystallites [17]. Monodispersity in nanocrystals is quite attractive because it leads to 2D organization of the nanocrystallites [18] into mesolattice, which finds applications in many optoelectronic [19] and other devices [20].

This chapter focuses on various MOP-based synthetic routes that have emerged in the literature to prepare nanocrystals of noble and seminoble transition metals, Au, Ag, Pd, Pt, Cu, Co, Ni, and Bi, and their binary compositions. To reveal the morphology, structural, and optical properties of various metal nano- and microcrystallites, a number of state-of-the-art tools such as transmission electron microscopy (TEM), selected area electron diffraction (SAED), ultraviolet-visible (UV) spectroscopy, and x-ray diffraction (XRD) techniques have been used. Producing nanocrystallites from a MOP involves reduction of the metal ions, followed by controlled growth for which the following methods have been developed:

1. Chemical reduction of metal ions in a reaction medium
2. Chemical reduction of metal ions at the liquid–liquid interface
3. Thermal decomposition of MOPs
    a. Direct heating of solid precursor in air or under various atmospheres
    b. Metal Organic Chemical Vapor Deposition (MOCVD)
    c. Aerosol-Assisted Chemical Vapor Deposition (AACVD)
    d. Inkjet printing

In what follows, each method is detailed out giving relevant examples from the literature.

9.1 CHEMICAL REDUCTION OF METAL IONS IN A REACTION MEDIUM

This method involves the reduction of metal ions using a reducing agent, with MOP and reducing agent dissolved in the same solution. Usually, solution-based methods have been widely used for preparing metal nanocrystallites due to the possibility of obtaining monodisperse nanocrystals from a homogeneous medium and for a possible tuning of their size and shape. In some cases, the reduction process can be accelerated by external heating. Using this method, Au, Ag, Pt, Ru, and other metal nanocrystals have been prepared.

9.1.1 Au Nanocrystals from Au(I) Complexes

Au is known for its excellent chemical and structural stability under rigorous conditions and is also biocompatible. Hence, it is being used for bio- [21] and optoelectronic applications [22,23]. Due to the finite size, Au nanocrystals show distinct optical properties compared to the bulk state [24]. To prepare Au nanocrystals, Au(III) derivatives are widely used, though Au(I) state is also known. For instance, $HAuCl_4$ is one of the highly used Au(III) complexes due to its high solubility in water, ease of handling, and stability [25]. However, it is found that Au(I) is an intermediate during the reduction of Au(III) to Au(0) [26]. It has been proposed that understanding the reduction of Au(I) to Au(0) is helpful in controlling the size and shape of the Au microcrystals [26]. Although Au(I) halides such as AuCl, AuBr, and AuI are readily available, their low solubility in various solvents limits their usage for preparing Au nanocrystals. The solubility in nonpolar solvents can be improved by introducing organic ligands such as thiol [27], amine [28], phosphine [29], and other complex molecules. Unlike Au(III) derivatives, decomposition of the Au(I) complexes needs milder conditions, due to the lesser stability of Au(I) halides under ambient conditions.

Recently, Xia et al. have reported a suitable Au(I)-organic precursor to obtain monodisperse Au nanocrystals, with well-controlled size, at moderate temperatures (60°C) (Equation 9.1) [16]. AuCl(oleylamine) complex was obtained by dissolving AuCl and oleylamine in chloroform and agitated for few minutes to form a clear solution [16]. The molecular formula of the complex, $AuCl(C_{18}H_{37}N)$, was confirmed using mass and IR spectroscopy techniques. During the reduction, the color of the solution turned to pink, indicating Au nanoparticle formation [30]. In the given reaction mixture, oleylamine acts as a reducing agent for Au(I) to Au(0) and also as a capping agent once Au nanocrystals form. The obtained Au nanocrystals are of ~12.7 nm with a standard deviation of 8% (see Figure 9.1a and b). The size of the nanocrystals could be well controlled from ~3 to 12.7 nm by halting the reaction at different times

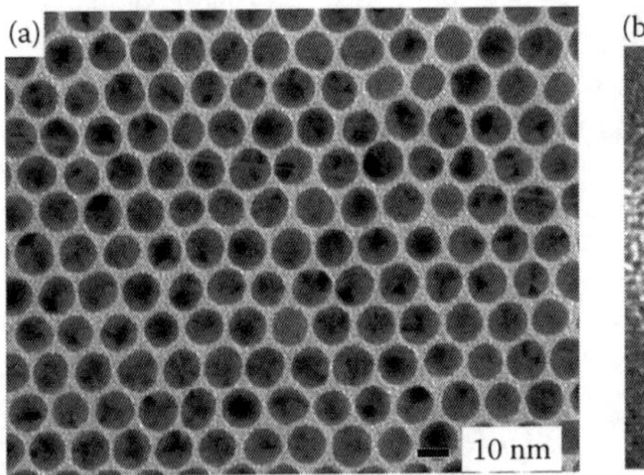

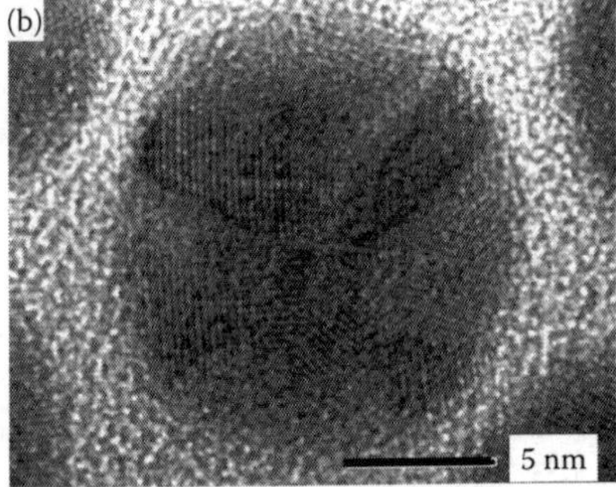

FIGURE 9.1 (a and b) Transmission electron microscopy (TEM) images of Au nanocrystallites obtained from Au(I)oleylamine complex. (Reprinted with permission from Lu, X. et al., *Chem. Eur. J.*, 14(5), 1584, 2008.)

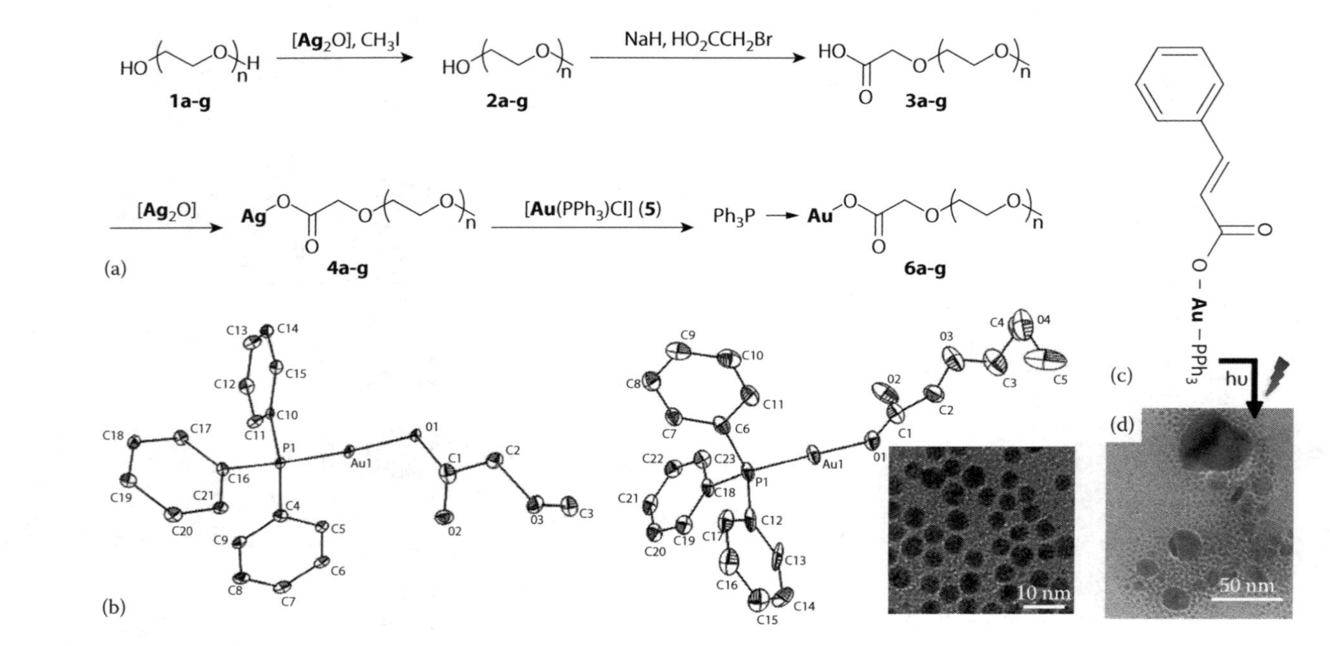

FIGURE 9.2 (a) Synthesis protocol for the preparation of [$(Ph_3P)AuO_2CCH_2(OCH_2CH_2)nOCH_3$]. (b) Oak Ridge Thermal Ellipsoid Plot (ORTEP) diagrams of [$(Ph_3P)AuO_2CCH_2(OCH_2CH_2)_nOCH_3$] (n = 0–1). Inset in Figure 9.2b shows TEM image of Au nanocrystals. (c) Molecular structure of Au(I)phosphine complex, and (d) TEM image of Au nanocrystals obtained by the photo reduction of Au(I) complex shown in (c). (b: Reprinted with permission from Tuchscherer, A. et al., *Dalton Trans.*, 41(9), 2738, 2012; d: Reprinted with permission from Schliebe, C. et al., *Chem. Commun.*, 49(38), 3991, 2013.)

of thermolysis. Besides thermolysis temperature, the nature of organic ligand and the mole ratio of ligand and AuCl also play important roles to control the size of the Au nanocrystals. With the ratio above 20, the crystallites size was uniform with <10% standard deviation, while at lower ratios, nanocrystals showed polydispersity with ~10%–30% variation. The proposed mechanism is that C=C in oleylamine forms a strong coordination bond with AuCl and slows down the reduction process, thereby controlling the size of the particle. When octadecylamine was used as stabilizing agent instead of oleylamine, the obtained crystallites were ~8 times larger.

$$\mathrm{AuCl} + \mathrm{Oleylamine} \rightarrow \mathrm{AuCl(oleylamine)} \rightarrow \mathrm{Au(0)} \tag{9.1}$$

Phosphine- [31] and carboxylate- [32] based Au(I) complexes are another set used for the synthesis of Au nanocrystals. Lang et al. have reported a bottom-up methodology to prepare Au nanocrystals using Au(I) carboxylates $[(Ph_3P)AuO_2CCH_2(OCH_2CH_2)_nOCH_3]$ (n=0–6) as precursors [32]. The precursor was obtained in a four-step procedure (Figure 9.2a). The precursor consists of Au source, reducing agent, and stabilizing agent and hence addressed as all-in-one precursor (Figure 9.2b). Under the given synthetic conditions polyols (ethylene glycol and its derivatives) and PPh_3 act as a reducing agent and capping agent respectively. The reported crystallite size is in the range of 3–4 nm and ~7 nm at higher concentrations. The same authors have used Au(I) phosphine compounds to synthesize Au nanocrystals via photo reduction (λ = 366 nm) (Figure 9.2c) [9]. The complex was dissolved in dimethylglyoxime (DMG), and the obtained solution was exposed to UV LED for 60 min. The synthesized nanocrystals were well below 5 nm (Figure 9.2d).

9.1.2 Ag Nanocrystals

Like Au, Ag nanocrystals are also of great interest due to their catalytic, electronic, and optical properties [33,34]. Ag nanocrystallites are commonly used in surface enhanced Raman spectroscopy (SERS) studies [35,36]. Instead of following the traditional approach, which involves the precursor of Ag(I) salts and reducing agent in a solvent, Lin et al. used a single-source precursor in an organic solvent [37a]. Narrowly dispersed Ag nanocrystals (7–11 nm) were obtained by the thermal reduction of Ag(I) trifluoroacetate in isoamyl ether in the presence of oleic acid (surfactant), without involving any size selection process. This method is versatile and easy to control the size of the crystallites. By varying the oleic acid-to-Ag(I) molar ratio, the diameter of the nanocrystals could be tuned [37a]. The TEM image clearly shows the monodispersed Ag nanocrystals (Figure 9.3a). Inset is the SAED of Ag nanocrystals. The polycrystalline nature of the Ag nanocrystals is evident from the broad ED rings and inhomogeneous contrast within individual crystallites [37a]. Vernal et al. have prepared Ag nanocrystallites by reduction of $[(PPh_3)_2Ag(O_2CC_{13}H_{27})]$ with azoisobutyronitrile (AIBN) at 80°C under N_2 atmosphere [38]. The average size of Ag nanocrystallites prepared from Ag myristate complex is ~4.4 nm [37b] (Figure 9.3b and c). In another report, monodispersed Ag nanocrystals of ~30 nm were obtained by reaction between Ag ligand clusters and borohydrides [39]. Gattorno et al. have used Ag(I)-ethylhexanoate in dimethyl sulfoxide (DMSO) to prepare Ag nanocrystallites

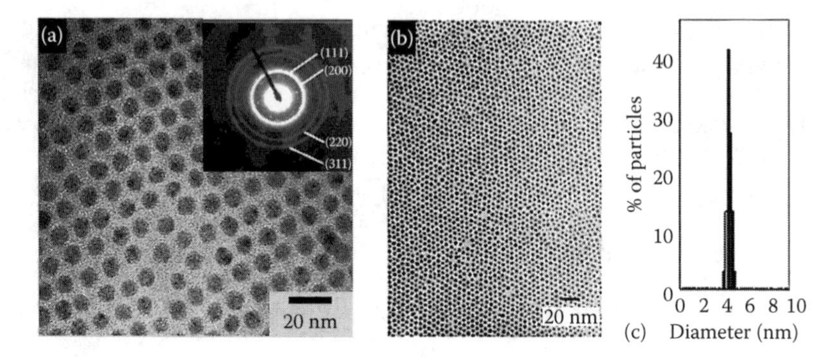

FIGURE 9.3 (a) TEM image of Ag nanocrystals prepared from Ag(I) trifluoroacetate. Inset is SAED of the Ag nanocrystals. (b and c) TEM image and particle size distribution of Ag nanocrystallites obtained from Ag myristate complex. (a: Reprinted with permission from Lin, X.Z., Teng, X., and Yang, H., *Langmuir*, 19(24), 10081. Copyright 2003 American Chemical Society; b and c: Reprinted with permission from Yamamoto, M. and Nakamoto, M., *J. Mater. Chem.*, 13(9), 2064, 2003.)

of 4.4 nm [40]. The stability of the colloidal Ag could be enhanced by adding sodium citrate as the capping agent.

9.1.3 Pt Nanocrystals

Pt with its high work function finds a wide usage in electrochemical applications, but is expensive. In nanoparticle form, the relative activity can be enhanced [41]. Besides the size effects, the crystallography of the nanocrystal faces also influences its activity. For O_2 reduction, {100} facets show better activity than {111} facets [42]. Pt-organic complexes are used extensively to prepare Pt nanocrystals of various shapes. Sun et al. have reported the synthesis of Pt nanocubes from $Pt(acac)_2$ by dissolving it in 1-octadecene along with oleic acid and oleylamine and heating the solution at 200°C in the presence of $Fe(CO)_5$ [43]. In the given mixture, oleic acid

FIGURE 9.4 (a and b) TEM images of Pt nanocubes obtained from $Pt(acac)_2$. (Reprinted with permission from Wang, C., Daimon, H., Lee, Y. et al., *J. Am. Chem. Soc.*, 129(22), 6974–6975. Copyright 2007 American Chemical Society.)

and oleylamine served as stabilizing and shape controlling agents, respectively. Thus, the obtained Pt nanocubes were ~8 nm sidewise (Figure 9.4a), enclosed with {200} facets as evident from the HRTEM image (Figure 9.4b). In the absence of $Fe(CO)_5$, spherical Pt nanocrystals were obtained. However, the decomposition of $Pt(acac)_2$ needs a higher temperature, which is a disadvantage. Recently, Chi et al. have used $Pt(hfac)_2$ instead of $Pt(acac)_2$ to prepare Pt nanocubes at ~140°C with size ranging from 2 to 16 nm depending on the solvent and surfactant [44].

9.1.4 Ru Nanocrystals

Among the transition metals, Ru is widely used in synthetic organic chemistry [45]. Ru nanocrystals have shown better catalytical activity than bulk Ru. Ru-organic precursors such as $Ru(acac)_3$ and Ru(Cod) (Cot) (Cod = 1,5-cyclooctadiene, Cot = 1,3,5 cyclooctatriene) have been used to prepare Ru nanocrystals of various shapes. In a typical synthesis, $Ru(acac)_3$ and PVP were dissolved in triethylene glycol (TEG) and the solution was heated at 200°C for 3 h [46]. In the given reaction medium, PVP and TEG act as stabilizing and reducing agents, respectively. The obtained nanocrystals were ~2.1 nm in size with FCC lattice (Figure 9.5a) [46], surprisingly deviating from the bulk Ru, which exists in HCP [47]. The size of the Ru nanocrystals was varied from 2 to 6.5 nm by changing the ratio of metal precursor to solvent. Though Ru nanocrystals have shown stable catalytical properties, their activity can be improved by disjoining the capping agent from the surface of Ru nanocrystals. Since the adsorbed ligands affect the physical and chemical properties of nanocrystals, it is indeed important to prepare ligand-free metal nanocrystals. To synthesize capping-free Ru nanocrystals, Colliere et al. have reported a route to obtain clean-surface nanocrystals for catalysis applications. In this route, Ru nanocrystals have been prepared by the decomposition of Ru(cod)(cot) in pure alcohol or a mixture of

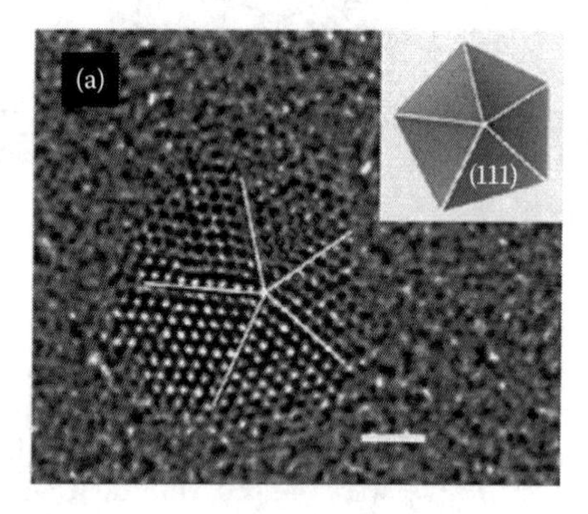

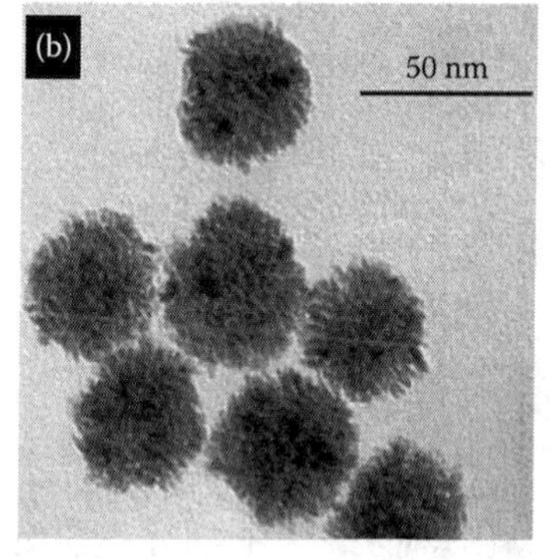

FIGURE 9.5 TEM images of Ru nanocrystals prepared by the reduction of (a) $Ru(acac)_3$ and (b) Ru(cod)(cot). Inset in (a) is a schematic illustration of a decahedral structure. ([a]: Reprinted with permission from Kusada, K., Kobayashi, H., Yamamoto, T. et al., *J. Am. Chem. Soc.*, 135(15), 5493–5496. Copyright 2013 American Chemical Society; [b]: Pelzer, K. et al., *Adv. Funct. Mater.*, 13(2), 118, 2003.)

alcohol and THF under H_2 atmosphere [48]. Here, the size and crystalline nature of the particles are influenced by the length of the alkyl chain of alcohol, volume ratio of alcohol to THF, and the reaction temperature. Single-crystalline Ru nanocrystals 2–3 nm in size were obtained in pentanol and the obtained nanocrystals were stable for few weeks. Ru nanocrystals prepared in other solvents were polycrystalline with size varying from 4 to 80 nm (Figure 9.5b). Unlike other synthetic routes, in this method solvent molecules (alcohols) stabilize the Ru nanocrystals. The stability of the sol was improved by taking mixture of methanol and THF instead of pure methanol. A sol prepared in a mixture of solvents (MeOH:THF = 5:95) was stable up to a year [48].

9.2 REDUCTION OF METAL-ORGANIC PRECURSOR AT THE LIQUID–LIQUID INTERFACE

Reduction of metal ions at the liquid–liquid interface is another way to prepare the nanocrystals. The liquid–liquid interface formed between two immiscible liquids shows a distinct density and viscosity from the bulk counterparts [49]. However, the liquid–liquid interface is not well understood, unlike the air–liquid interface, although the liquid–liquid interface has been effectively used in nanocrystal synthesis. The Brust method is a well-known example for the reduction of MOPs at the liquid–liquid interface [50]. It involves three major steps:

1. Transfer of metal anions to an organic layer or to dissolve the MOPs in an organic layer
2. Addition of reducing agent to the aqueous phase
3. Reduction of metal ions and growing metal nanocrystals at the interface

The reagent that transfers metal anion complex to the organic phase from aqueous is called phase transfer reagent; tetraoctylammonium bromide (ToABr) [50] and triphenylphosphine oxide (TPPO) [51] are well-known phase transfer reagents, used extensively for transferring the metal ions from aqueous to the organic layer. At the interface, polar groups (tetraoctylammonium or triphenylphosponium) bind to the metal anion center and transfer to the organic layer. Strong and mild reducing agents have been employed to reduce metal ions at the interface.

9.2.1 Au, Ag, and Their Alloy Nanocrystallites

Brust et al. have done a pioneer work on the preparation of Au nanocrystals at the water–toluene interface [50]. Briefly, $(AuCl_4)^-$ ions were transferred to the organic layer using ToABr named as AuToABr and reduced with sodium borohydride ($NaBH_4$). The formed Au nanocrystals in toluene were capped with dodecanethiol. Thus, obtained Au nanocrystals were in the range of ~1–3 nm [50]. The Brust method is a versatile method for producing metal nanocrystals due to its simplicity and its immediate extension for various noble metals such as Ag, Cu, Pd, and Pt. Since thiol addition rapidly quenches the growth of the nanocrystals, it may not be a suitable method to prepare anisotropic metal nanocrystals important in plasmonic applications. Recently, Kulkarni et al. have prepared various noble and reactive metal–ToABr complexes by transferring the metal anion to the organic layer with ToABr, henceforth the complexes named as MToABr (M = Au, Ag, Pd, Pt, Pb, Ga, In and Zn, and bimetal combination) [52].

Rao et al. have modified the Brust method; N_2H_4 and triphenylphosphine (PPh_3) have been used instead of $NaBH_4$ to reduce Au(III) to Au(0) and cap the nanocrystals with thiol [53]. In a typical synthesis, $(AuCl_4)^-$ ions were phase transferred to organic layer using ToABr as a phase transfer reagent, which produced red color in the organic layer. It was followed by the addition of PPh_3 to the organic layer resulting in a colorless solution, signifying the reduction of Au(III) to Au(I). Further, the desired quantity of N_2H_4 was added to the aqueous phase and allowed the reduction of Au(I) to metallic Au (Figure 9.6a). The film obtained at the liquid–liquid interface comprised of cauliflower-like units, with a diameter of ~700 nm (Figure 9.6b) [53]. The magnified image shown in Figure 9.6c shows the pentagonal nanorods projecting out of the core. In obtaining a flower-like morphology, PPh_3 plays a crucial role. In the absence of PPh_3, Au nanocrystals with no selective shape were obtained. Irregularly

FIGURE 9.6 (a) Digital photograph of Au nanocrystalline film formed at toluene–water interface in a glass beaker. (b) SEM images of Au cauliflower. The inset on the top right corner shows a high-resolution image of the cauliflower-like structures. The inset at the bottom shows the histogram of the size distribution of cauliflower-like structures. (c) High-resolution SEM image of the nanorods present in the cauliflower-like structures. Inset shows the end of one such nanorod, with a fivefold symmetry. SEM images of Au:Ag alloy nanostructures obtained by the reduction of Au and Ag organic precursors in the ratio of (d) 15:85, (e) 75:25. (a: Reprinted with permission from Rao, C.N.R., Kulkarni, G.U., Thomas, P.J.et al., *J. Phys. Chem. B*, 107(30), 7391–7395. Copyright 2003 American Chemical Society; b through e: Reprinted with permission from Agrawal, V.V., Kulkarni, G.U., and Rao, C.N.R., *J. Colloid Interface Sci.*, 318(2), 501, 2008.)

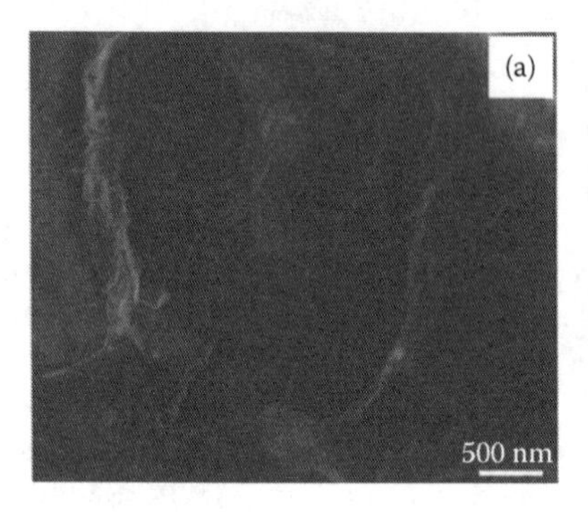

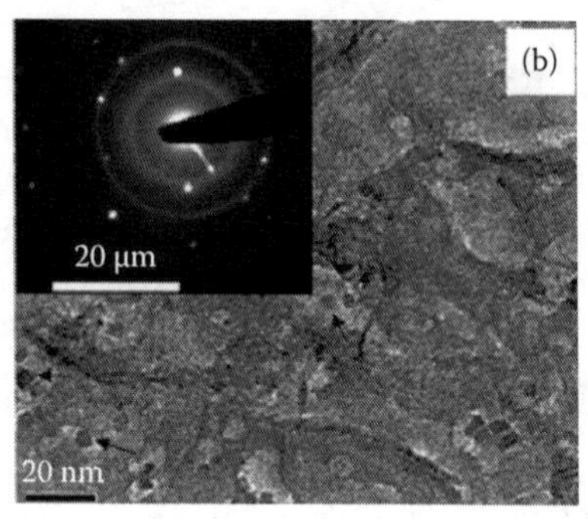

FIGURE 9.7 (a and b) Microscopy images of Ni films obtained at liquid–liquid interface. SAED pattern of Ni film is shown in inset. (Reprinted with permission from Varghese, N. and Rao, C.N.R., *Mater. Res. Bull.*, 46(9), 1500–1503, 2011.)

shaped Au nanocrystals were produced at higher concentrations of ToABr (18–91 μmol). Using this method, alloys and bimetallic nanocrystals can also be prepared. Au- and Ag-organic precursors were dissolved in toluene and reduced with N_2H_4 to obtain alloys of Au and Ag [53]. The morphology of the alloys changed with Ag concentration. Dendrite and mesoballs like crystallites were obtained at high (85%) and low (25%) concentrations of Ag, respectively (Figure 9.6d and e). In another case, $Au(PPh_3)_2Cl$ was used as Au source in toluene, instead of transferring the Au source from aqueous to organic layer [54]. Alkaline tetrakis(hydroxymethyl)phosphonium chloride (THPC) was used to reduce the Au(I) complex. Films obtained at the liquid–liquid interface after 24 h of reduction comprised tiny Au nanocrystals (~9 nm) [54].

9.2.2 Ni Nanocrystals

Ni nanocrystallites used as the catalyst to grow carbon nanotubes [55]. Unlike noble metals, there are only a few synthetic routes to prepare Ni crystallites from MOPs. Ni crystallites have been prepared by the reduction of nickel cupferronate with $NaBH_4$ at the toluene–water interface [56]. Similar to the synthesis of other metal nanocrystals, Ni source was dissolved in toluene and $NaBH_4$ in water. The obtained film at the interface was ~20 nm thick and consisted of nanocrystals of size, 2–3 nm (Figure 9.7a and b).

9.3 THERMOLYSIS OF METAL-ORGANIC PRECURSOR

Thermolysis of MOPs is a well-known method for the preparation of metal nanocrystallites. The advantages of thermolysis over solution routes are the following:

1. The obtained nanocrystals are free of surface capping agents.
2. The size of the crystallites can be tuned from nm to submillimeter, which is rather difficult to achieve with solution-based routes.

3. In situ growth studies during thermolysis, using a simple optical and electron microscopy.
4. Growth almost independent of the nature of the solvent and of the substrate.
5. Immobilization of the nano-/microcrystallites on a desired solid surface.

9.3.1 Direct Heating of Solid Precursors in Air or under Various Atmospheres

9.3.1.1 Au Nano and Microcrystals

Fukusumi and coworkers introduced a modified Brust method to prepare Au nanocrystals by the thermal decomposition of solid precursor of Au(I) thiolate instead of $NaBH_4$ reduction. The Au(I) complex was prepared by the addition of dodecanethiol to the organic layer containing AuToABr and the derived formula is $[C_{14}H_{29}(CH_3)_3N][Au(SC_{12}H_{25})_2]$. Au nanocrystallites were prepared by thermal reduction of Au(I) thiolate at 180°C under N_2 atmosphere (Equation 9.2). The obtained Au nanocrystals were spherical in shape (Figure 9.8a), and the average diameter was ~25 nm (Figure 9.8b) [8].

$$\left[R\left(CH_3\right)_3N\right]\left[Au\left(SC_{12}H_{25}\right)_2\right] \xrightarrow{\Delta} Au(0) + \left(CH_3\right)_3N + \left(C_{12}H_{25}S\right)_2 \qquad (9.2)$$

Other Au(I) complexes such as $[Au(C_{13}H_{27}COO)(PPh_3)]$ [57] and $Au(PPh_3)_2Cl$ [58] have also been used to prepare Au nanocrystals via the thermolysis route. Size of the Au nanocrystals obtained from these precursors ranges from 20 to 200 nm.

Due to thermodynamics restrictions, it is rather difficult to grow millimeter or submillimeter long single-crystalline metal crystallites, enclosed with ultrasmooth facets. Recently, Kulkarni et al. have modified the Brust method to produce submillimeter (~500 μm) Au crystallites using AuToABr complex (see Figure 9.9a) [59,60].

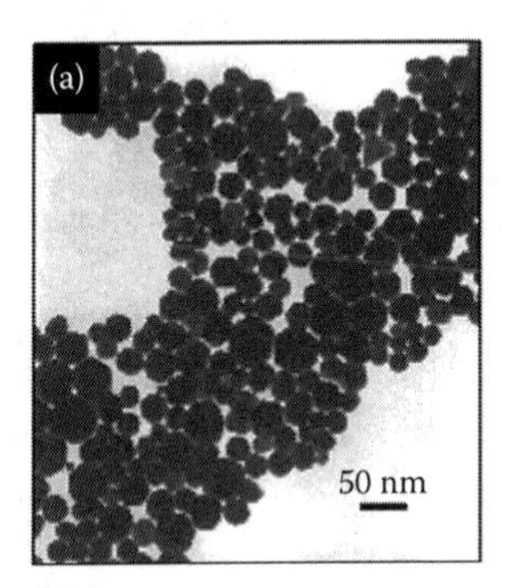

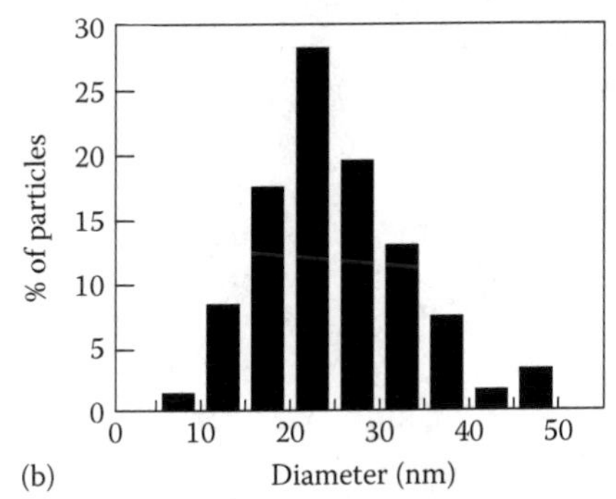

FIGURE 9.8 (a) TEM image of Au nanocrystals obtained by the thermolysis of $[C_{14}H_{29}(CH_2)_3N][Au(SC_{12}H_{25})_2]$ in N_2 atmosphere at 180°C. (b) Histogram depicts the yield of Au nanocrystals of different sizes. (Reprinted with permission from Nakamoto, M. et al., *Chem. Commun.* (15), 1622–1623, 2002.)

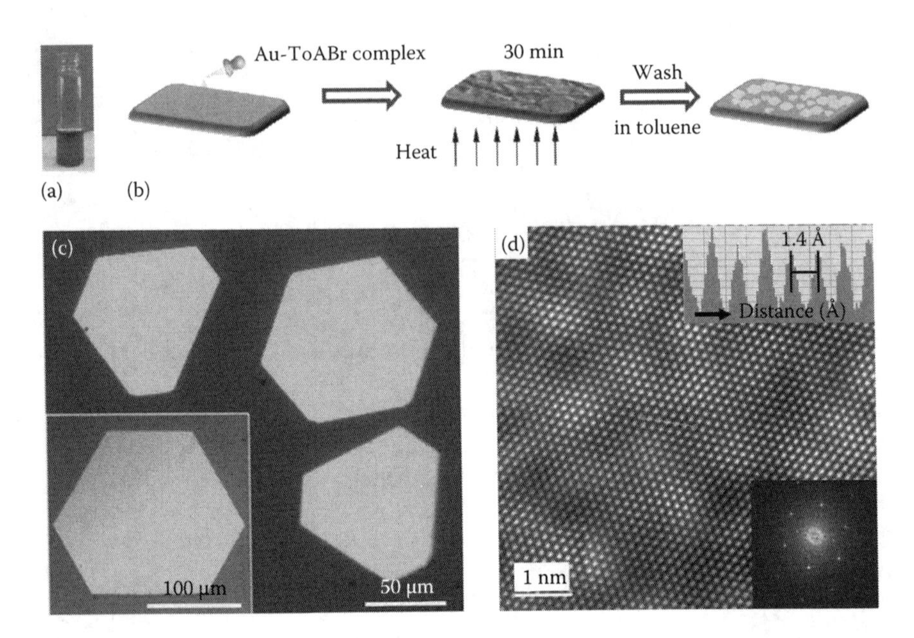

FIGURE 9.9 (a) Digital photograph image of AuToABr solution. (b) Schematic illustration of Au microplates preparation. (c) Optical microscope image of Au polygons. (d) HRTEM image collected from Au plate and the corresponding fast Fourier transform (FFT) is shown as an inset. (Reprinted with permission from Radha, B. and Kulkarni, G.U., *Nano Res.*, 3(8), 537, 2010.)

The derived formula of the complex is $N^+(C_8H_{17})_4AuCl_4^-.2N(C_8H_{17})_4Br$. The MOP was thermalized at 130°C, which is well below the decomposition temperature, for 24 h. Among the obtained Au crystallites, polygons such as hexagons and triangles were the major products (see Figure 9.9b). The edge length of the obtained crystallites was above 100 μm, while the thickness was more often, only sub μm. Most importantly, the polygonal top surface, typically the {111} facet (Figure 9.9c), was obtained as atomically smooth with an average roughness of ~1 nm. Side faces were enclosed with alternative {111} and {100} facets [61]. As shown in HRTEM image, the hexagonal arrangement atoms with d-spacing of 1.44 Å refer to the FCC{220} planes of Au (Figure 9.9d). Such microcrystallites can be ideal substrates for STM studies, as well as to grow semiconducting nanowires [62]. The in-situ studies have shown that the growth of the microcrystallite commenced with tiny Au nanocrystals that assembled to form nanotriangles where the intermediate stages could be captured due to the lower decomposition rates [11]. This method has received immense attention as it is a simple synthetic route involving mild thermolysis conditions where the nature of the substrate is not so important. Due to the larger dimensions of Au triangle and hexagons, these crystallites can be manipulated with a simple needle.

Metal 1D nanowires have gained importance due to the extensive optical properties and wide tunability of the absorption from visible to NIR region.

They possess two distinct surface plasmon absorption modes associated with their dimensionality. Such wires, typically of Ag, are prepared by the polyol reduction of $AgNO_3$ [63].

Tapered nanowires are a subclass of 1D nanowires where center is bulged out and tapered at the two ends [64]. Although the synthetic routes of nanowires are well documented, there is only limited literature on tapered metal nanowires. Bipyramids and its derivatives geometries are typical examples of tapered metal nanowires. These crystallites were obtained by adding Ag(I) ions to the AuToABr precursor so that Ag(I) binds to the Au{100} facets and the growth along the <110> and <112> lead to corrugated pentagonal bipyramids (CPB) of Au. The obtained crystallites were ~10 μm long with a central diameter of about 1 μm (Figure 9.10a) and the tip diameter of ~55 nm (Figure 9.10b) [15]. The surface is covered with highly corrugated nanosurfaces (Figure 9.10c). The cross-section view of μ-CPB shows the fivefold symmetry of the tips running along the length (Figure 9.10d). The mole ratio of

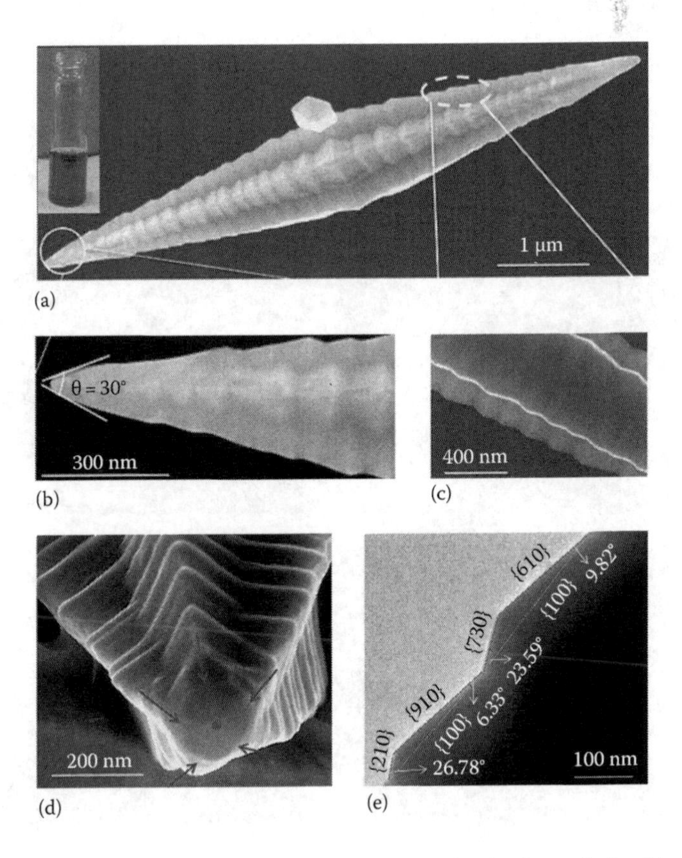

FIGURE 9.10 (a–c) SEM images of Au μ-CPBs obtained by the thermolysis of AuAgToABr. Digital camera photograph of AuAgToABr solution. (d) Cross-section image of Au μ-CPBs. (e) HRTEM image obtained corrugated surface. (Reprinted with permission from Mettela, G. et al., *Sci. Rep.*, 3, 1793, 2013.)

Au(III) to Ag(I) and the thermolysis temperature influence the length and surface corrugations of Au bipyramids. The nanocorrugations were assigned to high-index facets (Figure 9.10e) and these reactive facets are stabilized by a monolayer or sub-monolayer of Ag. At higher thermolysis temperatures (200°C–220°C), Au crystallized with tetragonal (BCT) and orthorhombic (BCO) lattice structures [17].

9.3.1.2 Ag Nano/Microcrystals

Ag nanocrystals with almost uniform SERS enhancement over the entire region should be ensured for practical usage along with its ability to measure ultralow volumes [65]. Reusability of the SERS active substrates would make the analytical process resource intensive and cost-effective. Easy handling of bigger-sized nanocrystals is another advantage. Recently, Kulkarni et al. have reported the synthesis of Ag microflowers by simple thermolysis of AgToABr complex at 250°C, followed by $NaBH_4$ reduction (see schematic in Figure 9.11a). The reported microflower size is in the range of 50–100 μm (Figure 9.11b through e) and consists of interconnected Ag nanocrystallites of 25–40 nm. These microflowers can be

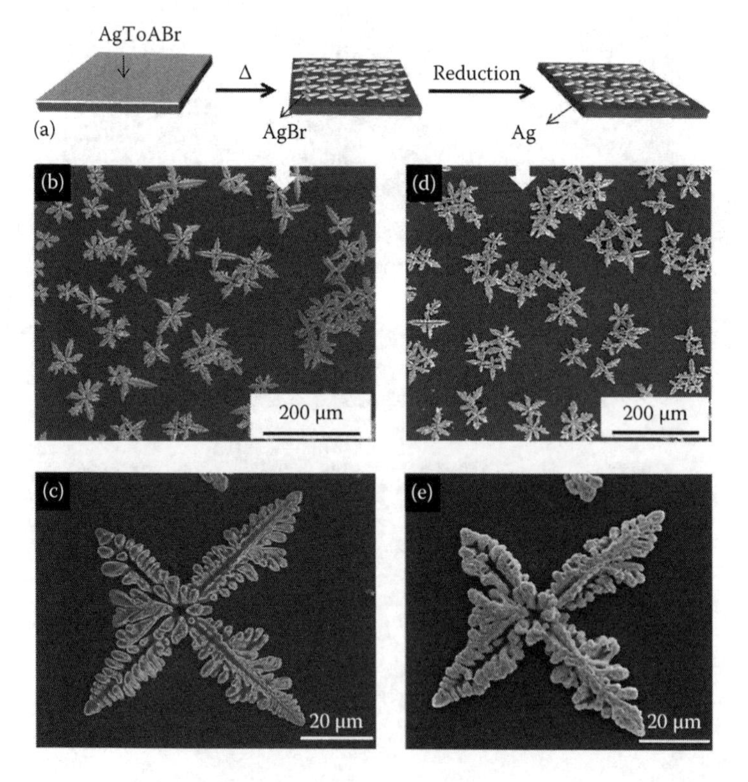

FIGURE 9.11 (a) Schematic illustrating the synthesis of Ag microflowers. FESEM images of (b and c) AgBr microflowers (as prepared) and (d and e) Ag microflowers after reduction with $NaBH_4$. (Mettela, G. et al., *Nanoscale*, 6(13), 7480, 2014. Reproduced by permission of The Royal Society of Chemistry.)

manipulated with a tiny needle. In this synthetic route, shape (cube to flower) and size (submicron to micron) of the microcrystal were well controlled. The obtained microflowers served as reusable SERS substrates for the detection of various bioactive molecules [66]. Various Ag(I) carboxylate complexes have also been used to prepare Ag nanocrystallites of sub 10 nm [67–70]. Chen et al. have reported the synthesis of disk-shaped Ag nanocrystallites from Ag thiolate by decomposition at 180°C–225°C under a N_2 atmosphere. The reported size and diameter are 16.1 and 2.3 nm, respectively [71].

9.3.1.3 Alloys and Bimetal Nanocrystallites of Noble and Seminoble Metals

Heterometallic systems like alloys and bimetals are quite interesting due to the enhanced optical [72], catalytic [73] and magnetic properties [74], and stability [75]. For example, AuCu alloy nanocrystallites show better oxidation stability [75], higher catalytical activity compared to the pristine Cu and Au nanocrystallites [76]. Kulkarni and coworkers have introduced a simple method to prepare submicron-sized AuCu alloy crystallites by the thermolysis of AuCuToABr complex at 375°C in N_2 atm [52a]. Similarly, AuPd bimetallic and alloy compositions were obtained from AuPdToABr complex by subjecting to moderate (250°C) and high thermolysis temperatures (500°C) in air (Figure 9.12a and b). This is noteworthy that the composition of the system can be tuned from bimetallic to alloy by simply varying the thermolysis temperature. Triangle- and hexagonal-shaped and submicron-sized crystallites were

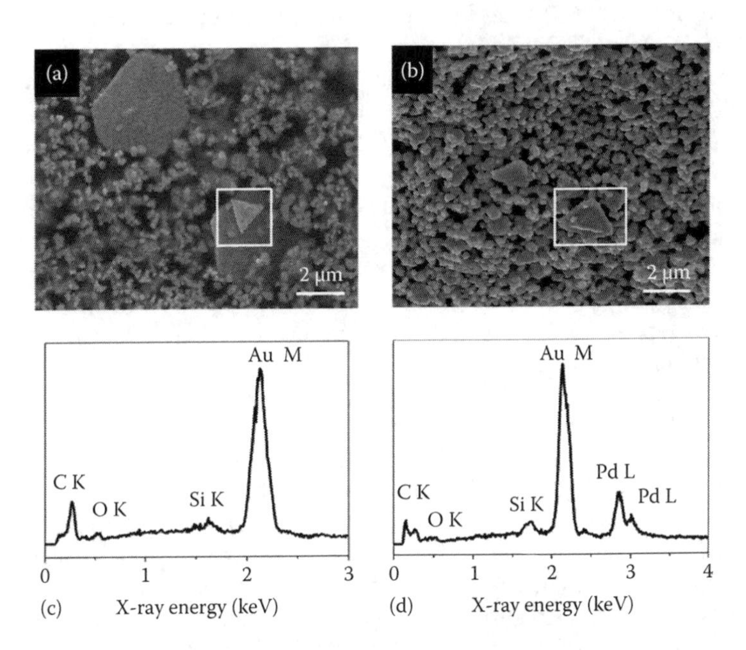

FIGURE 9.12 SEM images of Pd–Au bimetal (a) and alloy (b) nanocrystallites. EDS spectrum collected from the marked region in (a) and (b) are shown in (c) and (d), respectively. (From Kiruthika, S. et al., unpublished work.)

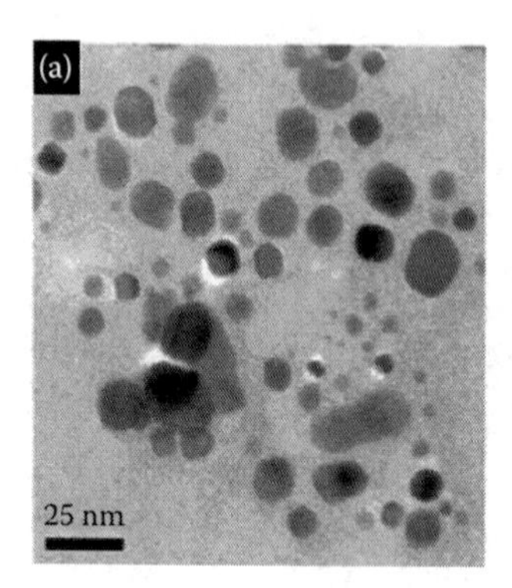

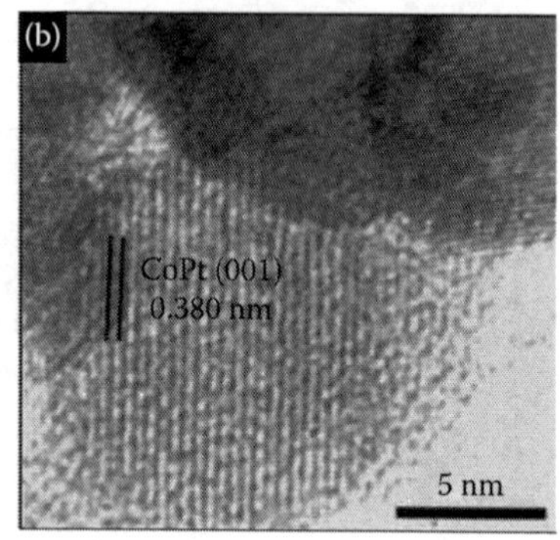

FIGURE 9.13 (a) TEM micrographs of as-prepared $L1_0$ CoPt nanopowder and high magnification image of CoPt particle is shown in (b). The d-spacing confirms the lattice (001) of $L1_0$ CoPt. (Wellons, M.S. et al., *J. Mater. Chem. C*, 1(37), 5976, 2013. 2013. Reproduced by permission of The Royal Society of Chemistry.)

obtained at both temperatures (Figure 9.12a and b), yet the composition of a given crystallite is completely different. As shown in Figure 9.12c, polygons prepared at 250°C are composed of only Au, while those prepared at 500°C are composed of Au and Pd with at% of 52:48 (Figure 9.12d) [52b]. Deivaraj and Lee have prepared carbon-supported PtNi nanocrystals by thermalizing a mixture of carbon matrix and $[(bipy)_3Ni](PtCl_6)]$ at 550°C in H_2/Ar atmosphere. The average size of the PtNi nanocrystallites is ~8 nm [77].

Nanoscale magnetic materials, especially of binary metal alloys, are quite important in the context of high-density data storage and also useful in understanding ferromagnetism at nanoscale. Ferromagnetic CoPt nanopowders with an average particle size of 11 nm have been obtained by reductive decomposition of $CoPt(CO)_4$(dppe)Me/NaCl composite powder at 650°C under getter gas (9:1 N_2/H_2) [78]. Aqueous dissolution of NaCl support helps in isolating the CoPt nanocrystals. TEM images of thus formed alloy composite are shown in Figure 9.13a with high magnification image showing (001) lattice fringes of $L1_0$ CoPt (Figure 9.13b). These atomically ordered $L1_0$-type lattice structures exhibit high coercivity, large magnetic anisotropy, and hard ferromagnetic properties [78].

9.3.1.4 Cu Nanocrystals

Cu nanocrystals have been used for various applications such as catalysis [79], transparent conducting electrodes [80], etc. However, the stability of Cu nanocrystals is a major challenge due to higher reactivity of Cu compared to Au and Ag. This can be overcome by preparing Cu nanocrystals in nonpolar solvents under the inert atmosphere and embedding in polymer matrices. Kim et al. have prepared nearly spherical–shaped Cu nanocrystals (average size, ~10 nm) by the thermal decomposition of Cu-oleate complex in vacuum at 295°C (Figure 9.14). The lattice spacing 1.99 Å corresponds to {111} facets of Cu [81].

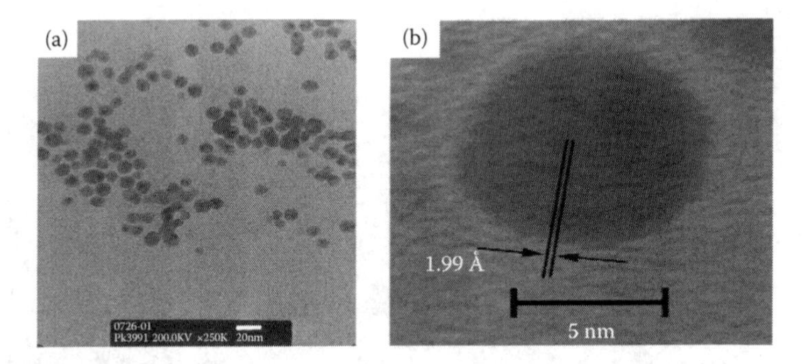

FIGURE 9.14 TEM image (a) and lattice pattern (b) of Cu nanocrystals obtained by the decomposition of Cu-oleate complex. (Reprinted with permission from Taylor & Francis, Kim, Y.H. et al., *Mol. Cryst. Liq. Cryst.*, 445(1), 231/[521].)

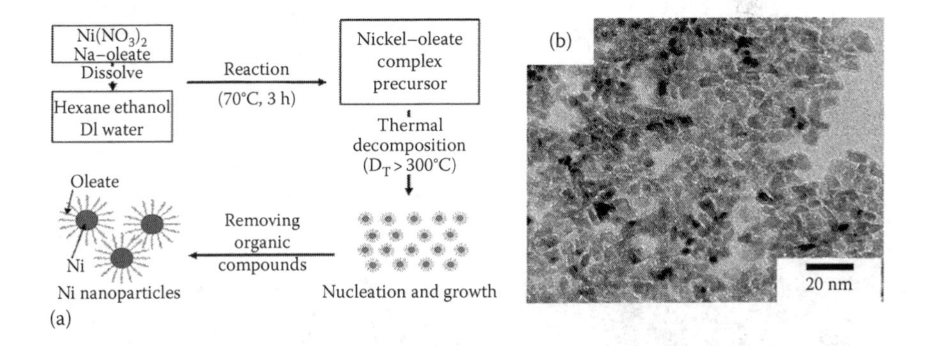

FIGURE 9.15 (a) Schematic illustrating the synthesis of Ni nanocrystals. (b) TEM images of Ni nanocrystals obtained from Ni–oleate complex. (Reprinted with permission from S.G. Kim et al., *Colloids Surf. A Physicochem. Eng. Asp.*, 337(1–3), 96, 2009.)

9.3.1.5 Ni Nanocrystals

Ni nanocrystals have been prepared by the thermal decomposition of Ni-oleate complex in N_2 atmosphere at 300°C and annealed at higher temperature to form continuous film composed of Ni crystallites of ~15 nm (Figure 9.15) [82]. The residual carbon formed due to the decomposition of oleate acts as a binder in obtaining continuous Ni films and also protects the Ni films from aerial oxidation.

9.3.1.6 Bi Nanocrystals

Bi is a semimetal with a small indirect band gap and converts to the semiconductor when the particle size is reduced below ~50 nm [83] and these crystallites are of interest due to their thermoelectric properties [67]. Various Bi nanostructures have been prepared by the thermal decomposition of Bi thiolate. Layered Bi nanorhombuses have been obtained by the decomposition of a layered precursor, $Bi(SC_{12}H_{25})_3$

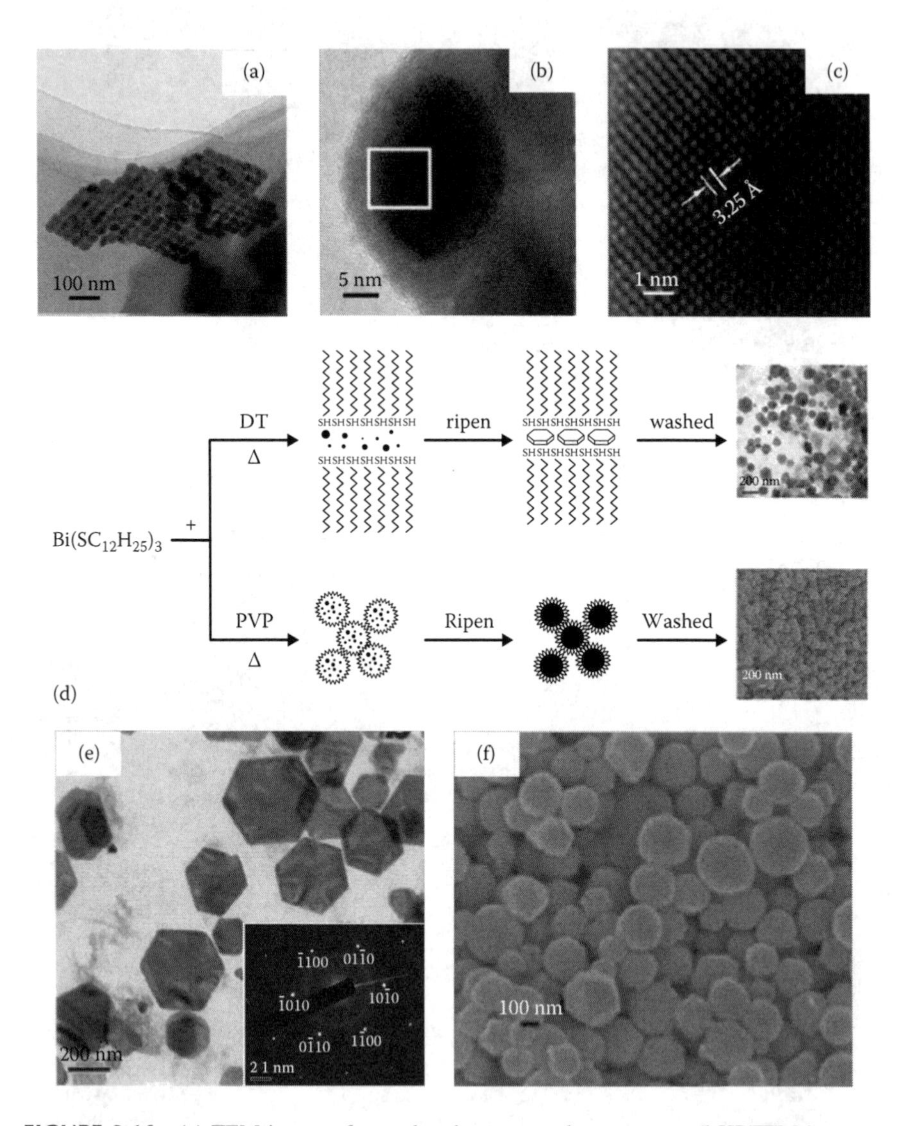

FIGURE 9.16 (a) TEM image of nanorhombuses monolayer array and HRTEM image of a single nanorhombus (b). An enlarged image of the selected part of panel in (b) is shown in (c). (d) Schematic showing possible mechanism of formation of nanodisks and nanospheres. (e) TEM image of Bi nanodisks prepared at 135°C for 10 h and inset is the SAED pattern of Bi disks. SEM image of Bi nanospheres formed in the presence of PVP capping agent is shown in (f). (a–c Reprinted with permission from Chen, J., Wu, L.M., and Chen, L. et al., *Inorg. Chem.*, 46(2), 586–591. Copyright 2007 American Chemical Society; d through f: Reprinted with permission from Wang, Y., Chen, J., Chen, L. et al., *Cryst. Growth Des.*, 10(4), 1578–1584. Copyright 2010 American Chemical Society.)

at 90°C (Figure 9.16a through c) [84]. Spherical Bi nanocrystals have also been prepared by Carotenuto et al. wherein the nanocrystals size and distribution could be controlled by the thermolysis temperature [85]. Small nanocrystals with narrow size distribution were obtained at working with elevated temperatures [86]. Wang et al. could tune the morphology of Bi nanocrystallites by varying the synthesis conditions (see Figure 9.16d). Bi nanodisks were obtained as a major product by the thermolysis of Bi thiolate at 120°C or 135°C for 10 h. The TEM image of thus formed nanodisks is shown in Figure 9.16e, with the SAED pattern in the inset. Nanospheres were obtained by annealing the thiolate precursor along with PVP capping agent at 105°C for 1 h. Figure 9.16f shows the SEM image of Bi nanospheres.

9.3.2 Metal Organic Chemical Vapor Deposition

CVD is a well-known chemical process to fabricate high-purity nanocrystalline metal films (Figure 9.17a). The MOPs used in the CVD process should have (1) high volatility, (2) an active component that remains stable during transport, and (3) appropriate decomposition temperature. Au is used as a conducting channel due to its high conductivity and chemical stability under varied conditions [87]. Various types of Au precursors such as Au(I) alkylphosphine, Au(I) alkyloxyphosphine, and Au(I) carboxylate complexes with tertiary phosphines as ligands have been used for CVD processes. Among these, the latter is considered as a promising material for the CVD

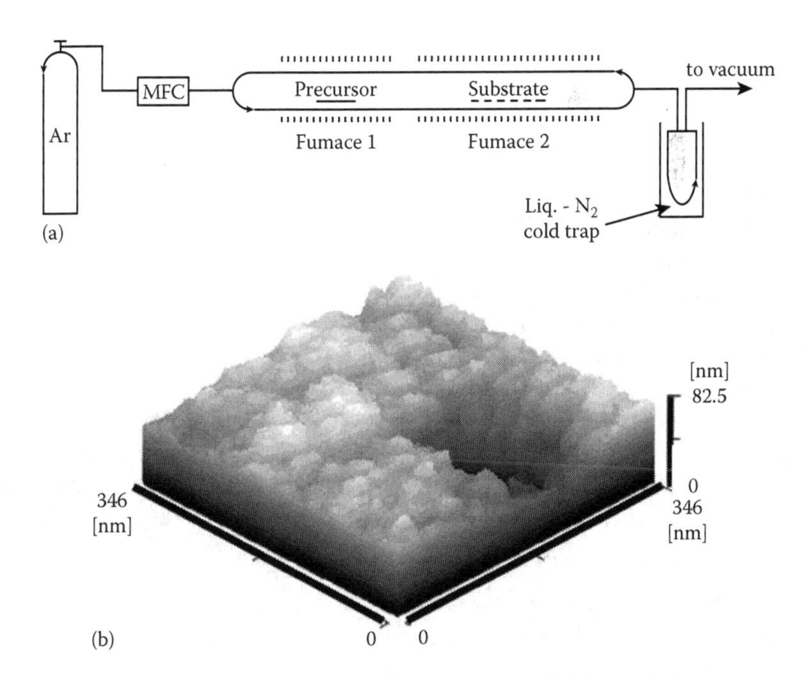

FIGURE 9.17 (a) Schematic illustration of reactor used for chemical vapor deposition (CVD) process. (b) STEM image of Au films obtained from $Au(OOCC_7F_{15})(PMe_3)$. (Reprinted with permission from Szłyk, E. et al., *Chem. Vap. Depos.*, 6(3), 105, 2000.)

process due to the tunability of the decomposition temperature by varying the length of the alkyl group. The decomposition temperature of the complex increases with the length of the alkyl chain. Thus, an Au film on Si(111) substrate was obtained by the decomposition of $Au(OOCC_7F_{15})(PMe_3)$ complex at 287°C (Figure 9.17b). The thickness of the film was in the range of 25–125 nm, comprising interconnected Au nanocrystallites of ~50 nm [88]. Au(III) complexes with S,S chelating ligands have also been employed to fabricate Au films in the H_2 atmosphere [89]. Crystallites size and their connectivity were modified by varying the deposition temperature between 210°C and 300°C. The CVD process was employed to form Ag and Cu films using precursors such as $C_3F_7COOAgPMe_3$[88] and $[Cu_2(oxalate)L_2]$ (L=$Me_3SiC{=}CSiMe_3$, $Me_3SiC{=}CSiBu^t{}_3$) [90].

9.3.3 Aerosol-Assisted Chemical Vapor Deposition

An Ag film consisting of the rod and spherical nanocrystallites has been deposited on a glass substrate using an AACVD technique (Figure 9.18a). Using ($[Ag\{(OPPh_2)_2N\}]_4{\cdot}2H_2O$) as precursor (for structure see Figure 9.18b) and Ar as carrier gas, Ag film was grown on glass held between 375°C and 475°C (Figure 9.18c). The feature size of rod and spherical nanocrystallites is in the range of 0.05–0.1 μm [91]. The other precursors such as $[Ag_3\{(SPiPr_2)_6N_3\}]_2$ and $AgO_2CR(PPh_3)_2$ have also been used for the deposition of Ag films by the AACVD method [92,93].

9.3.4 Inkjet Printing

9.3.4.1 Au Nanofilms

Inkjet printing is an inexpensive method for the fabrication of metal electrodes in circuitry and microelectronic devices from solubilized metal precursors on a desired substrate. It has gained interest among the researchers due to its low fabrication cost, printability on nonplanar substrates, short processing times, etc. Due to these advantages, inkjet printing is recognized as a simple, cost-effective, and environmental-friendly process [94–98]. Au(I)-diketonates [94] and phosphane Au(I)-carboxylate [95] complexes have been explored as inkjet "inks." Schoner et al. have reported the fabrication of nanocrystalline Au film from $[AuO_2CCH_2(OCH_2CH_2)_2OCH_3(nBu_3P)]$ complex. The obtained Au film was ~70 nm thick and consists of highly interconnected Au nanocrystals of ~30–40 nm (Figure 9.19a and b) [89]. The conductivity of the films measured using four-probe setup is of $1.9 \pm 0.2 \times 10^7$ $S{\cdot}m^{-1}$ which is 43% ± 4% less than bulk Au (4.4×10^7 $S{\cdot}m^{-1}$) [96]. Organic Au(III) and Au(I) compounds can be used for metallization process in various techniques such as CVD, spin coating, and drop coating [90].

9.3.4.2 Pd Microstripes and Dots

Pd thiolates have been explored as novel Pd-organic precursors to prepare Pd nanocrystallites taking advantages such as high solubility in various organic solvents and moderate decomposition temperatures [97]. These have been used as ink in various lithography techniques such as soft lithography, E-beam lithography (EBL),

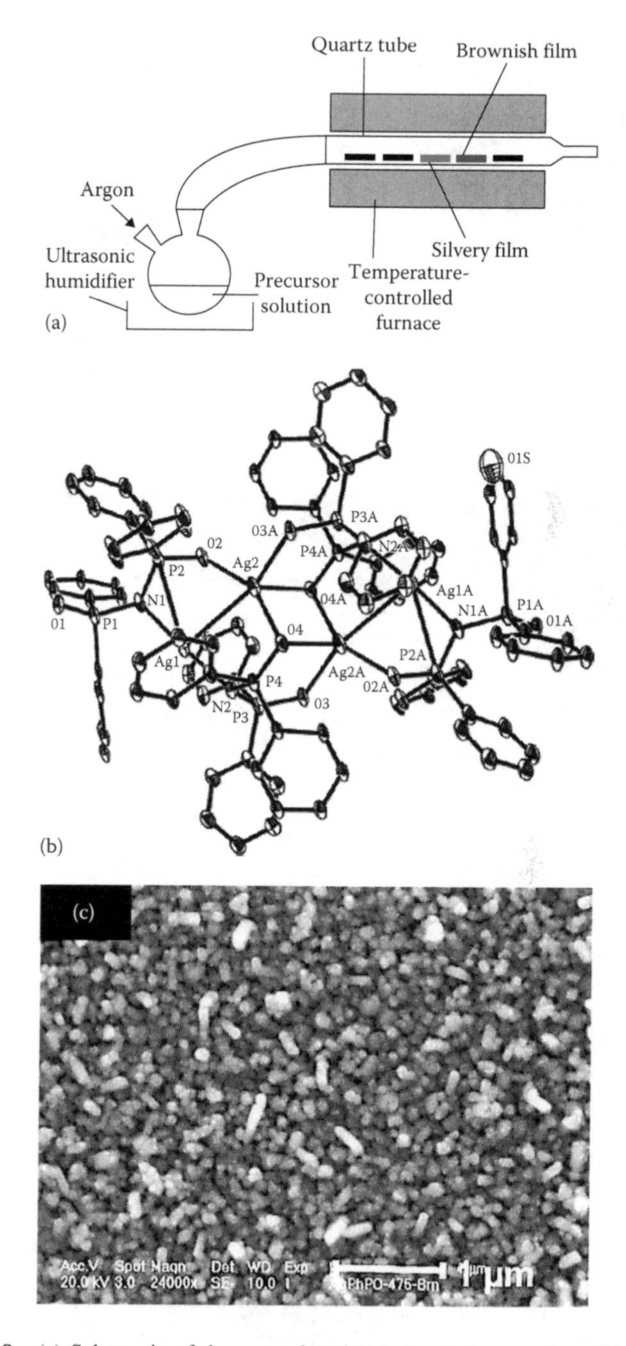

FIGURE 9.18 (a) Schematic of the aerosol-assisted chemical vapor deposition (AACVD) reactor. (b) The molecular structure of $[Ag\{(OPPh_2)_2N\}]_4.2H_2O$. For clarity, hydrogen atoms are omitted. (c) SEM images of AACVD-deposited Ag films at argon flow rate of 160 sccm on glass substrate at growth temperature of 475°C. (Reprinted with permission from Panneerselvam, A. et al., *Chem. Vap. Depos.*, 15(1–3), 57, 2009.)

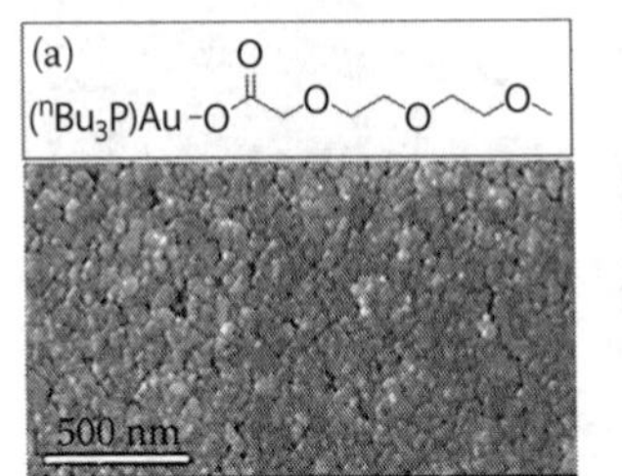

FIGURE 9.19 (a and b) SEM images of inkjet-printed Au films obtained by the decomposition of [$AuO_2CCH_2(OCH_2CH_2)_2OCH_3(nBu_3P)$]. Inset in (a) shows the molecular structure of the precursor. (Reprinted with permission from Schoner, C. et al., *Thin Solid Films*, 531, 147, 2013.)

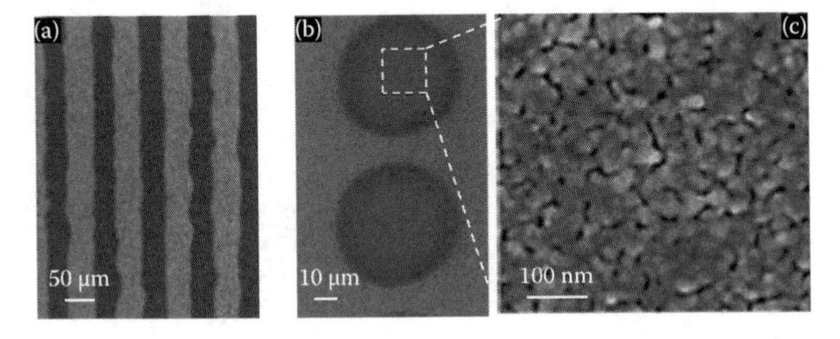

FIGURE 9.20 (a and b) SEM images of Pd microstripes and dots drawn using inkjet printing. (c) Magnified view of a dot shown in (b). (From the author's lab.)

nanoimprint lithography. Recently, Bhuvana et al. have used Pd hexadecanethiolate to fabricate Pd microstripes and dots using inkjet print technique [98]. The width of stripes is ~50–60 μm and composed of tightly packed Pd nanocrystals (~25 nm) as shown in Figure 9.20a through c. An inkjet printer used for this stripes fabrication typically consists of an HP TIPS thermal inkjet drop ejection system with 18 nozzles, which accurately delivers 1–220 pL from each nozzle, a high-precision XY motorized stage, a CCD imaging system to view the drop generation, and a laser registration system. The printing time depends upon the complexity and fidelity of the patterns. Pd thiolates have been important precursors for producing not only Pd nanocrystallites but also Pd-rich sulfides [59].

Besides the mentioned methods, radiation-induced reduction of MOPs has also been tried out to prepare metal nanocrystallites. However, the obtained metal nanocrystallites embedded in carbon matrices possess no specific shapes [12–14,99].

In conclusion, this chapter provides a nearly comprehensive survey of the literature related to synthesis routes to metal crystallites using MOPs. The literature activity as presented here demonstrates the importance of this research area and its possible implications in industry.

REFERENCES

1. K.J. Kilpin, W. Henderson, and B.K. Nicholson, *Polyhedron*, **26**(2), 2007, 434–447.
2. H. Westmijze, H. Kleijn, and P. Vermeer, *J. Organomet. Chem.*, **172**(3), 1979, 377–383.
3. H. Schmidbaur and A. Bayler, *PATAI'S Chemistry of Functional Groups*, John Wiley & Sons Ltd., Hoboken, New Jersey, 2009.
4. H.T. Liu, X.G. Xiong, P. Diem Dau et al., *Nat. Commun.*, **4**, 2201, 2013.
5. D. Benitez, N.D. Shapiro, E. Tkatchouk et al., *Nat. Chem.*, **1**(6), 482–486, 2009.
6. S. Seršen, J. Kljun, F. Požgan, B. Štefane, and I. Turel, *Organometallics*, **32**(2), 609–616, 2013.
7. D.G.H. Hetterscheid, J.M.M. Smits, and B. de Bruin, *Organometallics*, **23**(18), 4236–4246, 2004.
8. M. Nakamoto, M. Yamamoto, and M. Fukusumi, *Chem. Commun.,* (15), 1622–1623, 2002.
9. C. Schliebe, K. Jiang, S. Schulze et al., *Chem. Commun.*, **49**(38), 3991–3993, 2013.
10. N.S. John, P.J. Thomas, and G.U. Kulkarni, *J. Phys. Chem. B*, **107**(41), 11376–11381, 2003.
11. B. Radha, and G.U. Kulkarni, *Cryst. Growth Des.*, **11**(1), 320–327, 2011.
12. B. Radha, and G.U. Kulkarni, *Adv. Funct. Mater.*, **22**(13), 2837–2845, 2012.
13. T. Bhuvana, and G.U. Kulkarni, *ACS Nano*, **2**(3), 457–462, 2008.
14. Y. Yin, and X. Xu, *Chem. Commun.*(8), 941–942, 1998.
15. G. Mettela, R. Boya, D. Singh et al., *Sci. Rep.*, **3**, 1793, 2013.
16. X. Lu, H.Y. Tuan, B.A. Korgel, and Y. Xia, *Chem. Eur. J.*, **14** (5), 1584–1591, 2008.
17. G. Mettela, M. Bhogra, U.V. Waghmare, and G.U. Kulkarni, *J. Am. Chem. Soc.*, **137** (8), 3024–3030, 2015.
18. S.K. Eah, *J. Mater. Chem.*, **21**, (42), 16866–16868, 2011.
19. Y. Yang, W. Wang, T. Chen, and Z.R. Chen, *ACS Appl. Mater. Interfaces*, **6** (23), 21468–21473, 2014.
20. N. Lidgi-Guigui, C. Dablemont, D. Veautier et al., *Adv. Mater.*, **19** (13), 1729–1733, 2007.
21. P.K. Jain, I.H. El-Sayed, M.A. El-Sayed, *Nano Today*, **2**(1), 18–29, 2007.
22. Y.Y. Yu, S.S. Chang, C.L. Lee, and C.R.C. Wang, *J. Phys. Chem. B*, **101** (34), 6661–6664, 1997.
23. Y. Xia, Y. Xiong, B. Lim, and S.E. Skrabalak, *Angew. Chem. Int. Ed.*, **48**(1), 60–103, 2009.
24. K.L. Kelly, E. Coronado, L.L. Zhao, and G.C. Schatz, *J. Phys. Chem. B*, **107** (3), 668–677, 2003.
25. S. Chen, and K. Kimura, *Langmuir*, **15**(**4**), 1075–1082, 1999.
26. C. Li, K.L. Shuford, Q.H. Park et al., *Angew. Chem.*, **119**(18), 3328–3332, 2007.
27. P. Pyykkö, *Angew. Chem. Int. Ed.*, **43**(34), 4412–4456, 2004.
28. P.G. Jones, and B. Ahrens, *New J. Chem.*, **22**(10), 1041–1042, 1998.
29. A. Bayler, A. Schier, and, H. Schmidbaur, *Inorg. Chem.*, **37**(17), 4353–4359, 1998.
30. M.M.H. Khalil, E.H. Ismail, F. El-Magdoub, *Arab. J. Chem.*, **5**(4), 431–437, 2012.
31. Y. Yuan, A.P. Kozlova, K. Asakura et al., *J. Catal.* **170**(1), 191–199, 1997.
32. A. Tuchscherer, D. Schaarschmidt, S. Schulze et al., *Dalton Trans.*, **41**(9), 2738–2746, 2012.
33. R.L. Zong, J. Zhou, Q. Li et al., *J. Phys. Chem. B*, **108**(43), 16713–16716, 2004.
34. C.N.R. Rao, G.U. Kulkarni, P.J. Thomas, and P.P. Edwards, *Chem. Soc. Rev.*, **29**(1), 27–35, 2000.
35. G. Kumari and C. Narayana, *J. Phys. Chem. Lett.*, **3**(9), 1130–1135, 2012.
36. M. Rycenga, P.H.C. Camargo, W. Li et al., *J. Phys. Chem. Lett.*, **1**(4), 696–703, 2010.

37. (a) X.Z. Lin, X. Teng, and H. Yang, *Langmuir*, **19**(24), 10081–10085, 2003. (b) M. Yamamoto and M. Nakamoto, *J. Mater. Chem.*, **13**(9), 2064–2065, 2003.
38. V.N. Richards, N.P. Rath, and W.E. Buhro, *Chem. Mater.*, **22**(11), 3556–3567, 2010.
39. C.W. Liu, Y.R. Lin, C.S. Fang et al., *Inorg. Chem.*, **52**(4), 2070–2077, 2013.
40. G. Rodriguez-Gattorno, D. Diaz, L. Rendon, and G.O. Hernandez-Segura, *J. Phys. Chem. B*, **106**(10), 2482–2487, 2002.
41. R. Narayanan, M.A. El-Sayed, *Nano Lett.*, **4**(7), 1343–1348, 2004.
42. N.M. Markovic, H.A. Gasteiger, and P.N. Ross, *J. Phys. Chem.*, **99**(11), 3411–3415, 1995.
43. C. Wang, H. Daimon, Y. Lee et al., *J. Am. Chem. Soc.*, **129**(22), 6974–6975, 2007.
44. C.M. Wu, S.F. Song, and K.M. Chi, *J. Nanosci. Nanotechnol.*, **10**(9), 5715–5722, 2010.
45. F. Su, L. Lv, F.Y. Lee et al., *J. Am. Chem. Soc.*, **129** (46), 14213–14223, 2007.
46. K. Kusada, H. Kobayashi, T. Yamamoto et al., *J. Am. Chem. Soc.*, **135**(15), 5493–5496, 2013.
47. T. Zhang, G.Y. Guo, *Phys. Rev. B*, **71**(21), 214–442, 2005.
48. K. Pelzer, O. Vidoni, K. Philippot et al., *Adv. Funct. Mater.*, **13**(2), 118–126, 2003.
49. C.N.R. Rao and K.P. Kalyanikutty, *Acc. Chem. Res.* **41**(4), 489–499, 2008.
50. M. Brust, M. Walker, D. Bethell et al., *J. Chem. Soc. Chem. Commun.*, (7), 801–802, 1994.
51. S. Kondo, K. Furukawa, and K. Tsuda, *J. Polym. Sci., Part A: Polym. Chem.*, **30**(7), 1503–1506, 1992.
52. (a) B. Radha, S. Kiruthika, and G.U. Kulkarni, *J. Am. Chem. Soc.*, **133**(32), 12706–12713, 2011. (b) S. Kiruthika, B. Radha, and G.U. Kulkarni (Unpublished work).
53. V.V. Agrawal, G.U. Kulkarni, and C.N.R. Rao, *J. Colloid Interface Sci.*, **318**(2), 501–506, 2008.
54. C.N.R. Rao, G.U. Kulkarni, P.J. Thomas et al., *J. Phys. Chem. B*, **107**(30), 7391–7395, 2003.
55. M. Anna, G.N. Albert, and I.K. Esko, *J. Phys. Condens. Matter*, **15**(42), S3011, 2003.
56. N. Varghese, and C.N.R. Rao, *Mater. Res. Bull.*, **46**(9), 1500–1503, 2011.
57. M. Yamamoto, and M. Nakamoto, *Chem. Lett.*, **32**(5), 452–453, 2003.
58. B. Radha and G.U. Kulkarni, *Nano Res.*, **3**(8), 537–544, 2010.
59. B. Radha and G.U. Kulkarni, *Adv. Funct. Mater.*, **20**(6), 879–884, 2010.
60. B. Radha and G.U. Kulkarni, *Curr. Sci.*, **102**(1), 70–77, 2012.
61. G. Mettela and G.U. Kulkarni, *Nano Res.*, 2015, DOI 10.1007/s12274-015-0797-8.
62. J.H. Joo, K.J. Greenberg, M. Baram et al., *Cryst. Growth Des.*, **13**(3), 986–991, 2013.
63. Y. Sun, Y. Yin, B.T. Mayers et al., *Chem. Mater.*, **14**(11), 4736–4745, 2002.
64. X. Kou, W. Ni, C.K. Tsung et al., *Small*, **3**(12), 2103–2113, 2007.
65. F. De Angelis, F. Gentile, F. Mecarini et al., *Nat. Photonics*, **5**(11), 682–687, 2011.
66. G. Mettela, S. Siddhanta, C. Narayana, and G.U. Kulkarni, *Nanoscale*, **6**(13), 7480–7488, 2014.
67. S.R. Hostler, Y.Q. Qu, M.T. Demko et al., *Superlattice. Microst.* **43**(3), 195–207, 2008.
68. K. Abe, T. Hanada, Y. Yoshida et al., *Thin Solid Films*, **327–329**, 524–527, 1998.
69. D.K. Lee and Y.S. Kang, *ETRI J.*, **26**(3), 252–256, 2004.
70. M. Cavicchioli, L.C. Varanda, A.C. Massabni, and P. Melnikov, *Mater. Lett.*, **59**(28), 3585–3589, 2005.
71. Y.B. Chen, L. Chen and L.M. Wu, *Inorg. Chem.*, **44**(26), 9817–9822, 2005.
72. B. Rodriguez-Gonzalez, A. Burrows, M. Watanabe et al., *J. Mater. Chem.*, **15**(17), 1755–1759, 2005.
73. J. Llorca, A. Casanovas, M. Domínguez et al., *J. Nanopart. Res.*, **10**(3), 537–542, 2008.
74. H. She, Y. Chen, X. Chen et al., *J. Mater. Chem.*, **22**(6), 2757–2765, 2012.
75. Y. Yoshida, K. Uto, M. Hattori, and M. Tsuji, *CrystEngComm*, **16**(25), 5672–5680, 2014.

76. R. He, Y.C. Wang, X. Wang et al., *Nat. Commun.*, **5**, 4327, 2014.
77. T.C. Deivaraj and J.Y. Lee, *J. Electrochem. Soc.*, **151**(11), A1832–A1835, 2004.
78. M.S. Wellons, Z. Gai, J. Shen et al., *J. Mater. Chem. C*, **1**(37), 5976–5980, 2013.
79. M.R. Decan, S. Impellizzeri, M.L. Marin and J.C. Scaiano, *Nat. Commun.*, **5**, 4612, 2014.
80. H. Guo, N. Lin, Y. Chen, Z. Wang et al., *Sci. Rep.*, **3**, 2323, 2013.
81. Y.H. Kim, Y.S. Kang, W.J. Lee et al., *Mol. Cryst. Liq. Cryst.*, **445**(**1**), 231/[521]–238/[528].
82. S.G. Kim, Y. Terashi, A. Purwanto, and K. Okuyama, *Colloids Surf. A Physicochem. Eng. Asp.*, **337**(1–3), 96–101, 2009.
83. Y.M. Lin, X. Sun, and M.S. Dresselhaus, *Phys. Rev. B*, **62**(**7**), 4610–4623, 2000.
84. J. Chen, L.M. Wu, and L. Chen, *Inorg. Chem.*, **46**(2), 586–591, 2007.
85. G. Carotenuto, C. Hison, F. Capezzuto et al., *J. Nanopart. Res.*, **11**(7), 1729–1738, 2009.
86. Y. Wang, J. Chen, L. Chen et al., *Cryst. Growth Des.*, **10**(4), 1578–1584, 2010.
87. F. Maury, *J. Phys. IV France*, **05**(C5), C5-449–C5-463, 1995.
88. E. Szłyk, P. Piszczek, I. Łakomska et al., *Chem. Vap. Depos.*, **6**(3), 105–108, 2000.
89. A. Turgambaeva, R. Parkhomenko, V. Aniskin et al., *Phys. Procedia*, **46**, 167–173, 2013.
90. A. Grodzicki, I. Łakomska, P. Piszczek et al., *Coord. Chem. Rev.*, **249**(21–22), 2232–2258, 2005.
91. A. Panneerselvam, M.A. Malik, P. O'Brien, and M. Helliwell, *Chem. Vap. Depos.*, **15**(1–3), 57–63, 2009.
92. D.A. Edwards, M.F. Mahon, K.C. Molloy, and V.J. Ogrodnik, *Mater. Chem.*, **13**(3), 563–570, 2003.
93. A. Panneerselvam, M.A. Malik, P. O'Brien, and J. Raftery, *J. Mater. Chem.*, **18**(27), 3264–3269, 2008.
94. F.H. Brain and C.S. Gibson, *J. Chem. Soc.* (Resumed), 762–767, 1939.
95. P. Römbke, A. Schier, H. Schmidbaur et al., *Inorg. Chim. Acta*, **357**(1), 235–242, 2004.
96. C. Schoner, A. Tuchscherer, T. Blaudeck et al., *Thin Solid Films*, **531**, 147–151, 2013.
97. B. Radha, S.H. Lim, M.S.M. Saifullah, and G.U. Kulkarni, *Sci. Rep.*, **3**, 1078, 2013.
98. T. Bhuvana, W. Boley, B. Radha et al., *Micro Nano Lett.*, **5**(5), 296–299, 2010.
99. N. Kurra, S. Kiruthika, and G.U. Kulkarni, *RSC Adv.*, **4**(39), 20281–20289, 2014.

Section IV

Devices

10 Molecular Materials for Organic Field-Effect Transistors

Takehiko Mori

CONTENTS

10.1 INTRODUCTION

Organic electronic and photonic devices are attracting a great deal of attention due to their potential application to large-area, low-cost, and flexible electronic devices [1]. Together with organic light-emitting diodes and organic solar cells, organic field-effect transistors are recognized as one of the three major organic electronic devices [2]. In the former two, however, the current flows in the longitudinal direction of the device, namely, in the thickness direction, but in organic field-effect transistors, the current flows in the lateral direction. The former two stem from Tang's benchmark works in 1987 [3,4], while organic field-effect transistors were invented by Japanese researchers in the 1980s [5,6]. Amorphous or scrambled structure is essential in the former two, whereas highly crystalline molecular arrangement is desirable in organic transistors, and even organic single-crystal transistors have been investigated. Organic light-emitting diodes convert electric energy to light, whereas organic solar cells transform light to electric energy, so, roughly speaking, organic solar cells have the opposite function to organic light-emitting diodes. These aspects demonstrate that organic field-effect transistors are electronic devices that are very different from the other two. In addition to the practical applications, many physicists are interested in organic field-effect transistors in connection with graphene transistors and as a fundamental technique to realize quantum Hall effect. In this chapter, we first provide fundamentals of organic transistors, including the device physics and the materials chemistry. Subsequently, we describe the recent progress in ambipolar materials, the use of organic metals, and the analysis of temperature-dependent transistor characteristics.

10.2 OPERATION OF ORGANIC TRANSISTORS

The structure of a typical pentacene transistor is depicted in Figure 10.1a. To investigate materials' properties, it is convenient to use a silicon wafer as a substrate. A highly doped and highly conducting silicon wafer with thermally oxidized SiO_2 layer is commercially available and used as the gate electrode. On the SiO_2 layer, for instance, pentacene is vacuum evaporated, and gold source and drain (S/D) electrodes are evaporated through a metal mask.

Usually, the source level is assigned to the ground level, from which the gate voltage, V_G, and the drain voltage, V_D, are defined. When V_G is applied, both sides of the SiO_2 layer are regarded as a capacitor, where the pentacene layer and the S/D electrodes are recognized as a combined conductor. If a negative voltage is applied to V_G, positive charges are induced on the interface between the pentacene and SiO_2 layers due to the holes injected from the source electrode. Then, when V_D is applied, the pentacene layer is conductive, and finite drain current I_D is observed. To achieve the hole conduction, we apply negative V_D, with the same polarity as V_G, to induce hole flow from S to D. When V_G is turned off, the device is no more a capacitor, and I_D drops zero. Then we can switch I_D by V_G.

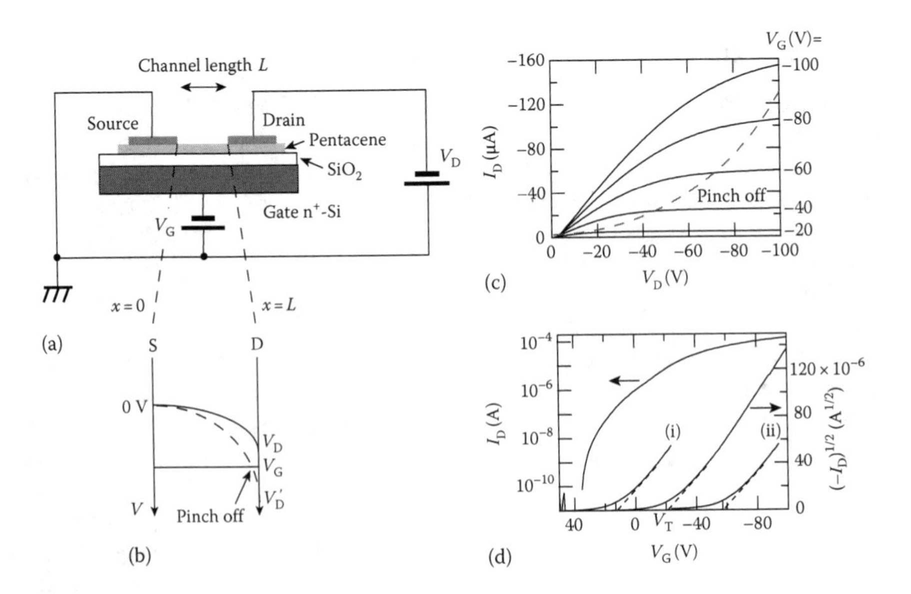

FIGURE 10.1 (a) Structure of a typical pentacene transistor and (b) the voltage distribution. (c) Output and (d) transfer characteristics of the resulting transistor.

In order to quantitatively estimate I_D, we have to consider that the channel voltage $V(x)$ drops gradually from the source ($x = 0$) to the drain ($x = L$) (Figure 10.1b). In this gradual channel approximation [7], the accumulated charge $Q(x)$ per area depends on the channel position as

$$Q(x) = C\left(V_G - V(x)\right) \tag{10.1}$$

where C is the capacitance per area. Conductivity is generally represented by $\sigma = ne\mu$, in which n is the carrier density, e is the elemental charge, and μ is the mobility. Here $Q(x) = ne$ and the electric field is $E = -dV(x)/dx$, so I_D is obtained from $I_D/WE = \sigma$ as

$$I_D = WQ(x)\mu\left(-\frac{dV(x)}{dx}\right) \tag{10.2}$$

where W is the channel width. Inserting Equation 10.1 into Equation 10.2, and integrating from $x = 0$ to $x = L$, we obtain

$$\int_0^L I_D\,dx = \int_0^{V_D} WC\mu\left(V_G - V(x)\right)dV. \tag{10.3}$$

Since I_D is constant from $x = 0$ to $x = L$, the integration affords

$$I_D = C\mu\frac{W}{L}\left[V_G V_D - \frac{1}{2}V_D^2\right]. \tag{10.4}$$

When I_D is plotted as a function of V_D for a given V_G (output characteristics), we obtain an upward parabola as shown in Figure 10.1c. In contrast to this linear region, when V_D exceeds V_G, the region near the drain is no more a capacitor and no charge accumulation occurs; this is called pinch off (Figure 10.1b). In this saturated region, the integration in the right-hand side of Equation 10.3 is limited by V_G to lead to

$$I_D = C\mu\frac{W}{2L}V_G^2. \tag{10.5}$$

Therefore, the output characteristics are constant with respect to V_D (Figure 10.1c). When I_D is plotted as a function of V_G for a relatively large constant V_D, we obtain the transfer characteristics as shown in Figure 10.1d. From Equation 10.5, square root I_D is proportional to V_G, so we can extract μ from the slope of such a plot. Note that we can use Equation 10.5 only in the saturated region at $V_G < V_D$. When V_G exceeds V_D, Equation 10.5 is not valid, and the transfer characteristics become comparatively flat [8].

Since the extrapolation does not cross zero exactly at $V_G = 0$ V, we have to consider the threshold voltage V_T, and Equations 10.4 and 10.5 are replaced by

$$I_{\mathrm{D}} = C\mu\frac{W}{L}\left[(V_{\mathrm{G}} - V_{\mathrm{T}})V_{\mathrm{D}} - \frac{1}{2}V_{\mathrm{D}}^{2}\right] \quad \text{(linear region)} \tag{10.6}$$

$$I_{\mathrm{D}} = C\mu\frac{W}{2L}(V_{\mathrm{G}} - V_{\mathrm{T}})^{2} \quad \text{(saturated region)} \tag{10.7}$$

The usual dielectric interface has somewhat surface charge $Q_s = CV_T$, so V_T is V_G, at which the original charge is eliminated. When holes exist even at $V_G = 0$ V ((i) in Figure 10.1d), we obtain a normally-on characteristics, but when there are considerable traps, I_D does not start to increase up to a large V_G, as shown in (ii) in Figure 10.1d.

The charge accumulation occurs due to the band bending φ, as shown in Figure 10.2 [8]. This figure is depicted for an n-channel transistor, and the Fermi energy E_F is located slightly below the conduction band edge E_c, where $E_c - E_F$ is typically 0.3–0.5 eV. When positive V_G is applied, the gate potential drops, and at the interface, E_c approaches to E_F. As a result, considerable charge carriers are accumulated at the interface. V_G is as large as 100 V, and the potential distribution in the channel $V(x)$ and V_D are in the same order [9], while the band bending φ is at most in the order of 0.5 eV. Such large V_G is necessary to accumulate $Q = CV_G$, because C is small.

It should be pointed out that the usual textbooks of device physics deal with single-crystal field-effect transistors of silicon, in which source/channel/drain are composed of n/p/n junctions [7]. For any V_D, the source (n/p) or the drain (p/n) is inversely biased, and the device is off. However, when we apply positive V_G, electrons are accumulated at the interface of the p-type channel, and this inversion layer conducts electrons from the n-channel source to the drain. Discussion on the formation of the inversion layer is not necessary in organic transistors as well as other thin-film transistors such as amorphous silicon transistors. In addition, in single-crystal transistors, the bulk semiconductor level is grounded, whereas in thin-film transistors the opposite end of the thin film is open, and the voltage is floating [10]. This considerably reduces the band bending and simplifies the analysis of the transistor operation [8].

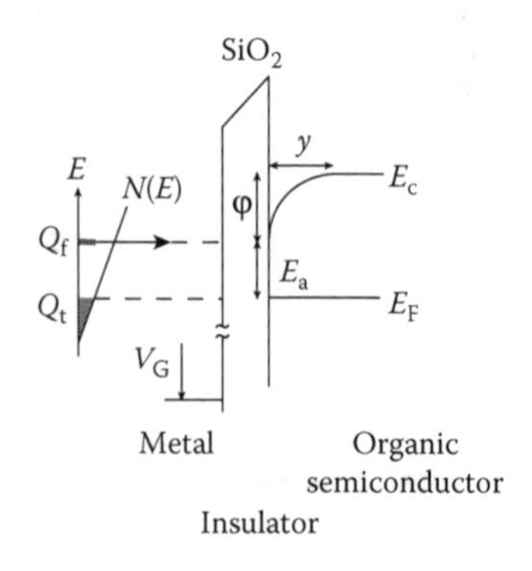

FIGURE 10.2 Band bending at the gate dielectric interface in an n-channel transistor.

10.3 MATERIALS

In order to conduct electricity, an organic molecule has to carry negative or positive charges even temporarily. A molecule carrying a negative charge is an anion with an extra electron, and an electron acceptor molecule receives such an electron. An electron acceptor is an easily reducible molecule, which is achieved in a molecule with a large electron affinity, or in a molecule with a low LUMO (lowest unoccupied molecular orbital) level (Figure 10.3). By contrast, a molecule carrying a positive charge is a cation with a hole, which is realized in an electron donor molecule. An electron donor is an easily oxidizable molecule, which is achieved in a molecule with a small ionization potential, or in a molecule with a high HOMO (highest occupied molecular orbital) level (Figure 10.3).

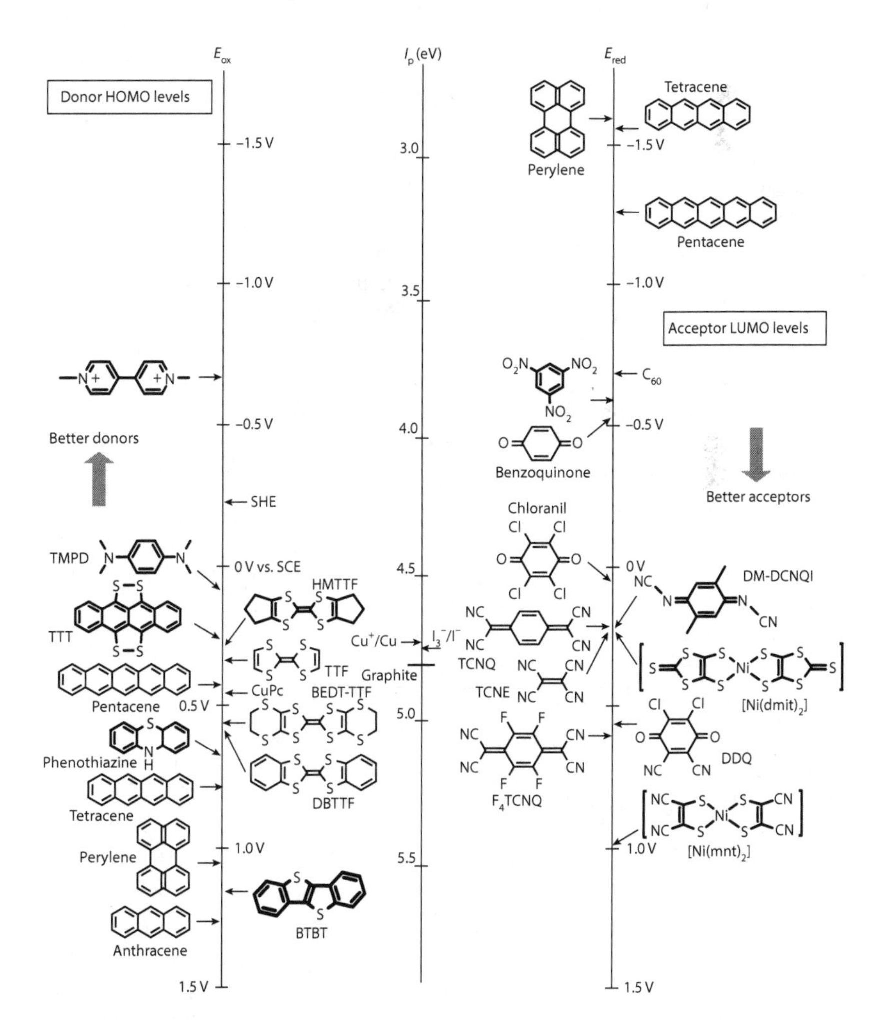

FIGURE 10.3 Redox potentials of representative organic donors and acceptors. (From Mori, T., *Electronic Properties of Organic Conductors*, Springer, Tokyo, Japan, 2016, p. 258.)

Figure 10.3 shows redox potentials of representative organic donors and acceptors that constitute organic conductors. The redox potentials are easily measured by cyclic voltammetry and represented by a potential versus a reference electrode; a conventional reference electrode is the standard calomer electrode (SCE), which is by 0.241 V more positive than the standard hydrogen electrode (SHE), used in thermodynamics. Another popular reference electrode Ag/AgCl is also by 0.222 V more positive than the SHE, but $Ag/AgNO_3$ is by 0.3–0.35 V more positive than the SCE. Since the potentials of these reference electrodes are not very certain, ferrocene (0.38 V vs. SCE) is sometimes used as a reference compound. These redox potentials are converted by

$$I_p = E_{redox}\ (\text{vs. SCE}) + 4.44\,\text{V} = E_{redox}\ (\text{vs. SHE}) + 4.2\ \text{V} \tag{10.8}$$

to energy levels or ionization potentials I_p, obtained from the photoelectron spectroscopy of solids [11]. Alternatively, E_{redox} is converted to I_p by assuming the energy level of ferrocene to be 4.8 eV [12]. The scale of I_p is depicted at the center of Figure 10.3.

Sometimes, we can measure only one of E_{ox} and E_{red}, when another is out of the electrochemical window. We can, however, estimate the HOMO–LUMO gap, $E_g = E_{LUMO} - E_{HOMO}$ from the absorption edge of the visible spectrum by using $E_g = (1240\ \text{nm eV})/\lambda$. In such a case, we can evaluate another energy level.

It has been recognized that organic molecules whose HOMO is located above 5.6 eV show p-channel transistor properties, whereas those with the LUMO level below 3.2 eV show n-channel conduction [12]. Here, we use the terms *p-channel* and *n-channel*, because organic semiconductors are generally used as an intrinsic form, namely, as a single-component material, so they are different from the doped p-type and n-type silicon. The transistor operation largely depends on the S/D metals from

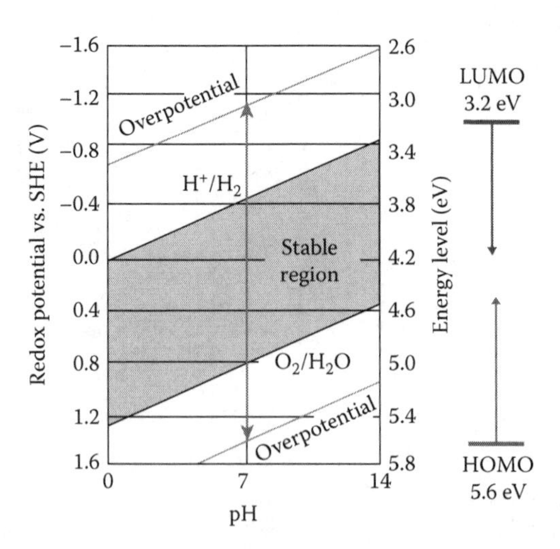

FIGURE 10.4 Redox potentials of water. (From Schriver, D.F. and Atkins, P.W., *Inorganic Chemistry*, 4th edn., Oxford University Press, 2006, Chapter 5.)

which the carriers are injected. This empirical relation is deduced from the devices with Au electrodes, where the work function of Au is located at 5.1 eV. This is rather close to the p-channel limit (5.6 eV), so the aforementioned limit is not mainly determined by the injection barriers.

Then we shall consider reaction with water (Figure 10.4) [13]. From the definition of SHE, H^+ is reduced to H_2 for $E_{redox} < 0$ V versus SHE. On the other hand, H_2O is oxidized to O_2 for $E_{redox} > 1.23$ V versus SHE. Accordingly, the region $0 \text{ V} < E_{redox} < 1.23$ V is called the stability field of water. These redox potentials are for pH = 1, but from the Nernst equation,

$$E = -\frac{\Delta G}{nF} = E^0 - \frac{RT}{nF}\ln K = E^0 + \frac{RT}{F}\ln\left[H^+\right] = E^0 - \left(0.059\text{V}\right)\text{pH}, \quad (10.9)$$

the potential is shifted to the negative direction by 0.413 V at pH = 7. Usually, we have to consider a somewhat overpotential (~0.6 V), so the resulting stability field is $-1.0 \text{ V} < E_{redox} < 1.4$ V, which corresponds to the energy level of $3.2 \text{ eV} < E < 5.6$ eV. Even when the transistor properties are measured in vacuum, the organic semiconductors have contact with SiO_2, and SiO_2 has many OH moieties and works like water. Accordingly, the operation of the material is limited by the reaction with water.

10.3.1 p-Channel Materials

Progress of mobility in p-channel organic transistors is shown in Figure 10.5, and the representative p-channel materials are depicted in Figure 10.6 [14,15]. The first organic transistors are fabricated by using such materials as merocyanine [5] and polythiophene [6], but in the 1990s considerable efforts have been devoted to oligothiophenes [16]. Oligothiophenes such as 6T (Figure 10.6) are vacuum evaporated to form well-ordered thin films and realize the mobility of the order of 0.01 cm²/V s [16]. Introduction of alkyl chains like DH6T improves the morphology, and the mobility

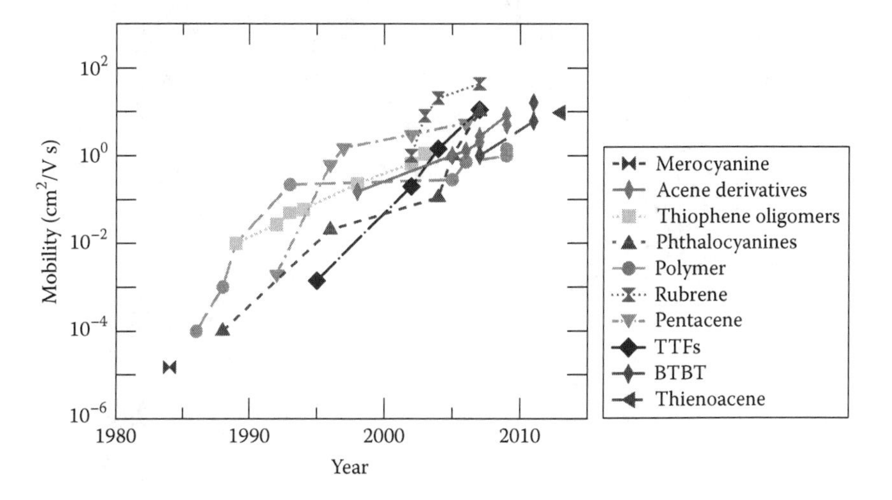

FIGURE 10.5 Progress of mobility in p-channel organic transistors.

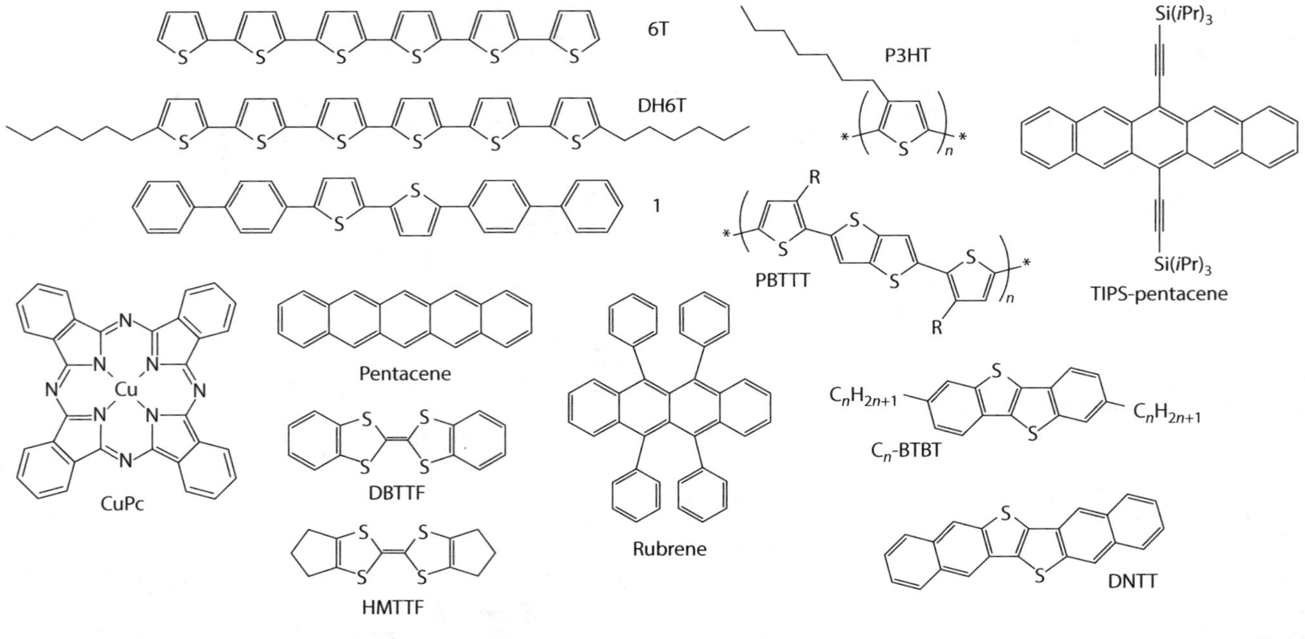

FIGURE 10.6 Representative p-channel materials. 6T, sexithiophene; DH6T, dihexylsexi-thiophene; CuPc, copper phthalocyanine; DBTTF, dibenzotetrathiafulvalene; HMTTF, hexamethylenetetrathiafulvalene; P3HT, poly-3-hexylthiopene; PBTTT, poly[2,5-bis(3-alkylthiophen-2-yl)thieno[3,2-b]thiophene; TIPS-pentacene, tri(isopropyl)silylethylylpentacene; C_n-BTBT, dialkylbenzothienobenzothiophene; and DNTT, dinaphthothienothiophene.

exceeding 0.1 cm^2/V s is achieved [17,18]. Co-oligomers of thiophene and phenylene (such as Compound 1 in Figure 10.6) have been extensively investigated, where many combinations are possible [19]. The phenyl groups reduce the donor ability and improve the device stability. Phthalocyanine is another conventional material [20], but the standard material in the 2000s is pentacene, in which mobility exceeding 1 cm^2/V s has been reported [21]. Pentacene is not soluble in organic solvents, and pentacene thin film is made by vacuum evaporation. Generally, these oligothiophenes and pentacene tend to have a herringbone structure [22]. Tetrathiafulvalene (TTF) is a representative electron donor and also shows p-channel transistor properties [15,23–26], among which DBTTF (Figure 10.6) is most investigated. Ordinary DBTTF (α-phase) has a herringbone structure with a large dihedral angle (132°), but there is a herringbone phase (β-phase) with an ordinary dihedral angle (52°) [22,27]. HMTTF exhibits better performance exceeding 1 cm^2/V s [15,28,29]. Since these TTF derivatives, as well as pentacene, are strong electron donors with the ionization energy smaller than 5.0 eV, it is somewhat susceptible to air oxidation, and carriers are slowly generated in the ambient conditions. In order to achieve more stable device operation, it is desirable to use materials with the ionization energy larger than 5.1 eV. However, the device stability is not solely determined by the ionization energy, and even the TTF derivatives show stable operation when substituted by bulky *t*-butyl groups [30–32]. In single-crystal rubrene, very high mobility of 40 cm^2/V s has been reported [33]. Similarly, single-crystal TTF shows excellent transistor performance with the mobility exceeding 1 cm^2/V s [34], but it is not easy to obtain the thin-film transistors by vacuum evaporation because of the small molecular weight and the resulting large vapor pressure [35].

For the solution process, soluble polythiophene such as P3HT (poly-3-hexylthiopene) is the conventional material, where the solubility is attained by the introduction of hexyl groups. The mobility is usually below 0.1 cm^2/V s [36], but recently improved values 0.2–0.6 cm^2/V s have been achieved in PBTTT [37]. Among small molecules, TIPS-pentacene is the well-known solution-process material [38]. Alkyl-BTBT is an excellent material [39–41]; first, the vacuum evaporated films are investigated [39], but highly crystalline films of C_n-BTBT have been obtained from the solution process, and mobility exceeding 10 cm^2/V s has been realized even in the solution process [42,43]. C_n-BTBT is a weak electron donor with an ionization potential of 5.6 eV, which is much larger than 4.85 eV in pentacene. This is a key to afford the long-term device stability in C_n-BTBT transistors. The naphthalene derivative DNTT is an excellent transistor material as well [44]. The monothieno derivative also shows excellent transistor performance (9.5 cm^2/V s) [45].

10.3.2 n-Channel Materials

The number of n-channel organic semiconductors (Figure 10.7) is limited in comparison with p-channel materials, and many are operated only in vacuum [46,47]. This is reasonable when we consider the relative stability of carboanion in comparison with carbocation. Perfluorophthalocyanine and oligothiophenes with perfluoroalkyl and perfluorophenyl groups show n-channel properties [48–50], and perfluoropentacene is another member of n-channel semiconductors [51]. Other

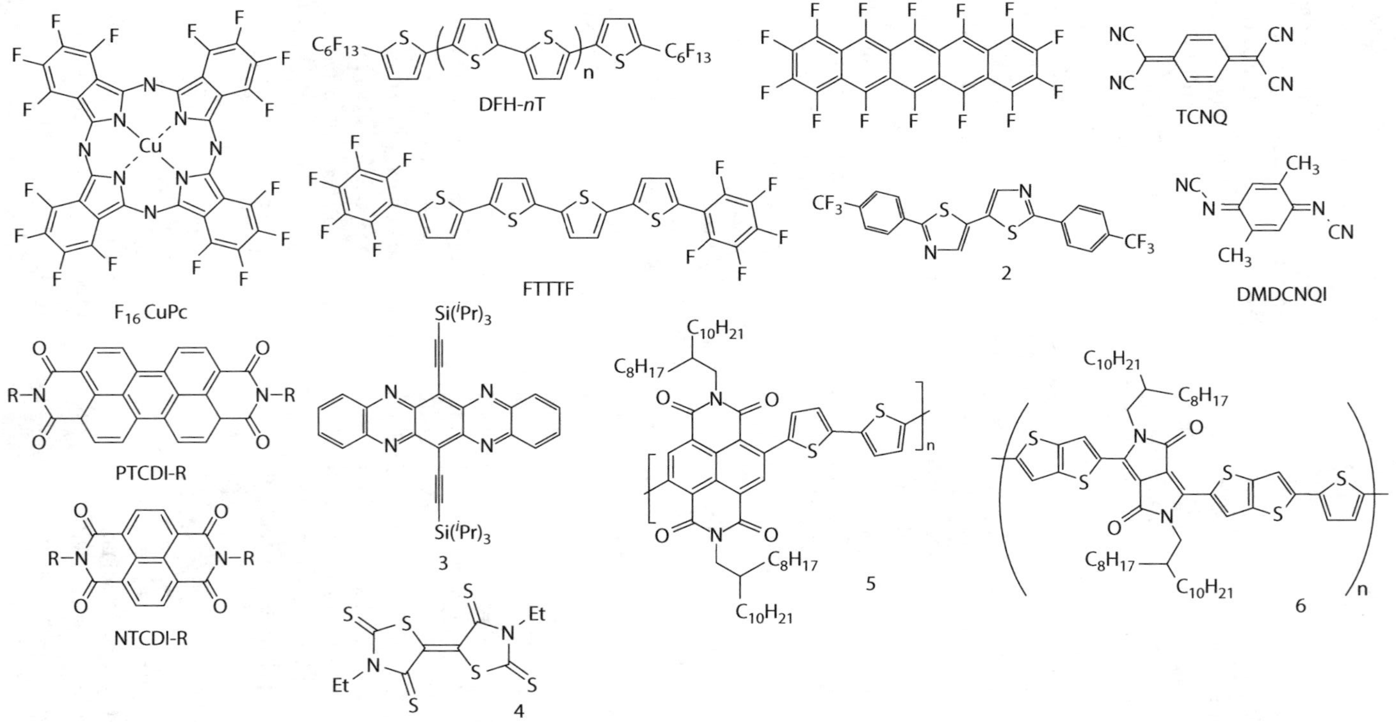

FIGURE 10.7 Representative n-channel materials. PTCDI, perylene tetracarboxylic diimides; NTCDI, naphthalene tetracarboxylic diimide; TCNQ, tetracyanoquinodimethane; and DMDCNQI, dimethyl dicyanoquinonediimine.

than perfluoro derivatives, thiazole is another source of an electronegative unit (Compound 2) [52,53]. C_{60} is a widely used n-channel organic semiconductor; an evaporated C_{60} transistor attains the mobility around 0.3 cm^2/V s [54]. Many derivatives of PTCDI and NTCDI have been extensively studied [47], among which mobility of 6.2 cm^2/V s has been reported in cyclohexyl NTCDI [55]. These molecules are compounds with wide π-framework, and particularly cyclohexyl NTCDI has a brickwork structure, where instead of the usual stacking structure, two molecules are equivalently located on the top of a molecule. Other NTCDI and PTCDI derivatives have a variety of structures in between the brickwork and stacking structures, but the mobility takes a maximum in the cyclohexyl derivative at the ideal brickwork structure [56]. Pyrazine-substituted TIPS-pentacene derivatives (e.g., Compound 3) show excellent n-channel characteristics exceeding 1 cm^2/V s [57–60]. Compound 4 is an electron acceptor, which realizes the single-crystal mobility of 0.22 cm^2/V s in air [61]. Excellent transistor performance has been reported in thiophene copolymers containing NTCDI (Compound 4, 0.45–0.85 cm^2/V s) and diketopyrrolopyrrole (Compound 5, 1.95 cm^2/V s) [62,63].

Some organic donors and acceptors that constitute organic conductors (Figure 10.3) are good transistor materials as well [23]. Single crystals of TCNQ show excellent transistor performance with the mobility exceeding 1 cm^2/V s [64–65], but it is not easy to obtain the thin-film transistors by vacuum evaporation because of the small molecular weight and the resulting large vapor pressure [35]. However, we can sometimes obtain the thin-film transistors by low-vacuum evaporation [35]. In particular, DMDCNQI makes air-stable n-channel transistors [35,66].

10.3.3 Ambipolar Materials

Ambipolar semiconductors are materials in which both electron and hole conduction is possible. Since this is a newly progressing field [67,68], here we describe the recent status of ambipolar semiconductors. Quinoidal oligothiophens (Compounds 7 and 8 in Figure 10.8) are materials that have been studied for a long time. The trimer (Compound 7) shows usually only electron transport (μ_e = 0.2 cm^2/V s), but when deposited above the substrate temperature of 230°C, ambipolar transport is observed, though the performance drops by several orders of magnitude (μ_h/μ_e = $10^{-4}/10^{-4}$ cm^2/V s) [69]. By contrast, the tetramer exhibits predominantly hole transport (μ_h/μ_e = 0.1/0.006 cm^2/V s) [70–73]. Upon oxygen exposure, C_{60} has been reported to show ambipolar transport, though the mobility lowers by several orders [74–76]. PCBM (methanofullerene [6,6]-phenyl-C_{61}-butyric acid methyl ester) and copper phthalocyanine are also reported to show ambipolar transport after annealing [77–80].

Recently, high-performance ambipolar transistors have been realized in polymers consisting of donor (D) and acceptor (A) parts [67,68,80–83]. A polymer consisting of thiophene (D) and diketopyrrolopyrrole (A: DPP) (Compound 9) shows the hole and electron field-effect mobilities of μ_h/μ_e = 0.9/0.1 cm^2/V s [84]. A polymer of thienothiophene and DPP (Compound 10) shows μ_h/μ_e = 1.36/1.56 cm^2/V s [85]. Several other D-A type polymers have been recently reported to show excellent ambipolar properties [80,81].

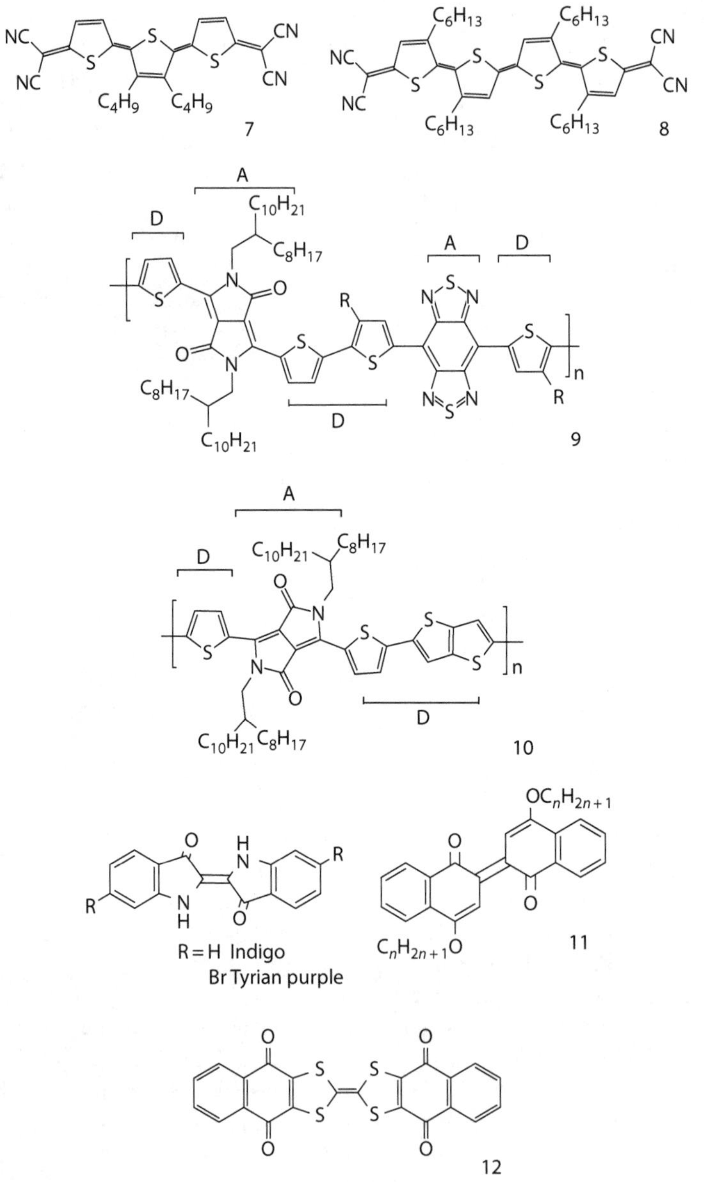

FIGURE 10.8 Representative ambipolar transistor materials.

Sometimes, self-assembled-monolayer (SAM) treatment is crucial to convert the carrier polarity [86,87]. In particular, it has been recently reported that tetratetracontane (TTC, $C_{44}H_{90}$) makes an extremely inert surface and is an excellent passivation material to achieve ambipolar transport [88]. For example, Compound 7 exhibits much improved and balanced ambipolar performance of μ_h/μ_e = 0.3/0.6 cm^2/V s on TTC [89]. When fabricated on TTC, materials such as CuPc, pentacene, TIPS-pentacene, and 6T, which usually show only p-channel properties, show ambipolar properties (Figure 10.9) [90]. For CuPc with a small HOMO–LUMO gap, ambipolar transport is achieved with Au electrodes, while for other ordinary p-channel materials, Al or Ag electrodes are necessary. Pentacene also shows ambipolar transport when Ca is used as the electrodes [91], and much improved ambipolar transport (μ_h/μ_e = 3/3 cm^2/V s) when Al electrodes are used together with PMMA (poly(methyl methacrylate)) treatment [92]. The energy levels of these materials are depicted in Figure 10.9, where approximately the energy levels have to be within the stability field of water. Since the interval of the stability field is 5.6 − 3.2 = 2.4 eV, the HOMO–LUMO gap has to be smaller than this value. Such a material has an optical absorption edge at a considerably long wavelength and should have a blue or violet color.

Along this line, it has been recently reported that indigo (Figure 10.8), a dye used in blue jeans, shows ambipolar transistor properties (μ_h/μ_e = 0.01/0.01 cm^2/V s) [88]. Dibromoindigo (Figure 10.8), which is another dye known as Tyrian purple from the ancient times, shows ambipolar properties (μ_h/μ_e = 0.4/0.4 cm^2/V s) [93,94]. Other halogenated and phenyl indigo derivatives also show ambipolar properties [95–97]. These molecules are of interest because of the minimal molecular structure consisting of electron-donating nitrogen atoms working as a donor and electron-withdrawing carbonyl groups working as an acceptor [98]. Compound 11 is a semiquinone with deep-blue color due to the small HOMO–LUMO gap and shows ambipolar transistor

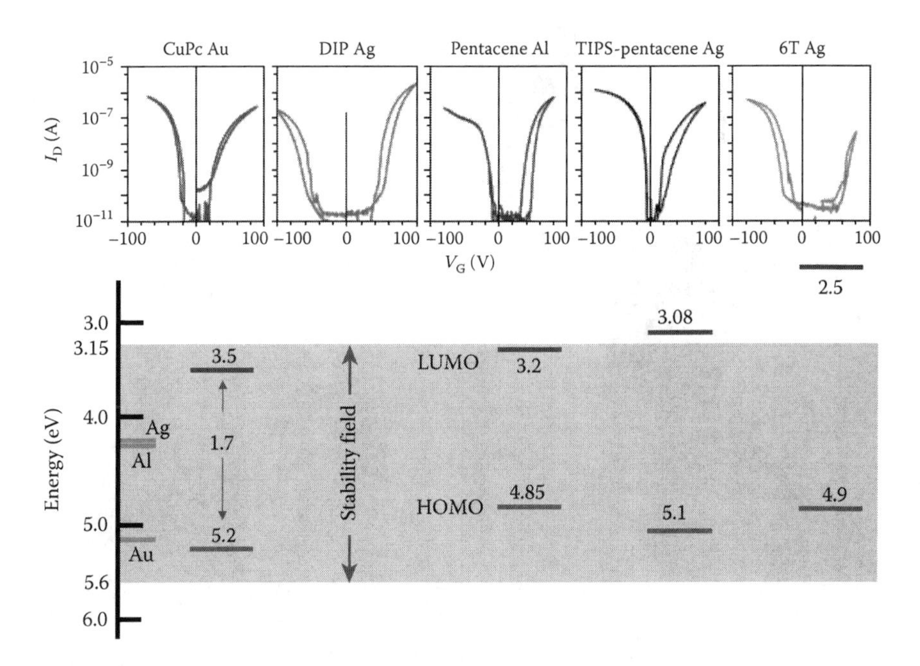

FIGURE 10.9 Ambipolar transfer characteristics of p-channel materials fabricated on TTC [90], together with the HOMO and LUMO levels.

properties [99]. Diketopyrrolopyrrole (DPP) is a representative "acceptor" unit in D-A polymers (A in Compounds 9 and 10), but diphenyl-DPP shows ambipolar transistor characteristics [100]. Ambipolar transistors are also realized on the basis of TTF derivatives fused with benzoquinone (Compound 12) and PTCDI [101,102].

10.3.4 Operation of Ambipolar Transistors

In ambipolar transistors, holes are injected from the source, and electrons are injected from the drain electrode at the same time. This strict ambipolar operation, however, occurs in a limited V_G and V_D region [67,68]. Here, we provide a few remarks on the operation of ambipolar transistors [96].

When the threshold voltages for electrons V_T and holes V_T' are nonzero, operation regions of ambipolar transistors are divided, as shown in Figure 10.10. This figure is depicted for $V_T' < 0 < V_T$, but V_T and V_T' may have the same sign; in this case, the whole figure is translated horizontally. The transistor characteristics are obtained assuming the gradual channel approximation [96]. When we follow the line (i) from right to left, the linear region (L) in $V_G - V_T > V_D$,

$$I_D = \frac{W\mu_e C}{L}\left((V_G - V_T)V_D - \frac{1}{2}V_D^2\right) \quad (V_D + V_T < V_G) \tag{10.10}$$

is followed by the unipolar electron-transporting saturated region (S)

$$I_D = \frac{W\mu_e C}{2L}(V_G - V_T)^2 \quad (V_D + V_T' < V_G < V_D + V_T) \tag{10.11}$$

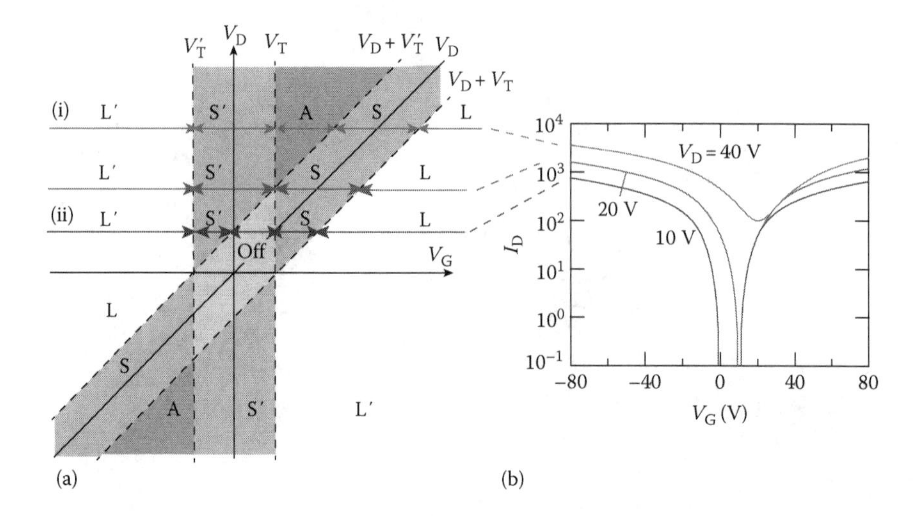

FIGURE 10.10 (a) Operation regions of ambipolar transistors considering nonzero threshold voltages. (b) Typical transfer characteristics for $V_T = 10$ V and $V_T' = -10$ V with $W\mu_e C/L = W\mu_e C/L = 1.0$.

at $V_G = V_D + V_T$. An ambipolar region (A) appears below $V_G = V_D + V'_T$,

$$I_D = \frac{W\mu_e C}{2L}(V_G - V_T)^2 + \frac{W\mu_h C}{2L}(V_D - V_G + V'_T)^2 \quad (V_T < V_G < V_D + V'_T), \tag{10.12}$$

which is certainly continuous to Equation 10.11 at $V_G = V_D + V'_T$. This is connected to the unipolar hole-transporting saturated region (S′) below $V_G < V_T$:

$$I_D = \frac{W\mu_h C}{2L}(V_D - V_G + V'_T)^2 \quad (V'_T < V_G < V_T) \tag{10.13}$$

where the first term of Equation 10.12 vanishes. Finally, a reversed hole-transporting linear region (L′) appears below $V_{GS} < V'_{TH}$:

$$I_D = \frac{W\mu_h C}{L}\left(\frac{1}{2}V_D^2 - (V_G - V'_T)V_D\right) \quad (V_G < V'_T), \tag{10.14}$$

which is continuous to Equation 10.13 at $V_G = V'_T$. Regions appear successively as L → S → A → S′ → L′, in which I_D follows V_{GS}^2 in the central S, A, and S′ regions, whereas I_D depends on V_G linear in the L and L′ regions. A typical transfer characteristics following this line is depicted in Figure 10.10b. When the A region appears, I_D does not drop zero even at the center of the A region.

Below $V_D < V_T - V'_T$, an off state appears, where I_D drops zero. When we trace the horizontal line (ii) in Figure 10.10a, regions appear successively as L →S → Off → S′ → L′. A transfer characteristics following this line is depicted in Figure 10.10b. Similar relations are obtained for hole transport, and the lower half of Figure 10.10a has a similar shape to the upper half. The off region is represented by the rhombic domain in Figure 10.10a, and the ambipolar region appears only above $V_D > V_T - V'_T$; this is the reason that ambipolar characteristics are observed only at comparatively large V_D in the actual ambipolar transistors. The off region appears in $V'_T < V_G < V_T$ at $V_D = 0$ V, but transistor characteristics are not observed at this $V_D = 0$ V limit, and the actual $I_D = 0$ region is always smaller than $V_T - V'_T$. The ambipolar region is sandwiched by the unipolar saturated regions, whose widths are again $V_T - V'_T$. It is therefore appropriate to extract the mobility from the formula of the unipolar saturated regions (Equations 10.11 and 10.13). Even in the conventional unipolar transistors, we should pay attention for V_G not to exceed V_D in order to keep the transfer characteristics in the saturated region [8]. In contrast, $V_T - V'_T$ is larger than V_D in many ambipolar transistors. It is also noteworthy that both the electron and hole sides have the same importance whatever is the V_D sign. We should be cautious to extract V_T and V'_T from the experimental data, because the apparent threshold voltages extracted from the transfer characteristics are dependent on V_D and different from the intrinsic threshold voltages, V_T and V'_T.

10.4 TRANSISTORS USING ORGANIC METALS

In order to achieve maximum performance of materials, device structure is an important factor. As described previously, it is not desirable that the organic semiconductor films are directly in contact with SiO_2, so it is usual to cover the SiO_2 surface by a self-assembled monolayer (SAM) using a silane coupling reagent. Sometimes, polymers such as PMMA, polystyrene, and CYTOP are used instead of SAMs [65]. TTC makes an extremely inert surface that is desirable to realize the ambipolar transport [90]. As an opposite end, it has been attempted to control the threshold voltage by using polar SAMs [103].

In Figure 10.1a, the S/D electrodes are located on the organic film, which is called the top-contact geometry. It seems natural to deposit organic semiconductors on the S/D electrodes; such a geometry is called bottom contact, but it has been known that bottom contact realizes by orders of magnitude lower performance than the top-contact geometry [104]. One of the reasons is ascribed to the morphological discontinuity (Figure 10.11); on the usual SiO_2 or SAMs, pentacene shows a characteristic dendritic pattern where the molecules are standing, whereas small domains are observed on Au, in which the face-on geometry is achieved [105]. Since pentacene conducts along the π-stacking, the face-on geometry is highly resistive and increases the contact resistance. Another reason is the interfacial potential generated on the metal/organic interface [106–109]. When an organic semiconductor, for instance, pentacene, is deposited on a metal, pentacene on the first monolayer is positively charged due to the donor ability. Then the interface polarizes, and the potential of the pentacene drops. This is reflected to the shift of the vacuum level Δ (Figure 10.11d). Although the level of pentacene (4.85 eV) is originally very close to the work function of Au (5.1 eV), the hole injection (Schottky) barrier φ_B increases by an amount of Δ. Usually, acceptors are charged negatively and increase the electron injection barrier. Accordingly, interfacial potential generally works to increase the carrier injection barrier.

In order to avoid the large contact resistance in the bottom-contact geometry, we have used conducting carbon as the S/D materials, expecting small interfacial dipole in carbon/organic interface [110]. Carbon electrodes in organic transistors have been attempted for chemical vapor deposition (CVD) carbon films [111], but we have used a dispersed carbon solution. We can achieve the patterning using the surface selective deposition (Figure 10.12a) [112], where self-assembled monolayers (SAMs) are selectively removed by irradiating ultraviolet light through a metal mask, and carbon

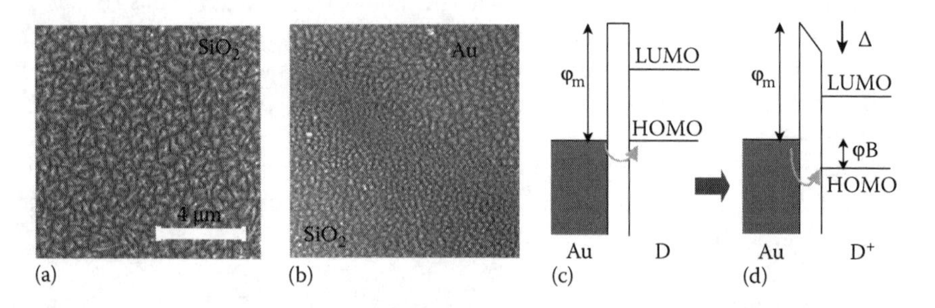

FIGURE 10.11 Atomic force microscope images of pentacene on (a) SiO_2 and (b) Au. [130] (c) Interfacial energy diagram of Au/pentacene, (d) after the contact, the hole injection barrier φ_B increases due to the potential energy shift Δ.

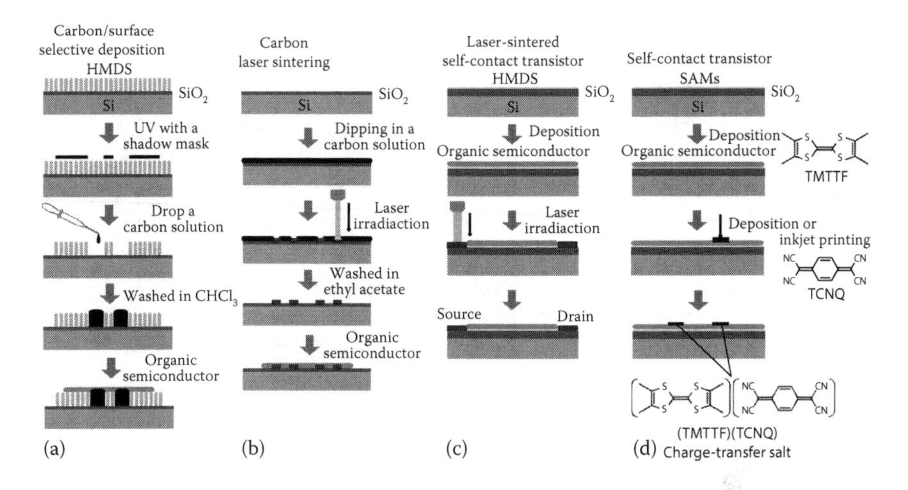

FIGURE 10.12 Fabrication of transistors with (a) carbon electrodes using the surface selective deposition, (b) laser sintering, (c) laser-sintered self-contact, and (d) chemically doped self-contact.

dispersion in polar ethyl acetate is deposited on the hydrophilic bare SiO_2 region. Various organic semiconductors are fabricated by evaporation or solution technique, and the resulting transistors have been investigated to exhibit excellent performance, even though this is a kind of bottom-contact transistor [110]. This technique is particularly important for materials with deep HOMO levels, which do not work easily in the bottom-contact geometry [113]. We have also achieved the patterning of the carbon film by selective laser irradiation (Figure 10.12b) [114]. The nonirradiated part is washed out with a polar solvent such as ethyl acetate, and laser-irradiated carbon film remains. By this laser-sintering method, we can obtain short-channel patterns down to 2 μm and relatively complicated patterns without using photolithography. The resulting carbon film is practically transparent because of the thin thickness of 60 nm, affording a transparent conducting film, which is interesting as a replacement of indium tin oxide (ITO). The laser sintering reminds us that many organic polymers are transformed to conducting carbon by laser irradiation [115,116]. Accordingly, films of pentacene or other organic semiconductors are selectively irradiated by laser and transformed to conducting carbon (Figure 10.12c). Using this part as S/D electrodes, "self-contact" organic transistors are constructed [117]. This is an ultimately easy way to fabricate organic transistors.

Another way to avoid the large contact resistance in the bottom-contact geometry is thiol treatments on the Au electrodes [118,119]. Alternatively, a thin layer of oppositely charging molecules between the electrode and the active layer improves the performance. For example, a thin layer of F_4TCNQ on bottom-contact metal electrodes has been used as a buffer layer of p-channel organic transistors [120–123]. When a thin layer of TCNQ is deposited on Ag or Cu electrodes before evaporating pentacene, the improvement of the performance has been reported [123,124]. However, metal TCNQ complexes are low-conducting insulators with 1:1 composition [125,126]. When DMDCNQI is deposited from a solution between Ag (or Cu) electrodes and pentacene, the resulting performance (0.3 cm^2/V s for Cu) is almost comparable to that of Au top-contact transistors (0.49 cm^2/V s), because the complex $Cu(DMDCNQI)_2$ is a highly conducting metal [127].

In order to reduce the contact resistance in bottom-contact transistors, a conducting polymer has been used as the electrode materials [128]. The electrode material has been also replaced by highly conducting organic charge-transfer salts such as (TTF)(TCNQ) [129], where the pentacene bottom-contact transistors show as good performance as the top-contact transistors [130]. (TTF)(TCNQ) electrodes work well in DBTTF transistors [105], dithiophene-TTF transistors [131], and n-channel transistors based on DMDCNQI [66] and $F_{16}CuPc$ [132]. In the DMDCNQI transistors, the expected order of the electrode metal dependence is Ag > Cu > Au, but the actually observed one is Au > (TTF) (TCNQ) > Ag > Cu. When we assume that the (TTF) (TCNQ) electrode is "null" of the interfacial potential shift Δ, Δ works in the direction to increase the injection barrier for Ag and Cu, but to decrease the barrier for Au. This is the reason that Au is generally a good electrode material even for n-channel transistors, though Au work function is relatively large.

We can easily obtain (TTF)(TCNQ) films by evaporating (TTF)(TCNQ) powder from a single crucible. This is based on similar sublimation temperatures of TTF and TCNQ (120°C). Even if one component is excess in the film, such a component is easily removed in the vacuum due to the high vapor pressure, and the resulting film contains only the 1:1 complex [35]. This method is, however, not applicable to organic conductors containing inorganic ions; the combination of an organic donor and an inorganic anion is called a radical cation salt, and a complex of an organic acceptor and an inorganic cation is called a radical anion salt.

However, nanoparticles of these salts have been prepared with the assistance of ionic liquid [133,134] or poly(vinylpyrrolidone) (PVP) [135]. Such nanoparticles are dispersed in common organic solvents and even water, and the patterns are obtained by the surface selective deposition similarly to carbon (Figure 10.11a). Then, pentacene or C_{60} is deposited; the resulting transistors show characteristics as shown in Figure 10.13. It is obvious that the characteristics of pentacene and C_{60} transistors do not largely depend on the electrode materials. This is somewhat surprising, because radical anion salts such as $Cu(DMDCNQI)_2$ and $TTF[Ni(dmit)_2]$ are electron-transporting materials, but holes are injected from these materials to pentacene. By contrast, electrons are injected from $(BEDT\text{-}TTF)_2I_3$ to C_{60}. We have to understand these observations considering that these organic conductors are organic metals with the Fermi surface, and either holes or electrons are injected from these materials.

In general, performance of organic transistors largely changes depending on the S/D electrode materials [120]. However, performance of these transistors does not depend on the electrode materials. This is partly due to the absence of undesirable interfacial potentials on the organic/organic interface. When we investigate the energy levels of these organic conductors, the HOMO levels of TTF and BEDT-TTF are located around 4.8–5.0 eV, whereas the LUMO levels of DMDCNQI and $[Ni(dmit)_2]$ are about 4.65 eV (Figure 10.12c). Upon band formation, the Fermi energy is shifted by a quarter of the bandwidth (~0.2 eV) because these CT salts are nearly quarter filled. Then the Fermi levels of the donors are pushed up by about 0.2 eV, and the Fermi levels of the acceptors are pushed down by the same amount.

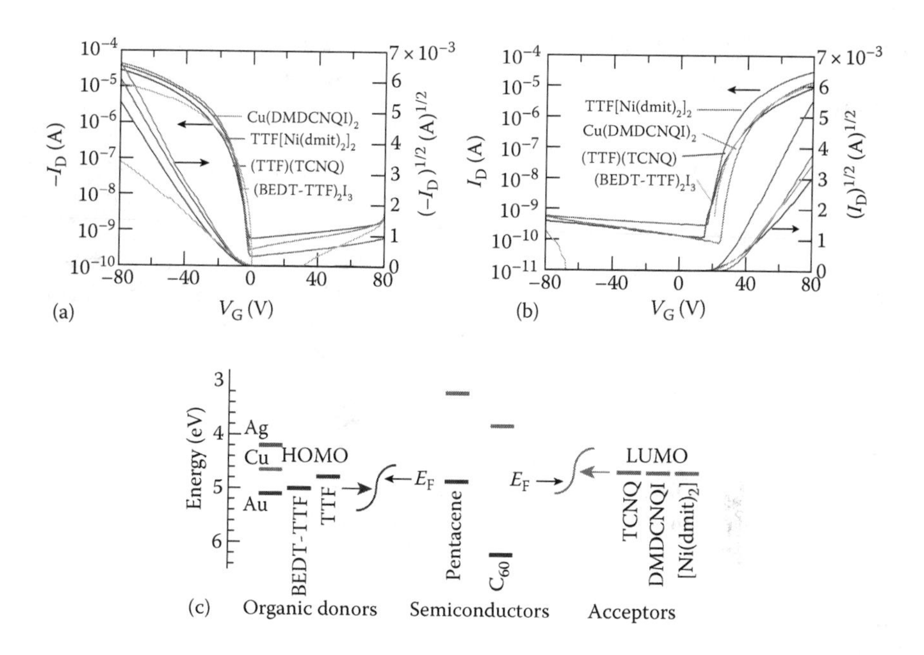

FIGURE 10.13 Transfer characteristics of (a) pentacene and (b) C_{60} transistors with organic conductor S/D electrodes. (c) Energy diagram of organic conductors and semiconductors [135].

As a consequence, all the Fermi levels of these organic metals are 4.8 eV. This is not entirely accidental, because conducting carbon has the same energy level as well; 4.8 eV seems to be the energy center for all organic compounds.

Since organic conductors work very well as the S/D materials in organic transistors, we have made transistors in which the organic semiconductor is converted to the conducting charge-transfer salt (Figure 10.12d). As an organic semiconductor, TMTTF is used, and the S/D parts are converted to (TMTTF)(TCNQ) by TCNQ evaporation [136]. Both TMTTF and (TMTTF)(TCNQ) have side-on geometry in the thin films, and TMTTF is easily converted to (TMTTF)(TCNQ) by annealing at 80°C for 2 h. Alternatively, we can use HMTTF (Figure 10.6) instead of TMTTF, and the mobility is improved to more than 1 cm^2/V s [137]. Since the S/D parts are composed of organic materials, we have replaced the substrate and the gate electrode to organic materials and achieved all-organic self-contact transistors [137]. Chemical doping is also achieved using inkjet printing, and n-channel self-contact transistors have been realized by reacting DMDCNQI with CuI [138]. It is very common in silicon technology to use p-doped and n-doped silicon, but organic semiconductors are generally used as an intrinsic form, namely, as a single component. This is partly because the usual organic semiconductors such as pentacene and oligothiophene do not make stable and high-conducting charge-transfer salts. The use of chemically doped charge-transfer salts is an essential technology in organic electronics.

As an opposite approach, we have explored charge-transfer salts based on representative organic semiconductors and found that BTBT (Figure 10.3) forms a conducting salt, $(BTBT)_2PF_6$ [139]. Although BTBT is a weak electron donor with the ionization potential of 5.6 eV, $(BTBT)_2PF_6$ exhibits as high a conductivity as 1500 S/cm at room temperature and very asymmetric Dysonian line shape in the ESR. The BTBT forms highly one-dimensional stacks and undergoes a resistivity jump around 150 K, followed by a true semiconducting state below 50 K. Other anions, AsF_6, SbF_6, and TaF_6, form isostructural salts, and the selenium analog $(BSBS)_2TaF_6$ shows similar properties [140].

It has been found that organic charge-transfer salts also work as an active layer of organic transistors [23]. (DBTTF)(TCNQ) is a highly resistive mixed-stack complex, and the carrier polarity sensitively changes when the Fermi level of the electrode (TTF)(TCNQ)-type salts is controlled by changing the combination of D and A [141,142]. In the thin-film transistors, (DBTTF)(TCNQ) usually shows n-channel properties [105]. In (BEDT-TTF)(TCNQ) transistors, ambipolar transport has been observed, and the mobility increases down to 240 K [143,144]. In TCNQ complexes of alkyl-BTBT, mainly n-channel transport has been observed, where the mobility of 0.4 cm^2/V s has been reported in the single-crystal transistors [145,146]. κ-(BEDT-TTF)$_2$Cu $[N(CN)_2]Br$ is an organic superconductor with the transition temperature of 11.8 K, but located on the border of the Mott insulator. When placed on a silicon substrate, the crystal exhibits an insulating behavior at low temperatures owing to the strain and shows an extremely high n-channel field-effect mobility of 94 cm^2/V s at 4 K [147,148]. Gate-induced and light-induced superconductivity has been achieved using this system [149,150].

10.5 TEMPERATURE DEPENDENCE OF TRANSISTOR CHARACTERISTICS

The definition of mobility μ (Equation 10.6) demonstrates that μ works like σ; to obtain μ, σ is divided by the carrier number $ne = Q$ because $Q = CV_G$ depends on V_G. Mobility of high-purity organic crystals measured by the time-of-flight method increases as the temperature is lowered [151]. In crystals such as naphthalene and anthracene, μ is around 1 cm^2/V s at room temperature, but increases to more than 100 cm^2/V s below 30 K; then, high-purity organic crystals show band transport. Increasing (metallic) mobility has been reported in single-crystal transistors [152–157], but thin-film transistors show decreasing (semiconducting) mobility.

Historically, analysis of temperature-dependent mobility in organic transistors dates back to Horowitz's work, in which the trap density of states (DOS) was obtained by analyzing temperature-dependent transfer characteristics [158–160]. When we investigate organic transistors following the conventional amorphous silicon (a-Si) transistors (Figure 10.2) [161–168], the depth of the accumulation layer (y) is smaller than the molecular length (1–3 nm), and it is appropriate to keep y constant; this is called interface approximation. Here, we discuss temperature-dependent characteristics according to the interface approximation.

In order to understand the temperature dependence, it is customary to assume the midgap localized states (Figure 10.2); we consider that the conduction band above

E_c is band like, but the midgap states below E_c are localized. The state density of the midgap states is usually approximated by an exponential function [8,165].

$$N(E) = N_G \exp\left[\frac{(E - E_c)}{kT_G}\right] \quad (10.15)$$

The total trapped charge Q_t (Figure 10.2) is obtained by integrating this equation up to $E_a = E_c - E_F - \varphi$,

$$Q_t = q\int_{-\infty}^{E_a} N_G \exp\left(-\frac{E}{kT_G}\right) dE = qN_G kT_G \exp\left(-\frac{E_a}{kT_G}\right) \quad (10.16)$$

where E_a is the activation energy at the interface. Although we should consider the Fermi distribution [162], the aforementioned integration is done for a step function at 0 K, because it does not alter the essential results [8]. The free charges Q_f are excited from E_F to E_c (Figure 10.2) by an activation energy of E_a.

$$Q_f = qN_c \exp\left(-\frac{E_a}{kT}\right) \quad (10.17)$$

The total charges are given by $CV_G = Q_t + Q_f$, but usually $Q_t \gg Q_f$, so E_a is obtained from Equation 10.16.

$$E_a = -kT_G \ln\frac{CV_G}{qN_G kT_G} \quad (10.18)$$

When we put this equation in Equation 10.17, we obtain

$$Q_f(V_G) = qN_c\left(\frac{CV_G}{qN_G kT_G}\right)^{T_G/T}. \quad (10.19)$$

We put this Q_f in Q in Equation 10.2 and integrate it according to the standard gradual channel approximation to lead to

$$I_D = \frac{W\mu_0 qN_c}{L}\left(\frac{C}{qN_G kT_G}\right)^{T_G/T}\frac{T}{T_G + T}\left[V_G^{\frac{T_G}{T}+1} - (V_G - V_D)^{\frac{T_G}{T}+1}\right]. \quad (10.20)$$

This equation affords the temperature-dependent output characteristics; when $T = T_G$, this equation reduces to the standard output characteristics (Equation 10.4). When the upper limit of the gradual channel approximation is replaced by V_G, Equation 10.20 is substituted by

$$I_D = \mu_0 \frac{WC}{L} \frac{qN_c}{C} \left(\frac{CV_G}{qN_G kT_G} \right)^{T_G/T} \frac{T}{T_G + T} V_G. \tag{10.21}$$

Although the observed I_D is reduced approximately by a factor of $CV_G/qN_GkT_G \sim Q_f/Q_t$, this is equivalent to Equation 10.5 at $T = T$, and affords the transfer characteristics in the saturated region.

To analyze temperature dependence in thin-film transistors, there are many methods developed for a-Si transistors, which assume variable y [161–168]. Kalb et al. have analyzed organic transistors based on these methods and proved that various methods afford approximately consistent exponential trap distribution [169–173]. However, Lang's method has been widely used in organic transistors [174,175], where the trap DOS is extracted from the Arrhenius plot of the drain current I_D [168,176,177]. As an example, the analysis of HMTTF transistor is shown in Figure 10.14 [8]. As shown in Figure 10.14a, when the temperature is lowered, the transfer characteristics (I_D) is gradually reduced together with the shift to the off (right) direction. At a given V_G, I_D decreases gradually, approximately following the Arrhenius law (Figure 10.14b). The slope affords E_a, and the V_G dependence of E_a is obtained as shown in

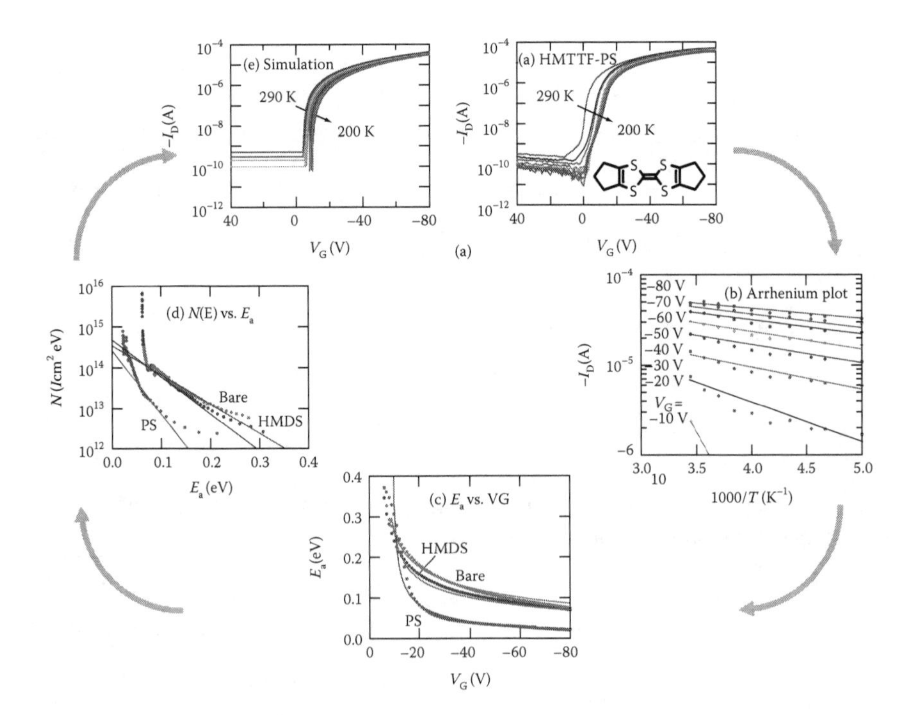

FIGURE 10.14 (a) Temperature-dependent transfer characteristics of an HMTTF transistor on polystylene. (b) The Arrhenium plot. (c) V_G dependence of E_a together with transistors with other SAM treatments. (d) Trap density of states. (e) Transistor characteristics calculated from Equation 10.21 with $qN_GkT_G/C = 80$ V, $T_G = 200$ K, and $qN_c/C = 1.0$ V. (From Akiyama, Y. and Mori, T., *AIP Adv.*, 4, 017126, 2014.)

Figure 10.14c. In Lang's method, all gate-induced charges are assumed to be trapped as $Q_t = CV_G$, and the trap DOS is obtained from

$$N(E) = \frac{C}{q}\left(\frac{dV_G}{dE_a}\right). \tag{10.22}$$

The results are shown in Figure 10.14d. Although strictly this is not a straight line following Equation 10.15, we can approximate the result by a single exponential, and we can calculate the theoretical temperature-dependent transfer characteristics from Equation 10.20, as shown in Figure 10.14e.

The actual observation in Figure 10.14a shows more detailed structure. The temperature-dependent shift of V_T is not very informative because it does not reflect the change of the trap states in the present framework. It is sometimes not very reproducible and probably depends on the temperature dependence of the surface charge. However, a small shoulder is obvious in the subthreshold region, and it is reproducible assuming the presence of an energetically discrete polaron-like state [8]. In general, the smaller the number of the trap states, the clearer the detailed structure. Similar localized states have been also predicted from the analysis of the ESR line width [178,179].

The observed transfer characteristics starts to deviate from the ideal theoretical equation in the large V_G region (Figure 10.15a); I_D becomes slightly flatter than the expectation from Equation 10.21. This is because the pinch off is lost at $V_G = V_D + V_{on}$, and the transistor goes into the linear region. Differentiating the transfer characteristics, we observe a clear inflection point as shown in Figure 10.15b. However, the inflection point appears at V_G lower than V_D. This is due to the contact voltage drop. As shown in the inset in Figure 10.15a, only the drain resistance influences the shift of the inflection point. The voltage drop due to the contact resistance is usually in the range of 10%–20% of the total voltage and approximately temperature independent. In general, the higher the channel mobility, the larger is the contribution of the contact resistance, because the channel resistance is small. In this method,

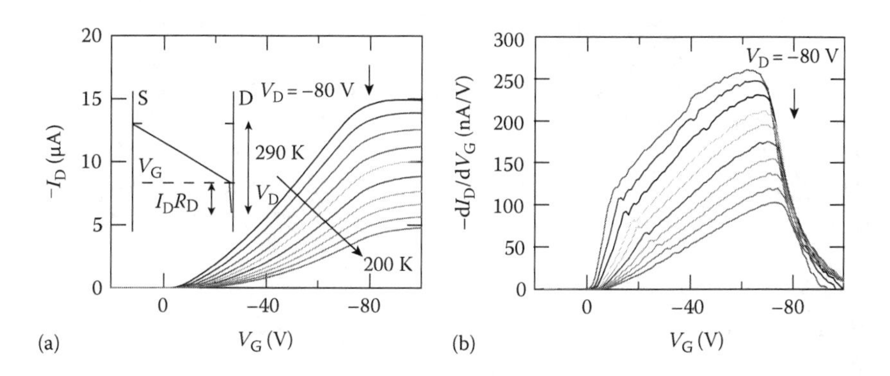

FIGURE 10.15 (a) Transfer characteristics of an HMDS-treated HMTTF transistor up to large V_G and (b) the derivative [8].

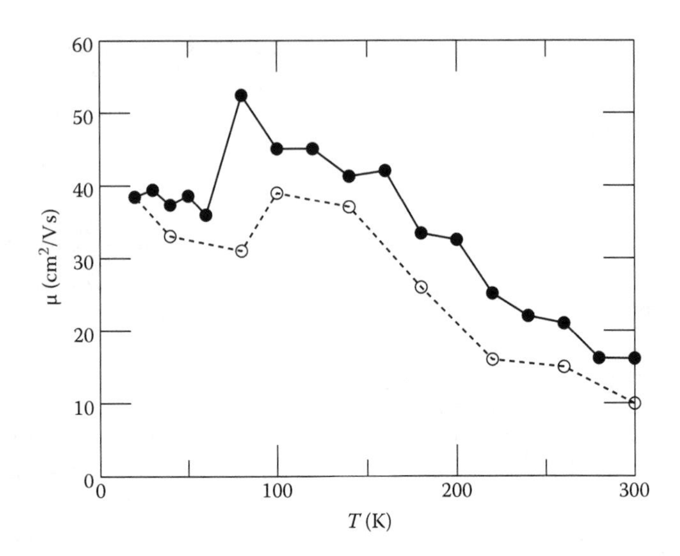

FIGURE 10.16 Temperature-dependent mobility of a C8-BTBT as-grown single-crystal transistor. (From Cho, J. and Mori, T., *Appl. Phys. Lett.*, 106, 193303, 2015.)

we can estimate contact resistance without using a special device such as the four-probe method, and only using a single device in contrast to the transfer-line method by measuring the transfer characteristics up to the linear region.

Equation 10.21 indicates $E_a = 0$ eV at $qN_GkT_G = CV_G$, where all trap states are full. Above this V_G, the gate-induced charge directly enters the conduction band, and band transport is expected [180]. More detailed calculation demonstrates that the crossover occurs at $CV_G = qN_GkT_G + qN_C$. In the usual organic transistors with SiO_2 gate dielectric (300 nm, $C = 13.7$ nF/cm^2), the molecular density corresponds to $V_G \sim 4000$ V, so at the usual $V_G \sim 40$ V, the carrier number is one hundredth of the whole molecules. The total trap number of thin-film transistors is usually in the order of $qN_GkT_G = 200$–500 V [8,180], but very clean thin-film transistors have as small number of traps as 40–80 V. When we apply V_G higher than this value, we may expect band-like transport, namely, I_D increases as lowering the temperature. Using as-grown single crystals [181], we can reduce the total trap number practically to 0 V. Here, the mobility increases with decreasing the temperature down to liquid helium temperatures [155], though still an anomaly appears around 80 K (Figure 10.16) [182]. Then band-like transport is attained in organic field-effect transistors as observed in the time-of-flight measurement [151].

10.6 CONCLUSION

It is characteristic of organic electronics that organic semiconductors are basically used as the single-component form. Then, we ask whether field-induced charge carriers in organic transistors are really the same as the chemically doped carriers in organic charge-transfer salts. Although studies of these two have been conducted almost independently, recent studies of self-contact transistors and field-induced superconductivity in organic metals have shown that something new exists in between these two, which is emphasized in this chapter. We also like to indicate that the study of organic transistors started from the application oriented interest, but it is now a basic technique to investigate molecular materials. The carrier polarity is clearly defined in transistors, and the carrier number is variable depending on V_G. We believe that it is an essential technique to understand charge carrier transport in molecular materials.

REFERENCES

1. H. Klauk ed., *Organic Electronics*, Wiley, Weinheim, Germany, 2006.
2. Z. Bao and J. Locklin ed., *Organic Field-Effect Transistors*, CRC Press, New York, 2007.
3. C. W. Tang and S. A. VanSlyke, *Appl. Phys. Lett.* 1987, **51**, 913.
4. C. W. Tang, *Appl. Phys. Lett.* 1987, **48**, 183.
5. K. Kudo, M. Yamashina, and T. Moriizumi, *Jpn. J. Appl. Phys.* 1984, **23**, 130.
6. H. Koezuka, A. Tsumura, and T. Ando, *Synth. Met.* 1987, **18**, 699.
7. J. P. Colinge and C. A. Colinge, *Physics of Semiconductor Devices*, Springer, New York, 2002, pp. 187–196.
8. Y. Akiyama and T. Mori, *AIP Adv.* 2014, **4**, 017126.
9. S. K. Passanner, K. Zojer, P. Pacher, E. Zojer, and F. Schurrer, *Adv. Funct. Mater.* 2009, **19**, 958.
10. Y. Xu, T. Monari, K. Tsukagoshi, R. Gwoziecki, R. Coppard, M. Benwadin, J. Chroboczek, F. Balestra, and G. Ghibaudo, *J. Appl. Phys.* 2011, **110**, 014510.
11. H. Meng, L. Zheng, A. J. Lovinger, B.-C. Wand, P. G. V. Patten, and Z. Bao, *Chem. Mater.* 2003, **15**, 1778.
12. M. L. Tang, A. D. Reicharrdt, P. Wei, and Z. Bao, *J. Am. Chem. Soc.* 2009, **131**, 5264.
13. D. F. Schriver and P. W. Atkins, in: Oxidation and reduction, *Inorganic Chemistry*, 4th edn., Oxford University Press, Oxford, U.K., 2006, Chapter 5.
14. A. R. Murphy and M. J. Fréchet, *Chem. Rev.* 2007, **107**, 1066.
15. T. Mori, *J. Phys. Condens. Matter.* 2008, **20**, 184010.
16. G. Horowitz, F. Garnier, A. Yassar, R. Hajlaoui, and F. Kouki, *Adv. Mater.* 1996, **8**, 52.
17. C. D. Dimitrakopoulos, B. K. Furman, T. Graham, S. Hegde, and S. Purushothaman, *Synth. Met.* 1998, **92**, 47.
18. H. E. Katz, A. J. Lovinger, and J. G. Laquindanum, *Chem. Mater.* 1998, **10**, 457.
19. A. Facchetti, *Handbook of Thiophene-Based Materials*, I. F. Perepichka and D. F. Perepichka eds., Wiley, Chichester, U.K., 2009, Chapter 16, Electroactive oligothiophenes and polythiophenes for organic field effect transistors.
20. Z. Bao, A. J. Lovinger, and A. Dodabalapur, *Appl. Phys. Lett.* 1996, **69**, 3066.
21. H. Klauk, M. Halik, U. Zscchieschang, G. Schmid, W. Radlik, and W. J. Weber, *Appl. Phys.* 2002, **92**, 5259.
22. H. Kojima and T. Mori, *Bull. Chem. Soc. Jpn.* 2011, **84**, 1049.
23. T. Mori, *Chem. Lett.* 2011, **40**, 428.
24. M. Mas-Torrent and C. Rovira, *J. Mater. Chem.* 2006, **16**, 433.
25. M. Mas-Torrent and C. Rovira, *Chem. Soc. Rev.* 2008, **37**, 827.
26. R. Pfattner, S. T. Bromley, C. Rovira, and M. Mas-Torrent, *Adv. Funct. Mater.* 2015, 2016, **26**, 2256, DOI: 10.1002/adfm.201502446.
27. A. Brillante, I. Bilotti, R. G. D. Valle, E. Venuti, S. Milita, C. Dionigi, F. Borgatti et al., *CrystalEngComm* 2008, **10**, 1899.

28. Y. Takahashi, T. Hasegawa, S. Horiuchi, R. Kumai, Y. Tokura, and G. Saito, *Chem. Mater.* 2007, **19**, 6382.
29. T. Yamada, R. Kumai, Y. Takahashi, and T. Hasegawa, *J. Mater. Chem.* 2010, **20**, 5810.
30. M. Kanno, Y. Bando, T. Shirakata, J. Inoue, H. Wada, and T. Mori, *J. Mater. Chem.* 2009, **19**, 6548.
31. T. Yoshino, K. Shibata, H. Wada, Y. Bando, K. Ishikawa, H. Takezoe, and T. Mori, *Chem. Lett.* 2009, **38**, 200.
32. J. Inoue, M. Kanno, M. Ashizawa, C. Seo, A. Tanioka, and T. Mori, *Chem. Lett.* 2010, **39**, 538.
33. J. Takeya, M. Yamagishi, Y. Tominari, R. Hirahara, Y. Nakazawa, T. Nishikawa, T. Kawase, T. Shimoda, and S. Ogawa, *Appl. Phys. Lett.* 2007, **90**, 102120.
34. H. Jiang, X. Yang, Z. Cui, Y. Liu, H. Li, W. Hu, Y. Liu, and D. Zhu, *Appl. Phys. Lett.* 2007, **91**, 123505.
35. T. Takahashi, S. Tamura, Y. Akiyama, T. Kadoya, T. Kawamoto, and T. Mori, *Appl. Phys. Exp.* 2012, **5**, 061601.
36. H. Sirringhaus, N. Tessler, and R. H. Friend, *Science* 1998, **280**, 1741.
37. I. Mcculloch, M. Heeney, C. Bailey, K. Genevivius, I. MacDonald, M. Shkunov, D. Sparrowe et al., *Nat. Mater.* 2006, **5**, 328.
38. J. E. Anthony, *Chem. Rev.* 2006, **106**, 5028.
39. H. Ebata, T. Izawa, E. Miyazaki, K. Takimiya, M. Ikeda, H. Kuwabara, and T. Yui, *J. Am. Chem. Soc.* 2007, **129**, 15732.
40. K. Takimiya, Y. Kunugi, and T. Otsubo, *Chem. Lett.* 2007, **36**, 578.
41. K. Takimiya, S. Shinamura, I. Osaka, and E. Miyazaki, *Adv. Mater.* 2011, **23**, 4347.
42. H. Minemawari, Y. Toshikazu, H. Matui, J. Tsutsumi, S. Haas, R. Chiba, R. Kumai, and T. Hasegawa, *Nature* 2011, **475**, 364.
43. J. Soeda, Y. Hirose, M. Yamagishi, A. Nakao, T. Uemura, K. Nakayama, M. Uno, Y. Nakazawa, K. Takimiya, and J. Takeya, *Adv. Mater.* 2011, **23**, 3309.
44. T. Yamamoto and K. Takimiya, *J. Am. Chem. Soc.* 2007, **129**, 2224.
45. T. Okamoto, C. Mitsui, M. Yamagishi, K. Nakahara, J. Soeda, Y. Hirose, K. Miwa et al., *Adv. Mater.* 2013, **25**, 6392.
46. C. R. Newman, C. D. Friebie, D. A. da Silva Filho, J. Bredas, P. C. Ewbank, and K. R. Mann, *Chem. Mater.* 2004, **46**, 4436.
47. X. Zhan, A. Facchetti, S. Barlow, T. J. Marks, M. A. Ratner, M. R. Wasielewski, and S. R. Marder, *Adv. Mater.* 2011, **23**, 268.
48. Z. Bao, A. J. Lovinger, and J. Brown, *J. Am. Chem. Soc.* 1998, **120**, 207.
49. A. Facchetti, M. Mushrush, M. H. Yoon, G. R. Hutchison, and M. A. Ratner, T. J. Marks, *J. Am. Chem. Soc.* 2004, **126**, 13859.
50. A. Facchetti, M. H. Yoon, C. L. Stern, H. E. Katz, and T. J. Marks, *Angew. Chem., Int. Ed.* 2003, **42**, 3900.
51. Y. Sakamoto, T. Suzuki, M. Kobayashi, Y. Gao, Y. Fukai, Y. Inoue, F. Sato, and S. Tokito, *J. Am. Chem. Soc.* 2004, **126**, 8138.
52. S. Ando, R. Murakami, J. Nishida, H. Tada, Y. Inoue, S. Tokito, and Y. Yamashita, *J. Am. Chem. Soc.* 2005, **127**, 14996.
53. Y. Yamashita, *Sci. Technol. Adv. Mater.* 2009, **10**, 024313.
54. R. C. Haddon, A. S. Perel, R. C. Morris, T. T. M. Palstra, A. F. Hebard, and R. M. Fleming, *Appl. Phys. Lett.* 1995, **67**, 121.
55. D. Shukla, S. F. Nelson, D. C. Freeman, M. Rajeswaran, W. G. Ahearn, D. M. Meyer, and J. T. Carey, *Chem. Mater.* 2008, **20**, 7486.
56. T. Kakinuma, H. Kojima, M. Ashizawa, H. Matsumoto, and T. Mori, *J. Mater. Chem. C* 2013, **1**, 5395.
57. Z. Liang, Q. Tang, J. Xu, and Q. Miao, *Adv. Mater.* 2011, **23**, 1535.
58. Z. Liang, Q. Tang, R. Mao, D. Liu, J. Xu, and Q. Miao, *Adv. Mater.* 2011, **23**, 5514.

59. C. Wang, Z. Liang, Y. Liu, X. Wang, N. Zhao, Q. Miao, W. Hu, and J. Xu, *J. Mater. Chem.* 2011, **21**, 15201.
60. Q. Miao, *Synlett* 2012, **23**, 326.
61. A. Filatre-Furcate, T. Higashino, D. Lorcy, and T. Mori, *J. Mater. Chem. C* 2015, **3**, 3569.
62. H. Yan, Z. Chen, Y. Zheng, C. Newman, J. R. Quinn, F. Dotz, M. Kastler, and A. Facchetti, *Nature* 2009, **457**, 679.
63. H. Bronstein, Z. Chen, R. S. Ashraf, W. Zhang, J. Du, J. R. Durrant, P. S. Tuladhar et al., *J. Am. Chem. Soc.* 2011, **133**, 3272.
64. E. Menard, V. Podzorov, S.-H. Hur, A. Gaur, M. E. Gershenson, and J. A. Rogers, *Adv. Mater.* 2004, **16**, 2097.
65. M. Yamagishi, Y. Tominari, T. Uemura, and J. Takeya, *Appl. Phys. Lett.* 2009, **94**, 053305.
66. H. Wada, K. Shibata, Y. Bando, and T. Mori, *J. Mater. Chem.* 2008, **18**, 4165.
67. J. Zaumseil and H. Sirringhaus, *Chem. Rev.* 2007, **107**, 1296.
68. S. Z. Bisri, C. Piliego, J. Gao, and M. A. Loi, *Adv. Mater.* 2014, **26**, 1176.
69. R. J. Chesterfield, C. R. Newman, T. M. Pappenfus, P. C. Ewbank, M. H. Haukaas, K. R. Mann, L. L. Miller, and C. D. Frisbie, *Adv. Mater.* 2003, **12**, 1278.
70. J. C. Ribierre, S. Watanabe, M. Matsumoto, T. Muto, and T. Aoyama, *Appl. Phys. Lett.* 2010, **96**, 083303.
71. J. C. Ribierre, T. Fujihara, T. Muto, and T. Aoyama, *Appl. Phys. Lett.* 2010, **96**, 233302.
72. J. C. Ribierre, T. Fujihara, T. Muto, and T. Aoyama, *Org. Electron.* 2010, **11**, 1469.
73. J. C. Ribierre, S. Ghosh, K. Takaishi, T. Muto, and T. Aoyama, *J. Phys. D* 2011, **44**, 205102.
74. A. Tapponnier, I. Biaggio, and P. Günter, *Appl. Phys. Lett.* 2005, **86**, 112114.
75. R. Könenkamp, G. Priebe, and B. Pietzak, *Phys. Rev. B* 1999, **60**, 11804.
76. T. Nishikawa, S. Kobayashi, T. Nakanowatari, T. Mitani, T. Shimoda, Y. Kubozono, G. Yamamoto, H. Ishii, M. Niwano, and Y. Iwasa, *J. Appl. Phys.* 2005, **97**, 104509.
77. T. D. Anthopoulos, D. M. de Leeuw, E. Cantatore, S. Setayesh, E. J. Meijer, C. Tanase, J. C. Hummelen, and P. W. M. Blom, *Appl. Phys. Lett.* 2004, **85**, 4205.
78. T. D. Anthopoulos, C. Tanase, S. Setayesh, E. J. Meijer, J. C. Hummelen, P. W. M. Blom, and D. M. deLeeuw, *Adv. Mater.* 2004, **16**, 2174.
79. F. Cicoira, N. Coppedé, S. Iannotta, and R. Martel, *Appl. Phys. Lett.* 2011, **98**, 183303.
80. J. D. Yuen and F. Wudl, *Energy Environ. Sci.* 2013, **6**, 392.
81. Y. Zhao, Y. Guo, and Y. Liu, *Adv. Mater.* 2013, **25**, 5372.
82. N. Olivier, D. Niedzialek, V. Lemaur, W. Pisula, K. Mullen, U. Koldemir, J. R. Reynolds, R. Lazzaroni, J. Cornil, and D. Beljonne, *Adv. Mater.* 2014, **26**, 2119.
83. E. Wang, W. Mammo, and M. R. Andersson, *Adv. Mater.* 2014, **26**, 1801.
84. J. D. Yuen, J. Fan, J. Seifter, B. Lim, R. Hufschmid, A. J. Heeger, and F. Wudl, *J. Am. Chem. Soc.* 2011, **133**, 20799.
85. Z. Chen, M. J. Lee, R. S. Ashraf, Y. Gu, S. Albert-Seifried, M. M. Nielsen, B. Schroeder, T. Anthopoulos, M. Heeney, I. McCulloch, and H. Sirringhaus, *Adv. Mater.* 2012, **24**, 647.
86. J. H. Seo, G. S. Chang, R. G. Wilks, C. N. Whang, K. H. Chae, S. Cho, K.-H. Yoo, and A. Moewes, *J. Phys. Chem. B* 2008, **112**, 16266.
87. M.-H. Yoon, C. Kim, A. Facchetti, and T. J. Marks, *J. Am. Chem. Soc.* 2006, **128**, 12851.
88. M. Irimia-Vlada, E. D. Glowacki, P. A. Troshin, G. Schwabegger, L. Leonat, D. K. Susarova, O. Krystal et al., *Adv. Mater.* 2012, **24**, 375.
89. T. Higashino, J. Cho, and T. Mori, *Appl. Phys. Exp.* 2014, **7**, 121602.
90. A. Opitz, M. Horlet, M. Kiwull, J. Wagner, M. Kraus, and W. Brutting, *Org. Electron.* 2012, **13**, 1614.
91. T. Yasuda, T. Goto, K. Fujita, and T. Tsutsui, *Appl. Phys. Lett.* 2004, **85**, 2098.

92. L.-Y. Chiu, H.-L. Cheng, H.-Y. Wang, W.-Y. Chou, and F.-C. Tang, *J. Mater. Chem. C*, 2014, **2**, 1823.
93. E. D. Glowacki, G. Voss, and N. S. Sariciftci, *Adv. Mater.* 2013, **25**, 6783.
94. E. D. Glowacki, L. Lenonat, G. Voss, M. A. Bodea, Z. Buzkurt, A. M. Ramil, M. Irimia-Vlada, S. Bauer, and N. S. Sariciftci, *AIP Adv.* 2011, **1**, 042132.
95. I. V. Klimovich, L. I. Leshanskaya, S. I. Troyanov, D. V. Anokhin, D. V. Novikov, A. A. Piryazev, D. A. Ivanov, N. N. Dremova, and P. A. Troshin, *J. Mater. Chem. C* 2014, **2**, 7621.
96. O. Pitayatanakul, T. Higashino, M. Tanaka, H. Kojima, M. Ashizawa, T. Kawamoto, H. Matsumoto, K. Ishikawa, and T. Mori, *J. Mater. Chem. C* 2014, **2**, 9311.
97. O. Pitayatanakul, K. Iijima, M. Ashizawa, T. Kawamoto, H. Matsumoto, and T. Mori, *J. Mater. Chem. C* 2014, **3**, 8612.
98. L. Serrano-Andres and B. O. Poos, *Chem. Eur. J.* 1997, **3**, 717.
99. T. Higashino, S. Kumeta, S. Tamura, Y. Ando, K. Ohmori, K. Suzuki, and T. Mori, *J. Mater. Chem. C* 2015, **3**, 1588.
100. E. D. Głowacki, H. Coskun, M. A. Blood-Forsythe, U. Monkowius, L. Leonat, M. Grzybowski, D. Gryko, M. S. White, A. Aspuru-Guzik, and N. S. Sariciftci, *Org. Electron.* 2014, **15**, 3521.
101. F. Oton, R. Pfattner, E. Pavilica, Y. Olivier, G. Bratina, J. Cornil, J. Puigdollers et al., *CrysEngComm* 2011, **13**, 6597.
102. R. Pfattner, E. Pavilica, M. Jaggi, S.-X. Liu, S. Decurtins, G. Bratina, J. Veciana, M. Mas-Torrent, and C. Rovira, *J. Mater. Chem. C* 2013, **1**, 3985.
103. S. Kobayashi, T. Nishikawa, T. Takenobu, S. Mori, T. Shimada, T. Mitani, H. Shimotani, N. Yoshimoto, S. Ogawa, and Y. Iwasa, *Nat. Mater.* 2004, **3**, 317.
104. M. J. Panzer and C. D. Frisbie, in *Organic Field-Effect Transistors*, Z. Bao and J. Locklin eds., CRC Press, New York, 2007, p. 139.
105. K. Shibata, K. Ishikawa, H. Takezoe, H. Wada, and T. Mori, *Appl. Phys. Lett.* 2008, **92**, 023305.
106. H. Ishii, K. Sugiyama, E. Ito, and K. Seki, *Adv. Mater.* 1999, **11**, 605.
107. I. G. Hill and A. Kahn, *Proc. SPIE*, 1998, **3476**, 168.
108. N. J. Watkins, Q. T. L. S. Zorba, L. Yan, Y. Gao, S. F. Nelson, C. S. Kuo, and T. N. Jackson, *Proc. SPIE* 2001, **4466**, 1.
109. R. J. Murdey and W. R. Salaneck, *Proc. SPIE* 2004, **5519**, 125.
110. H. Wada and T. Mori, *Appl. Phys. Lett.* 2008, **93**, 213303.
111. C. Di, D. Wei, G. Yu, Y. Liu, Y. Guo, and D. Zhu, *Adv. Mater.* 2008, **20**, 3289.
112. T. Minari, M. Kano, T. Miyadera, S.-D. Wang, Y. Aoyagi, M. Seto, T. Memoto, S. Isoda, and K. Tsukagoshi, *Appl. Phys. Lett.* 2008, **92**, 173301.
113. T. Kadoya, O. Pitayatanakul, and T. Mori, *Org. Electron.* 2015, **21**, 106.
114. H. Wada and T. Mori, *Appl. Phys. Lett.* 2009, **95**, 253307.
115. R. Srinivasan and B. Braren, *Chem. Rev.* 1989, **89**, 1303.
116. F. Raimondi, S. Abolhassani, R. Brütsch, F. Geiger, T. Lippert, J. Wambach, J. Wei, and A. Wokaun, *J. Appl. Phys.* 2000, **88**, 3659.
117. J. Inoue, H. Wada, and T. Mori, *Jpn. J. Appl. Phys.* 2010, **49**, 071605.
118. I. Kymissis, *IEEE Trans. Electron Dev.* 2001, **48**, 1060.
119. C. Bock, D. V. Pham, U. Kunze, D. Käfer, G. White, and Ch. Wöll, *J. Appl. Phys.* 2006, **100**, 114517.
120. K. Tsukagoshi, I. Yagi, K. Shigeto, K. Yanagisawa, J. Tanabe, and Y. Aoyagi, *Appl. Phys. Lett.* 2005, **87**, 183502.
121. C. Vanoni, S. Tsujino, and T. A. Jung, *Appl. Phys. Lett.* 2007, **90**, 193119.
122. T. Minari, T. Miyadera, K. Tsukagoshi, Y. Aoyagi, and H. Ito, *Appl. Phys. Lett.* 2007, **91**, 053508.
123. C. Di, G. Yu, Y. Liu, X. Xu, D. Wei, Y. Song, Y. Sun et al., *J. Am. Chem. Soc.* 2006, **128**, 16418.

124. C. Di, G. Yu, Y. Liu, Y. Guo, Y. Wang, W. Wu, and D. Zhu, *Adv. Mater.* 2008, **20**, 1286.
125. L. Shields, *J. Chem. Soc. Faraday Trans. 2* 1985, **81**, 1.
126. R. A. Heints, H. Zhao, X. Ouyang, G. Grandinetti, J. Cowan, and K. R. Dunbar, *Inorg. Chem.* 1999, **38**, 144.
127. Y. Yu, M. Kanno, H. Wada, Y. Bando, M. Ashizawa, A. Tanioka, and T. Mori, *Physica B* 2010, **405**, S378.
128. H. Sirringhaus, T. Kawase, R. H. Friend, T. Shimoda, M. Inbasekaran, W. Wu, and E. P. Woo, *Science* 2000, **290**, 2123.
129 R. Pfattner, C. Rovira, and M. Mas-Torrent, *Phys. Chem. Chem. Phys.* 2015, **17**, 26545.
130. K. Shibata, H. Wada, K. Ishikawa, H. Takezoe, and T. Mori, *Appl. Phys. Lett.* 2007, **90**, 193509.
131 R. Pfattner, M. Mas-Torrent, C. Moreno, J. Puigdollers, R. Alcubilla, I. Bilotti, E. Venuti et al., *J. Mater. Chem.* 2012, **22**, 16011.
132. K. Shibata, Y. Watakabe, K. Ishikawa, H. Takezoe, H. Wada, and T. Mori, *Appl. Phys. Exp.* 2008, **1**, 051801.
133. D. de Caro, K. Jacob, C. Faulmann, J.-P. Legros, F. Senocq, J. Fraxedas, and L. Valade, *Synth. Met.* 2010, **160**, 1223.
134. D. de Caro, K. Jacob, H. Hahioui, C. Faulmann, L. Valade, T. Kadoya, T. Mori, J. Fraxedas, and L. Valade, *New J. Chem.* 2011, **35**, 1315.
135. T. Kadoya, D. de Caro, K. Jacob, C. Faulmann, L. Valade, and T. Mori, *J. Mater. Chem.* 2011, **21**, 18421.
136. S. Tamura, T. Kadoya, T. Kawamoto, and T. Mori, *Appl. Phys. Lett.* 2013, **102**, 063305.
137. S. Tamura, T. Kadoya, and T. Mori, *Appl. Phys. Lett.* 2014, **105**, 023301.
138. T. Kadoya, S. Tamura, and T. Mori, *J. Phys. Chem. C* 2014, **118**, 23139.
139. T. Kadoya, M. Ashizawa, T. Higashino, T. Kawamoto, S. Kumeta, H. Matsumoto, and T. Mori, *Phys. Chem. Chem. Phys.* 2013, **15**, 17818.
140. T. Higashino, T. Kadoya, S. Kumeta, K. Kurata, T. Kawamoto, and T. Mori, *Eur. J. Inorg. Chem.* 2014, 3895.
141. Y. Takahashi, T. Hasegawa, Y. Abe, Y. Tokura, K. Nishimura, and G. Saito, *Appl. Phys. Lett.* 2005, **86**, 063504.
142. Y. Takahashi, T. Hasegawa, Y. Abe, Y. Tokura, and G. Saito, *Appl. Phys. Lett.* 2006, **88**, 073504.
143. M. Sakai, H. Sakuma, Y. Ito, A. Saito, M. Nakamura, and K. Kudo, *Phys. Rev. B* 2007, **76**, 045111.
144. M. Sakai, Y. Ito, T. Takahara, M. Ishiguro, M. Nakamura, and K. Kudo, *J. Appl. Phys.* 2010, **107**, 043711.
145. J. Tsutsumi, S. Matsuoka, S. Inoue, H. Minemawari, T. Yamada, and T. Hasegawa, *J. Mater. Chem. C* 2015, **3**, 1976.
146. Y. Shibata, J. Tsutsumi, S. Matsuoka, K. Matsubara, Y. Yoshida, M. Chikamatsu, and T. Hasegawa, *Appl. Phys. Lett.* 2015, **106**, 143303.
147. Y. Kawasugi, H. M. Yamamoto, M. Hosoda, N. Tajima, T. Fukunaga, K. Tsukagoshi, and R. Kato, *Appl. Phys. Lett.* 2008, **92**, 243508.
148. Y. Kawasugi, H. M. Yamamoto, M. Hosoda, N. Tajima, T. Fukunaga, K. Tsukagoshi, and R. Kato, *Phys. Rev. Lett.* 2009, **103**, 116801.
149. H. M. Yamamoto, M. Nakano, M. Suda, Y. Iwasa, M. Kawasaki, and R. Kato, *Nature Commun.* 2013, **4**, 2379.
150. M. Suda, R. Kato, and H. M. Yamamoto, *Science* 2015, **347**, 743.
151. W. Warta and N. Karl, *Phys. Rev. B* 1985, **32**, 1172.
152. V. Podzorov, E. Menard, A. Brissov, V. Kiryukhin, J. A. Rogers, and M. E. Gershenson, *Phys. Rev. Lett.* 2004, **93**, 086602.
153. V. Podzorov, E. Menard, J. A. Rogers, and M. E. Gershenson, *Phys. Rev. Lett.* 2005, **95**, 226601.

154. I. N. Hulea, S. Fratini, H. Xie, C. L. Mulder, N. N. Iossad, G. Rastelli, S. Ciuchi, and A. F. Morpurgo, *Nat. Mater.* 2005, **5**, 982.
155. T. Sakanoue and H. Sirringhaus, *Nat. Mater.* 2010, **9**, 736.
156. T. Uemura, M. Yamagishi, J. Soeda, Y. Takatsuki, Y. Okada, Y. Nakazawa, and J. Takeya, *Phys. Rev. B* 2012, **85**, 035313.
157. N. A. Minder, S. Ono, Z. Chen, A. Facchetti, and A. F. Morpurgo, *Adv. Mater.* 2012, **24**, 503.
158. G. Horowitz, R. Hajlaoui, and P. Delannoy, *J. Phys. III* 1995, **5**, 355.
159. G. Horowitz, M. E. Hajlaoui, and R. Hajlaoui, *J. Appl. Phys.* 2000, **87**, 4456.
160. G. Horowitz, *J. Mater. Res.* 2004, **19**, 1946.
161. S. Kishida, Y. Naruke, Y. Ushida, and M. Matsumura, *Jpn. J. Appl. Phys.* 1983, **22**, 511.
162. M. Shur and M. Hack, *J. Appl. Phys.* 1984, **55**, 3831.
163. M. Shur, M. Hack, and C. Hyun, *J. Appl. Phys.* 1984, **56**, 382.
164. M. Shur, C. Hyun, and M. Hack, *J. Appl. Phys.* 1986, **59**, 2488.
165. G. Fortunato and P. Migliorato, *J. Appl. Phys.* 1990, **68**, 2463.
166. M. Grünewald, P. Thoms, and D. Würtz, *Phys. Status Solidi B* 1980, **100**, K139.
167. M. Grünewald, K. Weber, W. Fuhs, and P. Thomas, *J. Phys. (France)* 1981, **42**, 523.
168. G. Fortunato, D. B. Meakin, P. Migliorato, and P. G. Le Comber, *Philos. Mag. B* 1988, **57**, 573.
169. W. L. Kalb, F. Meier, K. Mattenberger, and B. Batlogg, *Phys. Rev. B* 2007, **76**, 184112.
170. W. L. Kalb, K. Mattenberger, and B. Batlogg, *Phys. Rev. B* 2008, **78**, 035334.
171. W. L. Kalb and B. Batlogg, *Phys. Rev. B* 2010, **81**, 035327.
172. W. L. Kalb, S. Haas, C. Kreller, T. Mathis, and B. Batlogg, *Phys. Rev. B* 2007, **81**, 155315.
173. K. Willa, R. Häusermann, T. Mathis, A. Facchetti, Z. Chen, and B. Batlogg, *J. Appl. Phys.* 2013, **113**, 133707.
174. V. Y. Butko, X. Chi, D. V. Lang, and A. P. Ramirez, *Appl. Phys. Lett.* 2003, **83**, 4773.
175. D. V. Lang, X. Chi, T. Siegrist, A. M. Segrent, and A. P. Ramirez, *Phys. Rev. Lett.* 2004, **93**, 86802.
176. N. Kawasaki, T. Nagano, Y. Kubozono, Y. Sako, Y. Morimoto, Y. Takaguchi, A. Fujiwara, C.-C. Chu, and T. Imae, *Appl. Phys. Lett.* 2007, **91**, 243515.
177. J. Puigdollers, A. Marsal, S. Cheylan, C. Voz, and R. Alcubilla, *Org. Electron.* 2010, **11**, 1333.
178. A. S. Mishchenko, H. Matsui, and T. Hasegawa, *Phys. Rev. B* 2012, **85**, 085211.
179. H. Matsui, A. S. Mishchenko, and T. Hasegawa, *Phys. Rev. Lett.* 2010, **104**, 056602.
180. J. Cho, Y. Akiyama, T. Kakinuma, and T. Mori, *AIP Adv.* 2013, **3**, 102131.
181. C. Liu, T. Minari, X. Lu, A. Kumatani, K. Takimiya, and K. Tsukagoshi, *Adv. Mater.* 2011, **23**, 523.
182. J. Cho and T. Mori, *Appl. Phys. Lett.* 2015, **106**, 193303.

11 Biomedical Microdevices

Sharon Y. Wong, Mario Cabodi, and Catherine M. Klapperich

CONTENTS

In this chapter, we will review microfluidic devices with applications in biomedical sensing and diagnostics, with a particular emphasis on devices that can be used at the point of care (POC). We will discuss their manufacturing methods and compare the methods that are more accessible to academic labs and contrast them with methods that allow for mass production. With manufacturability in mind, we have focused on microfluidic systems that rely on geometry and other passive components (as opposed to electrically actuated systems) as the former can be more easily prototyped in academic labs and more readily mass produced at a low cost.

11.1 INTRODUCTION

The first examples of microfluidic systems were developed for the analysis of biochemical samples in chemically and mechanically robust devices. These devices were fabricated in silicon and glass, primarily because microfabrication techniques were already established for these materials from microelectronics applications. The first applications included gas/liquid chromatography systems in silicon wafers, in 1979.[1,2] Other systems used capillaries to replicate macroscale assays, especially analytical separations processes, and took advantage of shorter run times and smaller sample volumes. Mathies[3] used capillary array electrophoresis to massively parallelize existing separation schemes and obtained improvements in speed and throughput. As a first demonstration of the use of micromachining to obtain microfabricated devices, Harrison micromachined glass capillaries[4] for use in electrophoresis (Figure 11.1).

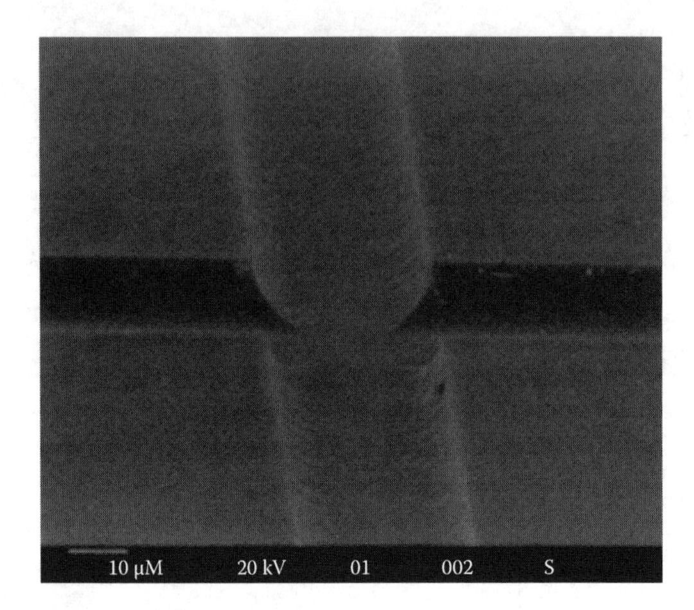

FIGURE 11.1 Electron micrograph of micromachined glass capillary channels for use in electrophoresis. (From Harrison, D.J. et al., *Science*, 261, 895, 1993.)

While some researchers initially focused on the miniaturization of individual aspects of bioanalysis, other groups sought to integrate multiple components onto one chip. Following the integration of PCR with electrophoresis[5] and separately, electrophoresis with electrochemical detection,[6] Manz introduced the concept of micro Total Analysis System (microTAS),[7] which would integrate both the functional separation unit and the detection system onto a single chip. These early examples took full advantage of the capabilities of microelectronic processes, resulting in devices that emphasized precise features and could include on-chip electronic components (e.g., for electrochemical detection). However, these systems were expensive, technically challenging to make, and most importantly required access to a cleanroom microfabrication facility for the fabrication step. This infrastructural requirement was a significant barrier to entry to most academic groups, since cleanrooms are expensive to set up and costly to operate. Thus, academic research groups had to balance the required resolution and desired device robustness with the cost and ease of manufacturing of the device itself.

In the 20 years following the original microfluidic demonstrations, many research groups found applications for microfluidic devices that did not require the high feature resolution, and attendant manufacturing processes, afforded by glass and silicon devices. With the shift in focus from high-precision, high-cost processes and materials to lower-cost, easier-to-manufacturer devices, researchers began exploring alternative materials.

Polymers emerged as an attractive and viable alternative to the conventional microelectronic materials.[8–11] Many commercially available polymers, such as polymethyl methacrylate (PMMA), polycarbonate (PC), and polystyrene (PS), offered properties similar to glass and silicon (e.g., optical clarity, mechanical strength, and durability). Some polymers [e.g., polydimethylsiloxane (PDMS)] could be surface treated (e.g., oxygen plasma treatment) to render them more hydrophilic and "glass-like," thereby improving their ability to promote protein and cell adhesion for biological applications.[12] Using these materials and the existing manufacturing and processing methods developed for commercial polymers, academic research groups began developing new methods to fabricate these polymers into microdevices. These methods allowed researchers to quickly and affordably make microdevices that did not require sub-100 μm features. More recently, the push for even lower-cost, easier-to-manufacture devices with greater applicability (e.g., for use in developing countries, at the POC) has made paper microfluidics an active area of research.

In the following sections, we will focus our discussion on polymer- and paper-based microfluidic systems and their applications. Within polymer-based microfluidic systems, we will consider two main categories of materials and their affiliated manufacturing methods: (1) polymers that can be scaled up to large device quantities (>1000 devices) and (2) polymers used for rapid prototyping in academic labs (1–100 devices).

11.2 POLYMER-BASED SYSTEMS

11.2.1 Large-Scale Manufacturing Processes and Materials

For microfluidics to find widespread use, both in high- and in low-resource settings, the eventual cost of the microdevices must be low enough to make disposable chips affordable.[13,14] This means that, after the initial prototyping stage, where designs and configurations may change, materials and methods to create large numbers of devices at low cost must be used. Polymers such as PMMA and PC were one of the first materials used in the development of microfluidics. These polymers were readily available commercially and easy to manufacture since industrial fabrication processes had already been developed, such as hot embossing and injection molding. These processes remain in use today.

In hot embossing, a patterned mold is used to stamp into a polymer layer that has been softened by raising its temperature slightly above its glass transition temperature. A variety of thermoplastic materials have been used for hot embossing, including PMMA,[11,15–18] PS,[11,19,20] PC,[11] and cyclic olefin polymer (COP).[21–23] One of the desired material properties for hot embossing is thus a low glass-transition temperature (T_g). During the embossing process, the "negative" of the desired pattern is transferred to the substrate. For microfluidics applications, this means that a ridge in the master mold becomes a trench in the substrate, needing only a "roof" to create a complete microchannel. Since it is straightforward to create patterns of lines/ridges in the original mold, this method is well suited to create in-plane features for microfluidic circuits. However, interconnects and other out-of-plane features, like valves, are challenging to produce. The resolution of this process is comparable to conventional photolithographic resolution, but structures with high aspect ratios are challenging to obtain. Some variations of embossing afford faster fabrication times or higher resolution—in hot roller embossing,[24] heated rollers are used to emboss into polymer foils under pressure; in nanoimprint lithography,[25] a UV-curable polymer is placed in contact with a nano-featured mold, then cured in place and removed from the nanomold. Nanoimprint lithography allows for replication of much smaller features,[26] but is relatively expensive due to the high initial cost of making the mold.

Injection molding, on the other hand, requires melting a thermoplastic polymer, then forcing it under pressure inside a master mold, where it is allowed to solidify as the temperature is lowered below its melting temperature. The molded material is then ejected and released from the mold. When injection molding is used to create microdevices, microscale features can be micromilled into master molds.[27] Unlike embossing, injection molding is inherently a 3-D process, which enables the creation of out-of-plane features, such as access ports and valves. Early examples of injection-molded microdevices were created in acrylic,[28] but COP has since become more common[29–31] due to its superior material qualities.

While these two large-scale manufacturing processes may afford high-volume production at a low cost, there are some disadvantages associated with each of these processes. Both hot embossing and injection molding require a high setup cost to manufacture molds that can withstand the high temperatures and pressures associated with each fabrication method. Often metal molds[32] are used, which add significant costs to the process, especially if they are micromachined. Alternatively, epoxy molds[18,33] are less expensive, quicker to manufacture, have smoother surfaces compared to micromachined metal molds, and can be easily integrated with existing soft lithography methods and workflow. However, their lower mechanical strength and durability imply that fewer devices per mold can be produced before mechanical failure of the mold.[18]

These manufacturing techniques also offer different solutions to the so-called world-to-chip interfacing issue—how to interface a microdevice with macroscale liquid handling instruments (e.g., pipettes, syringe pumps, tubing). Typically, inlet and outlet ports and microchannels are created to act as the interface. This process tends to be very time-consuming in traditional hot embossing techniques, since it is challenging to emboss features that are tall enough to pierce through the entire thickness of the plastic device. Embossed devices thus require subsequent processes (e.g., drill pressing or laser ablation) to create access ports. In such cases, injection molding becomes attractive and practical for large-scale product manufacturing.

Finally, whether created by injection molding or hot embossing, the different layers of polymers must be bonded together to create an enclosed microfluidic channel. One common method of sealing the layers together is through thermal bonding, where separate layers are brought together in a heated press, the temperature is raised above the T_g of the material, the layers are kept in contact under temperature and pressure for a certain amount of time, then the temperature is lowered and the bonded piece ejected from the press.[34] Since heat and pressure are applied to a material with microfeatures, this step can result in material reflow and loss of feature fidelity, depending on the material used. For example, polystyrene can deform considerably below its T_g, making it more difficult to maintain features during thermal bonding. Other materials, such as PMMA and PC, have better feature retention when thermally bonded.

11.2.2 Rapid Prototyping Materials

Although thermoplastic polymers enjoyed the advantage of inheriting well-established fabrication methods, initial setup costs generally remained too high for most academic labs looking to make a small number of prototype devices. Furthermore, miniaturization of biochemical assays often involves many rounds of optimization, which consequently requires the ability to quickly and inexpensively iterate through different designs and geometries. Thus, materials that could be hot embossed and injection molded were not ideal, since each new design required a new mold, making the process unaffordable for most academic labs. For this reason, researchers searched for new materials and methods that could be used to rapidly prototype microfluidic devices inexpensively.

Rapid prototyping of microfluidic devices rapidly increased after 1998 when Whitesides developed a low-cost technique using PDMS. Dubbed "soft lithography,"[35,36] this method involved creating a master mold using photolithography (and, later, high-resolution printing on transparencies) followed by a replica molding process in which the "negative" of the master mold is reproduced in PDMS (Figure 11.2). The required materials were inexpensive, the process itself was straightforward and, importantly, dust tolerant, which abrogated the need for cleanroom fabrication.

One of the main advantages to using PDMS for small batch fabrication is the possibility to make microdevices without specialized equipment, such as a hot embosser or injection molding equipment. Soft lithography enabled the creation of microfluidic devices by essentially any laboratory (by outsourcing the processes of printing the mask or creating the master mold). These attractive qualities of soft lithography in PDMS greatly increased access to microfluidics fabrication and the number of researchers in the field.[37]

While extremely successful and accessible, fabrication in PDMS has some disadvantages that arise from the properties of the elastomeric material itself: (1) channel deformation (PDMS is soft and channels can collapse for certain geometries), (2) evaporation of carrier fluid (PDMS is very permeable to gases including water vapor), (3) adsorption of solutes/analytes onto the microchannel walls (PDMS is inherently hydrophobic and proteins may adhere to nontreated surfaces), (4) absorption of solvents (some organic solvents can diffuse into PDMS causing it to swell), (5) leaching of uncross-linked PDMS chains into the channel after device fabrication is complete, and (6) hydrophobic recovery (even if is treated with plasma oxidation to be hydrophilic, its hydrophobicity may recover within hours after device assembly). These characteristics, especially the time-sensitive surface properties, imply a short shelf life for PDMS devices and the need to create devices immediately before use. As a result, widespread commercialization of PDMS-based devices has not yet been fully realized.

In addition to PDMS, thin polymeric films have also been used as the substrate for rapidly prototyped microfluidic devices. Materials such as COP[21,22,38] have excellent material and optical properties and can be obtained in various formats (e.g., rolls, films[39]) that are amenable to large-scale production. These films can be stacked together to obtain devices of different heights from standard film thicknesses. Lamination of these different layers can be achieved by thermal bonding,

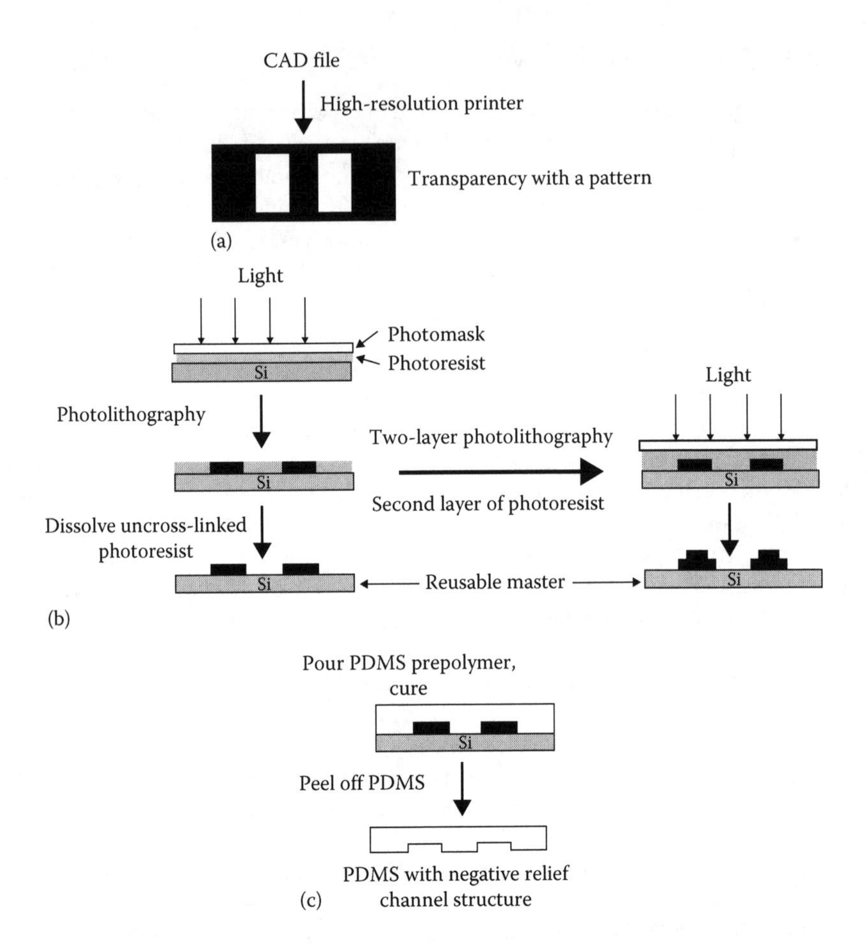

FIGURE 11.2 Schematic illustration for the soft lithography method for making PDMS microdevices. (a) A high-resolution transparency is used as a photomask. (b) The master mold is fabricated by spin coating photoresist onto a silicon wafer and cross-linked via UV exposure through the photomask. For 3-D structures, a second layer of photoresist is spin coated on and cross-linked through another photomask. The uncross-linked photoresist is dissolved to leave a positive relief of features. (c) Replica molds of microdevices are made by pouring PDMS onto the master mold, curing, and peeling the PDMS replica off the master. (Ng, J., Gitlin, I., Stroock, A., and Whitesides, G.: Components for integrated poly(dimethylsiloxane) microfluidic systems. *Electrophoresis*. 2002. 23. 3461–3473. Copyright Wiley-VCH Verlag GmbH & Co. KGaA. Reproduced with permission.)

where the layers are heated until polymer reflow occurs, or solvent-assisted bonding, where the surfaces to be bonded are first partially dissolved using a solvent, then brought into contact.[39]

In order to micropattern thin polymeric films for microfluidics, researchers have used direct-write methods, such as laser scribing and microcutting. Commercially available CO_2 lasers can be used to micromachine both plastics and silicon.[40–42] Either the laser itself or the substrate is mounted on an X–Y translation stage that is controlled using CAD software, and the laser beam is used to ablate material and create patterns into the substrate. The laser beam speed and the output power can be adjusted to achieve different channel depths. The main drawback of laser cutting lies primarily in the initial cost of the equipment and high maintenance costs. A simpler and more affordable fabrication method involves microcutting using a cutter plotter (Figure 11.3). A cutter plotter is similar to an ink plotter, but it cuts films with a knife

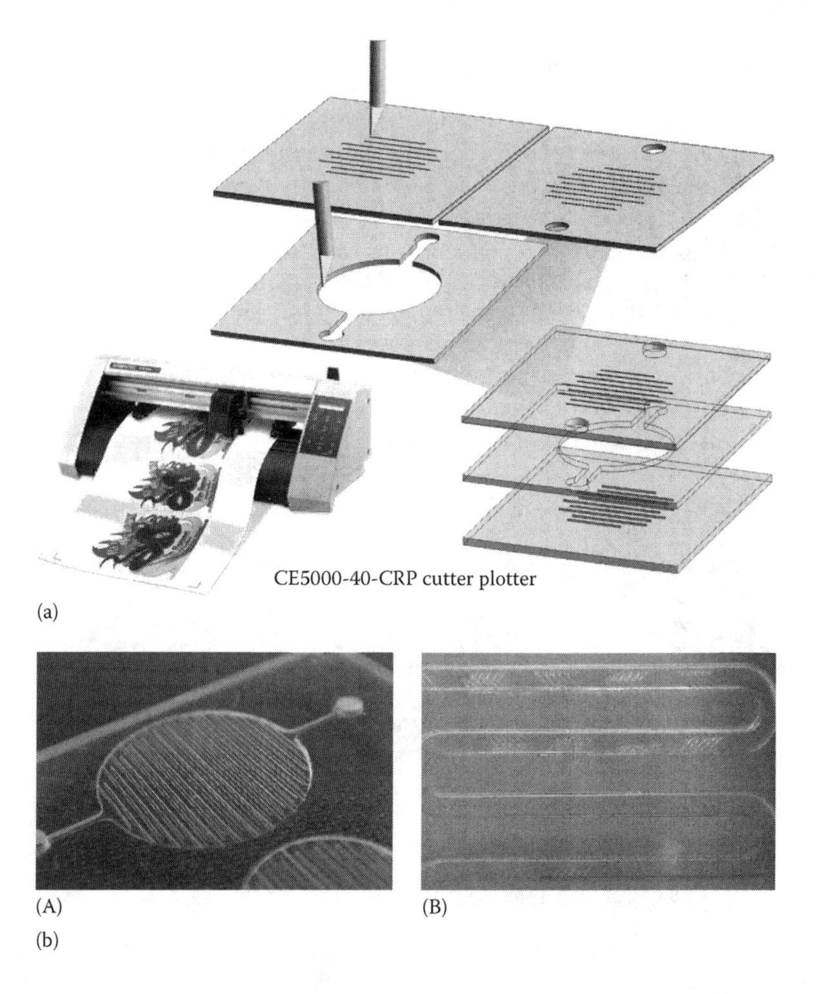

FIGURE 11.3 (a) Fabrication process of writing microfluidics using a cutter plotter showing both a through cut layer and scratched microfluidic patterned layers bonded to form a complete 3-D microfluidic channel. (b) Examples of fabricated microfluidic devices using the rapid prototyping technique: (A) a device for bubble-free filling of a low aspect ratio microfluidic chamber using surface tension flow guide and (B) a chaotic micromixer featuring raised features within a continuous flow channel.[45] (Reprinted from *Robot. Comput. Integr. Manuf.*, 27, Do, J., Zhang, J.Y., and Klapperich, C.M., Maskless writing of microfluidics: Rapid prototyping of 3D microfluidics using scratch on a polymer substrate, 245–248, Copyright 2011, with permission from Elsevier.)

blade instead of depositing ink onto them. As plotter technology has significantly improved, commercially available cutter plotters can now achieve a 10 μm resolution with high speed at a relatively low setup cost. Our group and others have demonstrated the use of cutter plotters to make microfluidic devices.[43,44] A major disadvantage of these fabrication methods is that they produce rough sidewalls, which can potentially interfere with device sealing, imaging across the device, and may cause sample loss to the wall due to increased roughness and an ill-defined surface.

11.3 PAPER-BASED SYSTEMS

An alternative substrate for microfluidic devices is paper. In paper-based microfluidics, the wicking properties of paper are used to drive fluid motion by capillary action. A familiar example of this principle is lateral-flow strips (e.g., home pregnancy tests). While the flow in these strips is one-dimensional, adding micropatterning to the paper substrate (e.g., by printing impermeable barriers) has allowed the creation of more sophisticated two-dimensional fluidic circuits. The printing of such barriers, generally made of wax, lessens concerns about any roughness of the sidewalls—now defined by the resolution of the printing process.

Whitesides demonstrated methods to fabricate microfluidics using paper and tape[45,46] (Figure 11.4), with a particular emphasis on applications in low-resource settings, including telemedicine approaches.[47,48] Yager demonstrated that simple fluidic sensors and filters can be translated into paper[49] and incorporated delays and timers into the fluidic circuit to further manipulate and control fluid movement in paper devices.[50] Whitesides[51,52] and others[53–56] also combined paperfluidic devices with electrochemical sensing, as a step toward integrating paperfluidics in standard sample analysis workflows.

These paperfluidic devices can perform simple assays and can easily be incinerated after use, obviating the need for biohazardous material disposal. Since the steps involved in the fabrication of these devices are straightforward (i.e., printing barriers, then cutting the paper substrates), this method has the potential to be mass manufactured at low cost. It is believed that scaling up paper-based diagnostics will become widespread.[57]

11.4 SPECIFIC APPLICATION—SAMPLE ANALYSIS AND DIAGNOSTICS

Microfluidic technologies have found a natural application in the sample analysis and diagnostics space, as they enable the miniaturization, integration, and automation of many biochemical assays.[58,59] On a laboratory scale, sample analysis generally entails three steps: (1) sample preparation—which includes all processes that extract a target (e.g., protein, nucleic acid) from a biological specimen (e.g., cell, biological fluid) for subsequent molecular assaying or detection, (2) the biochemical/molecular assay—which includes all processes used to prepare the target for subsequent detection, and (3) sample detection and readout—which includes all processes that enable detection or quantification of the target. By miniaturizing each step, microfluidics offers the possibility to integrate the entire bioanalytical process onto

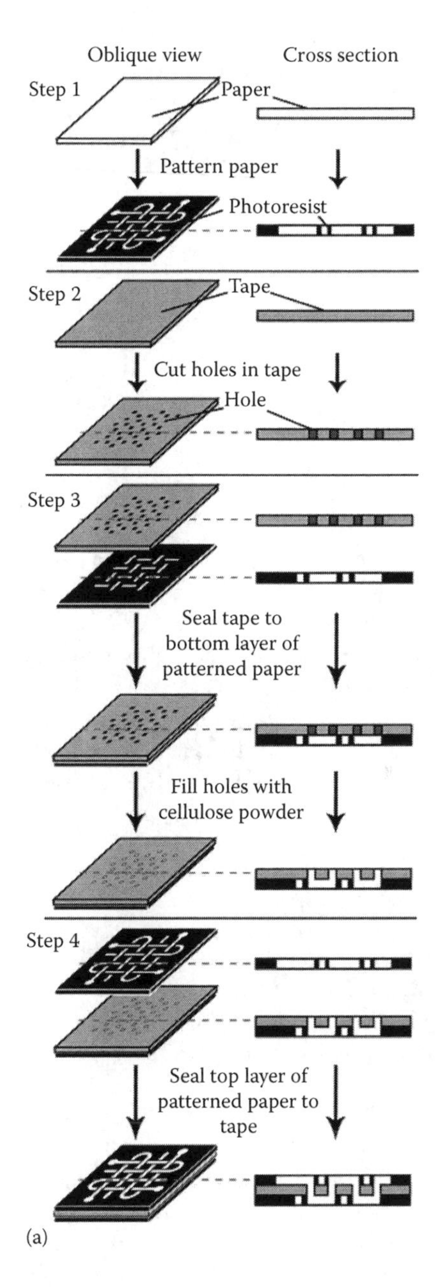

FIGURE 11.4 Preparation of a 3-D paper microfluidic device.[46] (From Martinez, A.W., Phillips, S.T., and Whitesides, G.M., Three-dimensional microfluidic devices fabricated in layered paper and tape, *Proc. Natl. Acad. Sci. USA*, 105, 19606–19611, and Copyright 2008 National Academy of Sciences, USA.)

a single device. One salient advantage of miniaturization is in point-of-care (POC) applications, where a microfluidic platform with a small footprint that works with inexpensive microfluidic cartridges may put fewer constraints (e.g., cost, resources) on the location where an assay is performed. In the following section, we will highlight recent technologies that have been developed to translate one or a combination of the three general steps in sample analysis onto a microfluidic platform.

11.4.1 Sample Preparation

Oftentimes, a number of preparatory steps are needed to render a sample compatible with downstream assaying. For nucleic acid (NA) analysis, for example, cell lysis, NA extraction, and amplification need to be carried out before actual NA manipulation (e.g., separation by size, detection) can occur. A few labs are working on miniaturizing these sample preparation steps onto microfluidic platforms so that they can be integrated into all-microfluidic systems. This approach would enable the creation of miniaturized platforms for complete analysis that can be portable, low cost, and easily distributed.

As an example of a disposable sample preparation system for different biological and diagnostic applications, we developed thermoplastic microfluidic modules for solid-phase extraction-based isolation of NAs from eukaryotic cells,[23,60] urine samples,[61] and whole blood.[62] The chips are fabricated in COP by hot embossing with a master mold. The solid phase consisted of a porous, monolithic polymer column impregnated with silica particles, which enabled isolation of NAs within the polymeric microdevice. To facilitate subsequent analysis of extracted NA samples, we also developed a portable, pressure-driven NA extraction system capable of storing NAs on-chip for over a week.[63]

11.4.2 Biochemical/Molecular Assays

Conventional assays have been miniaturized in both PDMS and thermoplastics, for applications that include manipulation and analysis of cells, nucleic acids and proteins, biosensors for biochemical and pathogen detection, and POC diagnostics.[64] Owing to the high surface-to-volume ratios in microfluidic devices, these miniaturized conventional assays enjoy reduced reaction times and increased yields; additionally, simply reducing the device footprint allows for large-scale parallelization of these assays.

In one of the most successful demonstrations of microfluidic parallelization, Quake used rapid prototyping in PDMS to develop high-density microfluidic chips with hundreds of individually addressable chambers. Their operation was based on networks of thousands of micromechanical PDMS valves that exploit its elastomeric properties and can be computer controlled for fast operation[65,66] (Figure 11.5). This platform was initially demonstrated to obtain and analyze DNA from bacterial cells and has since become widely used and commercialized (Fluidigm Corp.), with applications ranging from miRNA expression analysis[67] to genotyping.[68] This platform is

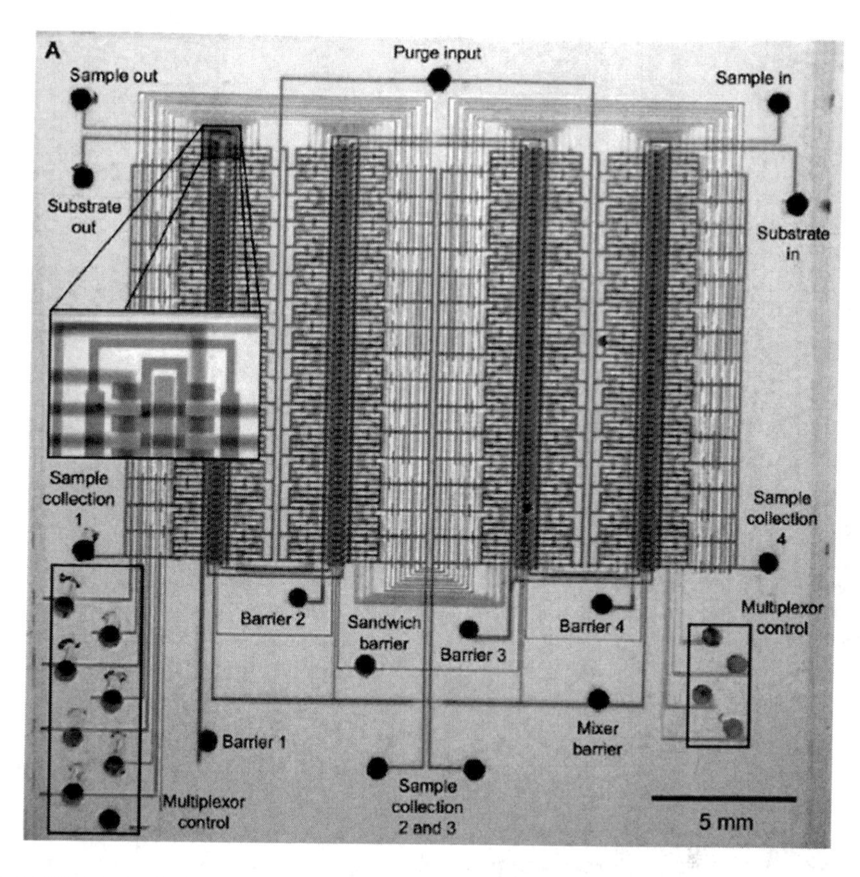

FIGURE 11.5 A high-density microfluidic chip with hundreds of individually addressable chambers. The device shown here is the microfluidic analog of an electronic comparator array. (From Thorsen, T. et al., *Science*, 298, 580, 2002.)

compatible with 96-well plate fluorescence readers and easily integrates into existing biochemical workflows.

Other examples of conventional, laboratory-scale assays that have been miniaturized include on-chip electrophoretic separation for gene analysis,[32] protein chip microarrays,[69,70] and nucleic acid amplification, which has been demonstrated by our group[18,33] and others.[71] In all cases, reductions in sample consumption and improvements in reaction times were observed.

Particularly in the field of cancer diagnostics, microfluidic technologies have been playing a larger role in both detection and therapy monitoring. In detection applications, cancer-specific protein biomarkers[69] can be identified directly in patient samples. In addition, microfluidic devices are very adept at physically separating and sorting circulating tumor cells (CTCs), which are shed by tumors and are larger than red and white blood cells. This ability to sort cells according to their size is realized due to laminar flow effects only seen at the microscale.[72,73] This strategy offers the

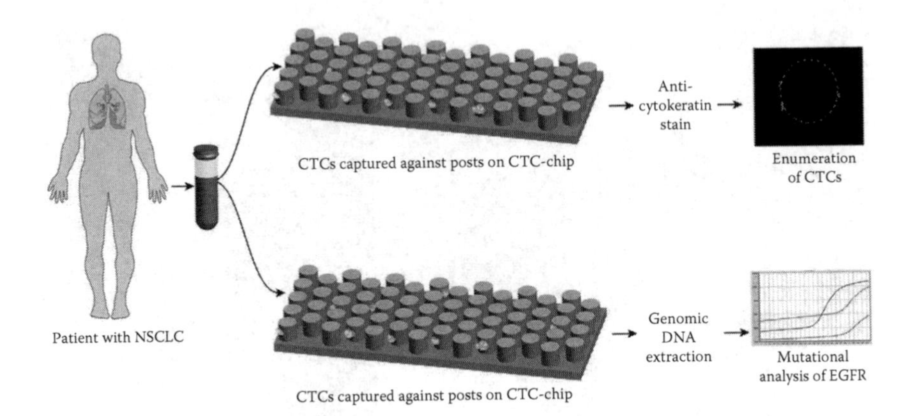

FIGURE 11.6 Illustrative cartoon demonstrating a patient with nonsmall cell lung cancer (NSCLC) donating a tube of peripheral blood, which is then processed in the circulating tumor cell (CTC)-chip immediately, without any required preprocessing. CTCs are captured against the sides of the anti-Ep-CAM-coated posts (epithelial cell adhesion molecule) and then can be stained with fluorescently labeled markers for enumeration or undergo genomic DNA extraction for epidermal growth factor receptor mutation or other molecular analysis.[82] (Reprinted from *J. Thorac. Oncol.*, 4, Sequist, L.V., Nagrath, S., Toner, M., Haber, D., and Lynch, T.J., The CTC-chip, 281–283, Copyright 2009, with permission from Elsevier.)

potential to obtain "liquid biopies" from cancer patients at different time points, to provide up-to-date molecular information about the cancer, and how patients respond to therapy[74–77] (Figure 11.6).

Toner et al. have pioneered this field with the "CTC-chip,"[78–81] with a micropillar-based cell sorter that is capable of processing large volumes of whole blood samples in a clinical setting. Cancer cells passing through the device can be captured on an array of microposts coated with monoclonal antibodies, allowing for the isolation of extremely rare cell subpopulations (1 target cell in 1 billion blood cells).

Other groups have used different strategies to capture CTCs, including the group of Di Carlo, which combined microscale vortices and inertial focusing[82,83] for high-purity extraction and enumeration of CTCs from the blood of patients with breast and lung cancer.[84,85] These technologies and similar approaches to enumerating CTCs can also be applied to real-time monitoring of chemotherapy response and have been fabricated in thermoplastics, showing that high-volume manufacturing of such devices is possible.[86,87]

11.4.3 Partially and Completely Integrated Systems

With the miniaturization of each step of sample analysis, it is now becoming feasible to create all-microscale platforms that replace laboratory-scale instruments. Some groups are addressing the formidable challenge of integrating the various steps to enable true sample-in, answer-out systems.

Our group and others have been developing minimally instrumented systems, where the sample preparation and assays steps are performed on a microfluidic platform, after which the samples are then detected or the results are read using standard laboratory instruments (e.g., fluorescent reader). An example of a partially integrated system developed by our group is a disposable microfluidic device that performs helicase-dependent isothermal amplification of DNA, starting with extraction from live bacteria in either liquid culture or stool samples collected from infected patients.[88,89] In both cases, the microfluidic extraction and amplification processes are followed by detection and quantification by gel capillary electrophoresis or polyacrylamide gel analysis. These systems do not require electricity or refrigeration and represent a major step forward in facilitating the use of molecular diagnostics in POC and resource-limited settings.

Progress toward a fully integrated microfluidic analytical system (microTAS) is slowly being realized with the development of, among others, "centrifugal microfluidics."[90,91] This technology exploits centrifugal forces in a spinning disc to drive fluid motion in microfluidic channels. Dubbed "Lab on a CD," one such disc-based system demonstrated the ability to fully automate plasma separation from whole blood and perform an enzyme-linked immunosorbent assay to detect either antigens or antibodies of the Hepatitis B Virus on its platform.[92,93]

Similarly, we have demonstrated a fully integrated microTAS for detecting bacteria via fluorescence detection.[94] The system, comprising a single-use disposable chip that interfaces with a tabletop instrument, performs DNA isolation from liquid samples containing whole *Bacillus subtilis* cells, PCR amplification of the isolated DNA, and subsequent fluorescent detection of the amplicons all within the platform. The instrument was designed to be compact, low cost, and manufacturable for POC applications.

11.5 CONCLUSION

In this chapter, we have focused on microfluidic devices used in biomedical applications (analysis and diagnostics), with an emphasis on devices developed in an academic research setting and usable at the POC. We have discussed how the choice of starting material has a fundamental influence on both the potential applications and the ease with which microfluidic devices can be fabricated in a lab setting. We have considered three fabrication processes: (1) the processes inherited from the microelectronics industry to shape, add, or remove material via lithography, deposition, or etching, respectively; (2) the molding processes adapted from industrial plastics manufacturing; and (3) the rapid prototyping strategies that use lithography and direct writing methods. These processes in turn have implications for feature fidelity, resolution, and equipment costs. We also briefly discussed the exciting potential of paper-based microfluidics.

In addition to the manufacturability issues that one must consider when fabricating microdevices, other considerations must also be taken into account. Beebe[19] reviewed the advantages and disadvantages of different classes of materials, with a particular focus on the question of why PDMS has become the material of choice for engineering groups (because of its ease of fabrication), while other plastics, especially polystyrene, remain the preferred substrate for biologists (because of their well-known surface properties and interactions with biological systems). Thus, adoption by the end user is another consideration that is orthogonal to manufacturability, but affects material selection and eventual applications.

An additional consideration for most bio-related diagnostic applications is the typical requirement for single-use chips that avoid cross-contamination from different patient samples and can be easily (and safely) disposed of. This aspect is particularly important for infectious disease diagnostics, where devices necessarily come in contact with biohazardous samples. In these cases, plastic and, in particular, paper devices can easily be incinerated, unlike silicon- and glass-based devices.

Another implication of the chosen material is the effect of the solvent on the integrity of the devices. While most applications with biological samples (cell, proteins, DNA) are in aqueous solutions, some applications require organic solvents, and one must ensure that these solvents will not compromise the structural integrity of the microfluidic device. In the case of PDMS, device damage occurs by excessive swelling of the material and breaking of the bonded device. For other polymers, such as thermoplastics, organic solvents cannot be used as working fluids since they may dissolve the device.

Finally, any material choice must also take into consideration the interaction of biological samples with the exposed surfaces of the device—often termed *biofouling*. This problem can be mitigated by using coatings on the exposed device surfaces; however, this step introduces complexity and cost, and one would ideally want to avoid it. Any method for manufacturing microfluidics devices must include a strategy for passivating or otherwise preparing the surface.

REFERENCES

1. Terry, S. C., Jerman, J. H., and Angell, J. B. A gas chromatographic air analyzer fabricated on a silicon wafer. *IEEE Trans. Electron. Dev.* **26**, 1880–1886 (1979).
2. Wohltjen, H. Chemical microsensors and microinstrumentation. *Anal. Chem.* **56**, 87A–103A (1984).
3. Mathies, R. A. and Huang, X. C. Capillary array electrophoresis: An approach to high-speed, high-throughput DNA sequencing. *Nature* **359**, 167–169 (1992).
4. Harrison, D. J. et al. Micromachining a miniaturized capillary electrophoresis-based chemical analysis system on a chip. *Science* **261**, 895–897 (1993).
5. Woolley, A. T. et al. Functional integration of PCR amplification and capillary electrophoresis in a microfabricated DNA analysis device. *Anal. Chem.* **68**, 4081–4086 (1996).
6. Woolley, A. T., Lao, K., Glazer, A. N., and Mathies, R. A. Capillary electrophoresis chips with integrated electrochemical detection. *Anal. Chem.* **70**, 684–688 (1998).
7. Manz, A., Graber, N., and Widmer, H. M. Miniaturized total chemical analysis systems: A novel concept for chemical sensing. *Sens. Actuators B Chem.* **1**, 244–248 (1990).
8. Becker, H. and Heim, U. Polymer hot embossing with silicon master structures. *Sens. Mater.* **11**, 297–304 (1999).
9. Becker, H. and Gartner, C. Polymer microfabrication methods for microfluidic analytical applications. *Electrophoresis* **21**, 12–26 (2000).
10. Becker, H. Polymer microfluidic devices. *Talanta* **56**, 267–287 (2002).
11. Becker, H. and Gärtner, C. Polymer microfabrication technologies for microfluidic systems. *Anal. Bioanal. Chem.* **390**, 89–111 (2008).
12. Makamba, H., Kim, J. H., Lim, K., Park, N., and Hahn, J. H. Surface modification of poly(dimethylsiloxane) microchannels. *Electrophoresis* **24**, 3607–3619 (2003).

13. Fiorini, G. S. and Chiu, D. T. Disposable microfluidic devices: Fabrication, function, and application. *Biotechniques* **38**, 429–446 (2005).
14. Kuo, J. S. and Chiu, D. T. Disposable microfluidic substrates: Transitioning from the research laboratory into the clinic. *Lab Chip* **11**, 2656–2665 (2011).
15. Mathur, A. et al. Characterisation of PMMA microfluidic channels and devices fabricated by hot embossing and sealed by direct bonding. *Curr. Appl. Phys.* **9**, 1199–1202 (2009).
16. Lee, G.-B., Chen, S.-H., Huang, G.-R., Sung, W.-C., and Lin, Y.-H. Microfabricated plastic chips by hot embossing methods and their applications for DNA separation and detection. *Sens. Actuators B Chem.* **75**, 142–148 (2001).
17. Wang, Y. et al. Microarrays assembled in microfluidic chips fabricated from poly(methyl methacrylate) for the detection of low-abundant DNA mutations. *Anal. Chem.* **75**, 1130–1140 (2003).
18. Cao, Q. et al. Microfluidic chip for molecular amplification of influenza A RNA in human respiratory specimens. *PLoS One* **7**, e33176 (2012).
19. Berthier, E., Young, E. W. K., and Beebe, D. Engineers are from PDMS-land, biologists are from polystyrenia. *Lab Chip* **12**, 1224–1237 (2012).
20. Huang, Y., Mather, E. L., Bell, J. L., and Madou, M. MEMS-based sample preparation for molecular diagnostics. *Anal. Bioanal. Chem.* **372**, 49–65 (2002).
21. Kameoka, J. et al. An electrospray Ionization source for integration with microfluidics. *Anal. Chem.* **74**, 5897–5901 (2002).
22. Nunes, P. S., Ohlsson, P. D., Ordeig, O., and Kutter, J. P. Cyclic olefin polymers: Emerging materials for lab-on-a-chip applications. *Microfluid. Nanofluid.* **9**, 145–161 (2010).
23. Chatterjee, A. et al. RNA isolation from mammalian cells using porous polymer monoliths: An approach for high-throughput automation. *Anal. Chem.* **82**, 4344–4456 (2010).
24. Yeo, L. P. et al. Investigation of hot roller embossing for microfluidic devices. *J. Micromech. Microeng.* **20**, 015017 (2010).
25. Austin, M. D. et al. Fabrication of 5 nm linewidth and 14 nm pitch features by nanoimprint lithography. *Appl. Phys. Lett.* **84**, 5299 (2004).
26. Guo, L. J. Nanoimprint lithography: Methods and material requirements. *Adv. Mater.* **19**, 495–513 (2007).
27. Attia, U. M., Marson, S., and Alcock, J. R. Micro-injection moulding of polymer microfluidic devices. *Microfluid. Nanofluid.* **7**, 1–28 (2009).
28. McCormick, R. M., Nelson, R. J., Alonso-Amigo, M. G., Benvegnu, D. J., and Hooper, H. H. Microchannel electrophoretic separations of DNA in injection-molded plastic substrates. *Anal. Chem.* **69**, 2626–2630 (1997).
29. Mair, D. A., Geiger, E., Pisano, A. P., Fréchet, J. M. J., and Svec, F. Injection molded microfluidic chips featuring integrated interconnects. *Lab Chip* **6**, 1346–1354 (2006).
30. Steigert, J. et al. Rapid prototyping of microfluidic chips in COC. *J. Micromech. Microeng.* **17**, 333–341 (2007).
31. Kim, D. S., Lee, S. H., Ahn, C. H., Lee, J. Y., and Kwon, T. H. Disposable integrated microfluidic biochip for blood typing by plastic microinjection moulding. *Lab Chip* **6**, 794–802 (2006).
32. Hupert, M. L. et al. Evaluation of micromilled metal mold masters for the replication of microchip electrophoresis devices. *Microfluid. Nanofluid.* **3**, 1–11 (2006).
33. Cao, Q., Kim, M.-C., and Klapperich, C. Plastic microfluidic chip for continuous-flow polymerase chain reaction: Simulations and experiments. *Biotechnol. J.* **6**, 177–184 (2011).
34. Tsao, C.-W. and DeVoe, D. L. Bonding of thermoplastic polymer microfluidics. *Microfluid. Nanofluid.* **6**, 1–16 (2008).

35. Ayon, A. Molding of deep polydimethylsiloxane microstructures for microfluidics and biological applications. *J. Biomech. Eng.* **121**, 28 (2007).
36. Ng, J., Gitlin, I., Stroock, A., and Whitesides, G. Components for integrated poly(dimethylsiloxane) microfluidic systems. *Electrophoresis* **23**, 3461–3473 (2002).
37. Whitesides, G. M. The origins and the future of microfluidics. *Nature* **442**, 368–373 (2006).
38. Bhattacharyya, A. and Klapperich, C. M. Microfluidics-based extraction of viral RNA from infected mammalian cells for disposable molecular diagnostics. *Sens. Actuators B Chem.* **129**, 693–698 (2008).
39. Paul, D., Pallandre, A., Miserere, S., Weber, J., and Viovy, J.-L. Lamination-based rapid prototyping of microfluidic devices using flexible thermoplastic substrates. *Electrophoresis* **28**, 1115–1122 (2007).
40. Klank, H., Kutter, J. P., and Geschke, O. CO(2)-laser micromachining and back-end processing for rapid production of PMMA-based microfluidic systems. *Lab Chip* **2**, 242–246 (2002).
41. Neils, C., Tyree, Z., Finlayson, B., and Folch, A. Combinatorial mixing of microfluidic streams. *Lab Chip* **4**, 342–350 (2004).
42. Munson, M. S., Hasenbank, M. S., Fu, E., and Yager, P. Suppression of non-specific adsorption using sheath flow. *Lab Chip* **4**, 438–445 (2004).
43. Bartholomeusz, D. A., Boutte, R. W., and Andrade, J. D. Xurography: Rapid prototyping of microstructures using a cutting plotter. *J. Microelectromech. Syst.* **14**, 1364–1374 (2005).
44. Do, J., Zhang, J. Y., and Klapperich, C. M. Maskless writing of microfluidics: Rapid prototyping of 3D microfluidics using scratch on a polymer substrate. *Robot. Comput. Integr. Manuf.* **27**, 245–248 (2011).
45. Martinez, A. W., Phillips, S. T., and Whitesides, G. M. Three-dimensional microfluidic devices fabricated in layered paper and tape. *Proc. Natl. Acad. Sci. USA* **105**, 19606–19611 (2008).
46. Martinez, A. W., Phillips, S. T., Wiley, B. J., Gupta, M., and Whitesides, G. M. FLASH: A rapid method for prototyping paper-based microfluidic devices. *Lab Chip* **8**, 2146–2150 (2008).
47. Martinez, A. W. et al. Simple telemedicine for developing regions: Camera phones and paper-based microfluidic devices for real-time, off-site diagnosis. *Anal. Chem.* **80**, 3699–3707 (2008).
48. Martinez, A. W., Phillips, S. T., Whitesides, G. M., and Carrilho, E. Diagnostics for the developing world: Microfluidic paper-based analytical devices. *Anal. Chem.* **82**, 3–10 (2010).
49. Osborn, J. L. et al. Microfluidics without pumps: Reinventing the T-sensor and H-filter in paper networks. *Lab Chip* **10**, 2659–2665 (2010).
50. Toley, B. J. et al. Tunable-delay shunts for paper microfluidic devices. *Anal. Chem.* **85**, 11545–11552 (2013).
51. Nie, Z. et al. Electrochemical sensing in paper-based microfluidic devices. *Lab Chip* **10**, 477–483 (2010).
52. Nie, Z., Deiss, F., Liu, X., Akbulut, O., and Whitesides, G. M. Integration of paper-based microfluidic devices with commercial electrochemical readers. *Lab Chip* **10**, 3163–3169 (2010).
53. Dungchai, W., Chailapakul, O., and Henry, C. S. Electrochemical detection for paper-based microfluidics. *Anal. Chem.* **81**, 5821–5826 (2009).
54. Abe, K., Suzuki, K., and Citterio, D. Inkjet-printed microfluidic multianalyte chemical sensing paper. *Anal. Chem.* **80**, 6928–6934 (2008).
55. Lu, Y., Shi, W., Jiang, L., Qin, J., and Lin, B. Rapid prototyping of paper-based microfluidics with wax for low-cost, portable bioassay. *Electrophoresis* **30**, 1497–1500 (2009).

56. Li, X., Ballerini, D. R., and Shen, W. A perspective on paper-based microfluidics: Current status and future trends. *Biomicrofluidics* **6**, (2012).
57. Mace, C. R. and Deraney, R. N. Manufacturing prototypes for paper-based diagnostic devices. *Microfluid. Nanofluid.* **16**, 801–809 (2013).
58. Haeberle, S. and Zengerle, R. Microfluidic platforms for lab-on-a-chip applications. *Lab Chip* **7**, 1094–1110 (2007).
59. Mark, D., Haeberle, S., Roth, G., von Stetten, F., and Zengerle, R. Microfluidic lab-on-a-chip platforms: Requirements, characteristics and applications. *Chem. Soc. Rev.* **39**, 1153–1182 (2010).
60. Bhattacharyya, A. and Klapperich, C. M. Thermoplastic microfluidic device for on-chip purification of nucleic acids for disposable diagnostics. *Anal. Chem.* **78**, 788–792 (2006).
61. Kulinski, M. D. et al. Sample preparation module for bacterial lysis and isolation of DNA from human urine. *Biomed. Microdevices* **11**, 671–678 (2009).
62. Mahalanabis, M., Al-Muayad, H., Kulinski, M. D., Altman, D., and Klapperich, C. M. Cell lysis and DNA extraction of gram-positive and gram-negative bacteria from whole blood in a disposable microfluidic chip. *Lab Chip* **9**, 2811–2817 (2009).
63. Byrnes, S. et al. A portable, pressure driven, room temperature nucleic acid extraction and storage system for point of care molecular diagnostics. *Anal. Methods* **5**, 3177–3184 (2013).
64. Yeo, L. Y., Chang, H.-C., Chan, P. P. Y., and Friend, J. R. Microfluidic devices for bioapplications. *Small* **7**, 12–48 (2011).
65. Melin, J. and Quake, S. R. Microfluidic large-scale integration: The evolution of design rules for biological automation. *Annu. Rev. Biophys. Biomol. Struct.* **36**, 213–231 (2007).
66. Thorsen, T., Maerkl, S. J., and Quake, S. R. Microfluidic large-scale integration. *Science* **298**, 580–584 (2002).
67. Jang, J. S. et al. Quantitative miRNA expression analysis using fluidigm microfluidics dynamic arrays. *BMC Genomics* **12**, 144 (2011).
68. Chan, M. et al. Evaluation of nanofluidics technology for high-throughput SNP genotyping in a clinical setting. *J. Mol. Diagn.* **13**, 305–312 (2011).
69. Hu, M. et al. Ultrasensitive, multiplexed detection of cancer biomarkers directly in serum by using a quantum dot-based microfluidic protein chip. *ACS Nano* **4**, 488–494 (2010).
70. Chang-Yen, D. A., Myszka, D. G., and Gale, B. K. A novel PDMS microfluidic spotter for fabrication of protein chips and microarrays. *J. Microelectromech. Syst.* **15**, 1145–1151 (2006).
71. Zhang, C., Xu, J., Ma, W., and Zheng, W. PCR microfluidic devices for DNA amplification. *Biotechnol. Adv.* **24**, 243–284 (2006).
72. Squires, T. and Quake, S. Microfluidics: Fluid physics at the nanoliter scale. *Rev. Mod. Phys.* **77**, 977–1026 (2005).
73. Pamme, N. Continuous flow separations in microfluidic devices. *Lab Chip* **7**, 1644–1659 (2007).
74. Pratt, E. D., Huang, C., Hawkins, B. G., Gleghorn, J. P., and Kirby, B. J. Rare cell capture in microfluidic devices. *Chem. Eng. Sci.* **66**, 1508–1522 (2011).
75. Chen, J., Li, J., and Sun, Y. Microfluidic approaches for cancer cell detection, characterization, and separation. *Lab Chip* **12**, 1753–1767 (2012).
76. Wlodkowic, D. and Cooper, J. M. Tumors on chips: Oncology meets microfluidics. *Curr. Opin. Chem. Biol.* **14**, 556–567 (2010).
77. Cima, I. et al. Label-free isolation of circulating tumor cells in microfluidic devices: Current research and perspectives. *Biomicrofluidics* **7**, 11810 (2013).
78. Toner, M. and Irimia, D. Blood-on-a-chip. *Annu. Rev. Biomed. Eng.* **7**, 77–103 (2005).

79. Nagrath, S. et al. Isolation of rare circulating tumour cells in cancer patients by microchip technology. *Nature* **450**, 1235–1239 (2007).
80. Wong, I. Y., Bhatia, S. N., and Toner, M. Nanotechnology: Emerging tools for biology and medicine. *Genes Dev.* **27**, 2397–2408 (2013).
81. Sequist, L. V., Nagrath, S., Toner, M., Haber, D., and Lynch, T. J. The CTC-chip. *J. Thorac. Oncol.* **4**, 281–283 (2009).
82. Di Carlo, D., Irimia, D., Tompkins, R. G., and Toner, M. Continuous inertial focusing, ordering, and separation of particles in microchannels. *Proc. Natl. Acad. Sci. U.S.A.* **104**, 18892–18897 (2007).
83. Di Carlo, D. Inertial microfluidics. *Lab Chip* **9**, 3038–3046 (2009).
84. Mach, A. J., Kim, J. H., Arshi, A., Hur, S. C., and Di Carlo, D. Automated cellular sample preparation using a Centrifuge-on-a-Chip. *Lab Chip* **11**, 2827–2834 (2011).
85. Sollier, E. et al. Size-selective collection of circulating tumor cells using Vortex technology. *Lab Chip* **14**, 63–77 (2014).
86. Jackson, J. M. et al. UV activation of polymeric high aspect ratio microstructures: Ramifications in antibody surface loading for circulating tumor cell selection. *Lab Chip* **14**, 106–117 (2014).
87. Adams, A. A. et al. Highly efficient circulating tumor cell isolation from whole blood and label-free enumeration using polymer-based microfluidics with an integrated conductivity sensor. *J. Am. Chem. Soc.* **130**, 8633–8641 (2008).
88. Mahalanabis, M., Do, J., ALMuayad, H., Zhang, J. Y., and Klapperich, C. M. An integrated disposable device for DNA extraction and helicase dependent amplification. *Biomed. Microdevices* **12**, 353–359 (2010).
89. Huang, S. et al. Low cost extraction and isothermal amplification of DNA for infectious diarrhea diagnosis. *PLoS One* **8**, e60059 (2013).
90. Gorkin, R. et al. Centrifugal microfluidics for biomedical applications. *Lab Chip* **10**, 1758–1773 (2010).
91. Cho, Y.-K. et al. One-step pathogen specific DNA extraction from whole blood on a centrifugal microfluidic device. *Lab Chip* **7**, 565–573 (2007).
92. Madou, M. et al. Lab on a CD. *Annu. Rev. Biomed. Eng.* **8**, 601–628 (2006).
93. Lee, B. S. et al. A fully automated immunoassay from whole blood on a disc. *Lab Chip* **9**, 1548–1555 (2009).
94. Sauer-Budge, A. F. et al. Low cost and manufacturable complete microTAS for detecting bacteria. *Lab Chip* **9**, 2803–2810 (2009).

12 Nanotemplated Materials for Advanced Drug Delivery Systems

Erica Schlesinger, Daniel A. Bernards, Rachel Gamson, and Tejal A. Desai

CONTENTS

12.1 INTRODUCTION

As healthcare costs rise, demand for reduced hospital or clinic visits and improved efficacy, safety, and user acceptability grow. Among solutions, companies and researchers seek to develop and commercialize long-acting controlled release drug delivery systems. The global revenue for these systems in 2013 was estimated at $181.9 billion with projected revenue growth to $212.8 billion by 2018 (Dewan, 2014). Enhanced drug uptake or permeability, advantageous pharmacokinetic profiles, and less frequent dosing aim to address issues of patient compliance, bioavailability, safety, and efficacy. Demand for such systems is broad and encompasses applications in the delivery of small molecules, peptides and proteins, and genetic material. The most common drug delivery routes of administration include oral, subcutaneous, ocular, and pulmonary. Therapeutic durations range from days to years, and systems may comprise of inorganic materials or polymers, some of which are biodegradable. Advancements in this area rely on the development of reproducible and controllable fabrication technologies and the availability of materials that are biocompatible and appropriate for system duration and design. Furthermore, therapeutic devices benefit from an understanding of the underlying mechanisms of controlled release as well as cellular and tissue responses to these devices and systems.

In the past three decades, medicine has witnessed the growth of nanotechnology and its translation to therapies, which has been made possible by advances in fabrication techniques, materials science, and analytical and imaging modalities. Nanotechnology-based drug delivery can enhance delivery systems by improving control of drug release, enhancing drug permeability across biological membranes, and reducing fibrotic responses to implants. Nanoscale features confer unique benefits to surface properties and specific interactions. Surface area-to-volume ratios are markedly increased, resulting in a significantly greater potential for surface interactions. Being on the same size scale as proteins, nanofeatures allow for interaction with, manipulation of, or even direct control over biomolecules. Additionally, nanoscale features may impact tissues on the cellular level through direct interactions with cellular proteins.

In this chapter, we introduce the use of nanoscale features in drug delivery systems. We will discuss systems that include both polymeric and inorganic materials fabricated with a variety of techniques. The applications we will cover include nanoporous membranes to control drug release, nanofeatures to improve adhesion, nanopillars and nanotopography to impact drug uptake and membrane permeability, and surface nanotopography on implants to reduce fibrotic response.

12.2 NANOPORES

In drug delivery systems, porous membranes are often employed as a means of modulating diffusion kinetics to control drug release. Drugs in solution pass through membranes with molecular diffusion driven by a concentration or osmotic gradient that is restricted by the tortuosity and constrictivity of the porous structure. Tortuosity is a measure of the actual diffusion path a molecule must transverse compared to a hypothetical direct path of diffusion in the absence of a membrane and accounts for the circuitous nature of the porous material. Constrictivity is a function of the ratio of the hydrodynamic diameter of the molecule-to-pore diameter. Diffusion rates can be tuned by altering the porosity, tortuosity, and constrictivity of the porous structure, which are defined by the density, diameter, volume, and interconnectivity of pores. There are a number of examples of continuous-release devices that use porous membranes to control diffusion of a therapeutic from a reservoir. Release rates for therapeutics are controlled in these systems by the size of the pore relative to the drug molecule, the length and structure of the pore or channel, and properties of the membrane material such as surface charge, polarity, and hydrophobicity. In some cases, release behavior can be further modulated with surface coatings. When the pore size is large compared to the molecule, release of drug from a reservoir through the porous membrane can be described by Fickian diffusion; however, when pore size is on the same order as the molecule diameter, diffusion through the pore is constrained, which can result in deviations from Fickian diffusion. When the pore is similar to the diameter of the molecule, "single-file" diffusion may occur, where molecular diffusion is influenced by the confinement within the pore and the presence of other molecules: the consequence is a constant release rate. In recent years, advances in micro and nanofabrication techniques have allowed for membranes with monodisperse pores and channels on the nanoscale, making it possible to achieve concentration-independent diffusion of proteins.

Fabrication of nanoporous membranes can be categorized as "top-down" (pores are introduced into a bulk material by substractive process) or "bottom-up" (pores are grown or introduced during the formation of the bulk material). Implementing top-down or bottom-up approaches will depend greatly on the specific material, and the processes and resulting porous structures both have their pros and cons. In this section, we will cover examples of nanoporous membranes fabricated using these approaches, focusing on materials, fabrication techniques, and resulting pore structure. Many of the nanoporous membranes discussed here are relevant to filtration applications as well as drug delivery systems. We will briefly discuss the use of these nanoporous membranes in various drug delivery applications.

12.2.1 Inorganic Nanoporous Materials

Inorganic materials are used to fabricate nanoporous membranes in drug delivery with good antifouling properties, well-controlled linear pores, and scalable manufacturing. Inorganic nanoporous materials are prevalent in biofiltration applications and are being applied to drug delivery as diffusion barriers in long-acting implants as well as drug-eluting coatings on implant surfaces. As discussed, by tuning pore diameter to be on the order of the hydrodynamic diameter of the therapeutic, single-file diffusion may be achieved and consequently drug release is concentration independent. In long-acting implants where a reservoir is loaded with a soluble therapeutic (such as a biologic), nanoporous membranes can be utilized as a diffusion barrier that results in constant-rate release over extended durations. Alternatively, inorganic nanoporous surface coatings may provide both drug loading and a means for controlled release. In these coatings, drugs may be loaded directly into the nanopores themselves, and constrained diffusion out of the pores results in sustained release of drug, which can be achieved with both small and large molecules. For such nanoporous coatings, pore size will control drug release rate and pore volume and density will determine drug-loading capacity (Figure 12.1).

Among inorganic materials, top-down approaches have primarily focused on nanoporous silicon, leveraging a breadth of expertise from the semiconductor industry. Two primary fabrication approaches predominate nanoporous membrane fabrication—photolithography and etching. Ferrari et al. developed several fabrication protocols for producing silicon membranes with nanopores ranging from 7 to 50 nm (Desai et al., 2000; Martin et al., 2005). Controlled release of several biologics using these silicon nanochannel membranes has been demonstrated both *in vitro* and *in vivo* (Martin et al., 2005). Furthermore, this and a similar technology have been integrated as the rate-limiting diffusion barrier for drug release from a subcutaneous implant systems such as NanoGATE (Gardner, 2006; Walczak et al., 2005), Debiotech's DebioStar™, Delpor's™ NANOPOR, and Alza's DUROS®.

An alternative approach to top-down fabrication of silicon nanoporous membranes used focused ion beam (FIB) etching combined with low-pressure chemical

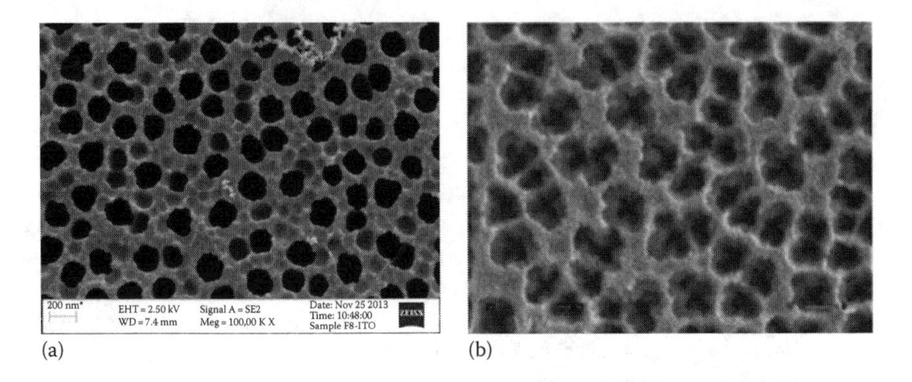

FIGURE 12.1 Inorganic nanoporous membranes. (a) Scanning electron microscope image of 20 nm anodized aluminum oxide (AAO) membrane. 200 nm scale bar. (b) Scanning electron microscope image of porous silicon. 1 μm scale bar. ([a]: From GE Healthcare, Piscataway, NJ; SEM image provided by Cade Fox; [b]: Moxon et al., *IEEE Trans. Biomed. Eng.*, 51(6), 881, 2004.)

vapor deposition (CVD). In this approach, nanopores were etched in SiN with a FIB and eventual pore sizes can be further tuned by chemical vapor deposition of SiN thin films. Pore sizes as small as 10 nm are achieved using this technique (Tong et al., 2004). These top-down approaches are highly controlled, allowing for precise control over pore features, including shape, orientation, size, density, and position.

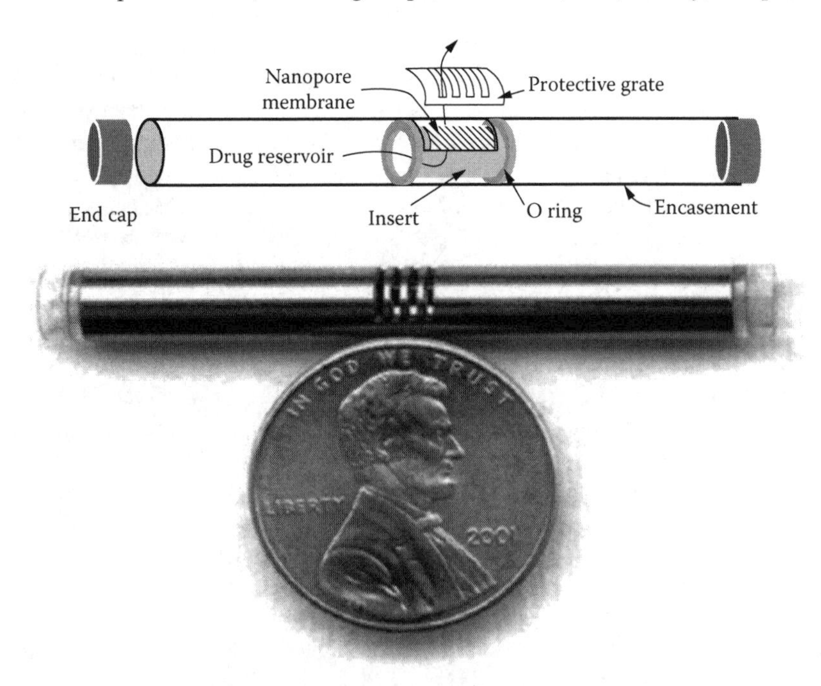

FIGURE 12.2 NanoGATE implant device. Nanoporous silicon membrane for controlled drug release from a titanium reservoir. (From Martin, F. et al., *J. Control. Release*, 102, 123, 2005.)

Unfortunately, FIB is a serial manufacturing technique, where each pore must be ablated individually and lacks any potential throughput enhancements associated with scale. Consequently, this approach is generally limited to relatively small surface areas or low pore densities, where precision and user-specified design is required (Figure 12.2).

The most common and versatile approach for bottom-up fabrication of inorganic nanoporous surfaces is anodization, where electrochemical oxidation generates nanostructures in the material. A range of metals, including aluminum, titanium, niobium, zirconium, and tungsten, can be anodized to generate membranes, and aluminum and titanium are the most common metals developed for therapeutic uses. In general, pore diameter and length are controlled by anodization process parameters, such as temperature, voltage, and anodization duration. In aluminum, anodization can create uniform hexagonal patterns of aluminum oxide pores that extended into the material perpendicular to the surface (Lee and Park, 2014; Masuda et al., 1997). Pore diameters less than 10 nm and up to hundreds of nanometers can be achieved with densities ranging from 10^{10} to10^{12} pores/cm^2. Anodic aluminum oxide (AAO) nanoporous membranes in 20, 100, and 200 nm pore diameters are commercially available (*Whatman, GE Healthcare*). AAO nanoporous materials have been used both in controlled-release drug delivery devices as diffusion-limiting membranes and in drug-eluting coatings on implants (Briggs et al., 2004; Gultepe et al., 2010). The bottom-up fabrication of anodized membranes allows for the processing of larger surface areas, uniform sizes, and high pore densities.

12.2.2 Polymeric Nanoporous Materials

While comparatively softer and with less mechanical strength than inorganic membranes, polymeric membranes tend to be more flexible and pliable and present their own advantageous material properties. A broad spectrum of polymers is available today from both synthetic and naturally derived sources; furthermore, continuous development generates novel polymers and improves upon existing ones. For the application of drug delivery, biocompatible and biodegradable polymers are typically the focus of development. Common biodegradable polymers degrade over timescales ranging from days to years. In most cases, degradation rate can be tuned through adjustments to the polymer's physical structure, such as molecular weight, degree of cross-linking, or incorporation of secondary copolymers. An additional benefit of polymeric nanoporous materials is the ability to fabricate as functional thin films, which is challenging to implement in common inorganic materials. Various fabrication approaches for nanoporous polymeric membranes have been developed for nano- and ultrafiltration applications, but more recently polymeric nanoporous materials have been developed as diffusion-limiting membranes in drug delivery systems (Bernards and Desai, 2010; Figure 12.3).

Many commercially available nanoporous polymer membranes utilize the top-down fabrication approach of track etching, where the bulk polymer is irradiated with uranium or californium and chemical etching is employed to generate porosity (Martin, 2005; Ulbricht, 2006). Each through pore corresponds to the path of an

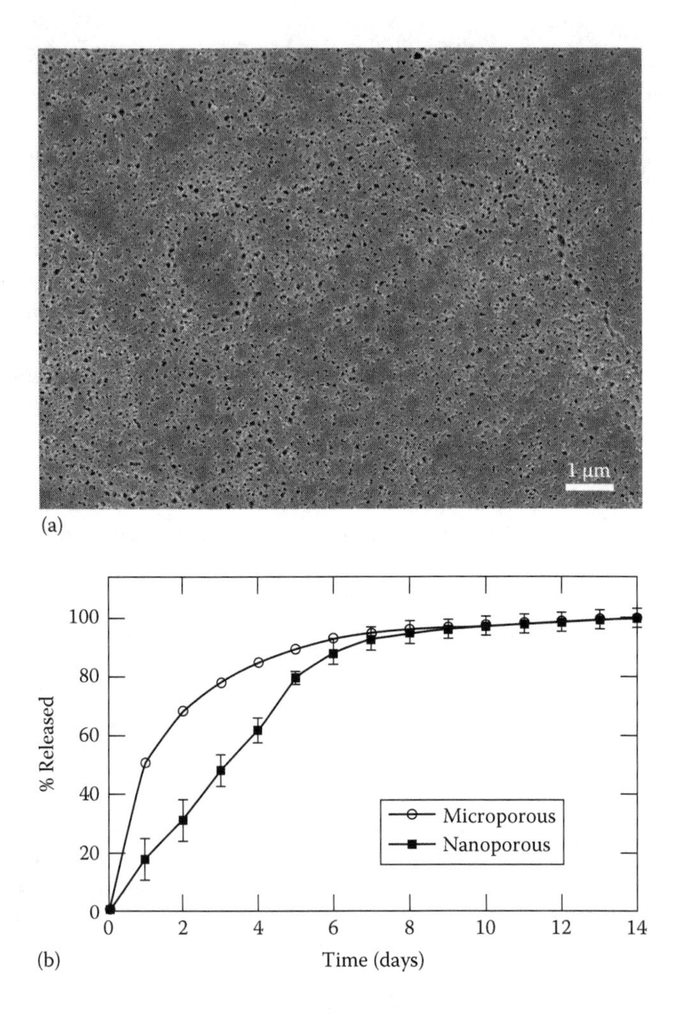

FIGURE 12.3 Polymeric nanoporous membrane. (a) Scanning electron microscope image of polycaprolactone thin film nanoporous membrane with 20–30 nm pore size. Scale bar 1 µm. (b) FITC-BSA release profiles from nanoporous polycaprolactone thin film (linear) and microporous polycaprolactone (nonlinear). Nanoporous membranes are used to achieve constant-rate release of protein therapeutics. ([a]: Daniel Bernards, unpublished, [b]: Adapted from Bernards, D. and Desai, T., *Soft Matter*, 6, 1621, 2010.)

irradiated particle, which results in uniform pores with diameters tunable to as small as 10 nm and as large as several microns. Pore size and shape depend on the polymer and etchant of choice as well as process parameters, such as etching temperature and level of irradiation energy. Pore densities in track-etched membranes range from 10^6 to 10^9 pores/cm^2, significantly lower than those achieved in anodized membranes (Martin et al., 2005). For filtration applications, commercially available track-etched polymer membranes come in a variety of polymers, including polycarbonate, polyethylene terephthalate, polypropylene, polyvinylidene fluoride, and polyether sulfone. In drug delivery, Rao et al. (2003) demonstrated the sustained release of ciprofloxacine hydrochloride (an antibiotic) from transdermal patches applied after medical procedures that employed track-etched polycarbonate membranes for drug delivery. Track-etched membranes are also commonly used as a component in more complex stimuli-responsive systems (Rattan and Sehgal, 2012).

In another approach for fabricating polymeric nanoporous thin films, the Desai lab uses a sacrificial zinc oxide template to form nanopores in polycaprolactone (PCL) films. Similar to the anodization process described, uniform zinc oxide nanorods can be grown perpendicular to a surface coated with a thin zinc oxide film. Activated by a thermal rather than an electrical driving force, density, diameter, and length of hydrothermally grown rods are affected by a variety of processing parameters, including growth temperature, duration, and precursor chemistry. Onto this nanorod template a polymer, such as PCL, is solvent cast, which results in a thin film on the nanorod template that partially covers the rods. The zinc oxide rods are subsequently etched, leaving nanopores in the PCL thin film. Constant-rate release of biologics has been demonstrated from reservoir drug delivery devices comprising thin films fabricated using this approach with a typical pore size of 20–30 nm. Such devices are being developed for long-acting controlled release of ocular therapeutics.

The most prevalent bottom-up approach for forming nanoporous polymer membranes is the self-assembly of block copolymers to create pores in polymer thin films. Block copolymers consist of two or more homopolymers covalently linked together in alternating patterns of polymer blocks. For a block copolymer composed of two homopolymers that are thermodynamically incompatible, phase separation can occur on the nanoscale creating phases rich in one of the polymer constituents. Because the homopolymers are covalently linked, rearrangement and organization of polymer chains are limited during rearrangement and these regions are confined to the nanoscale and defined by homopolymer block lengths and thermodynamic properties. The bulk film can then be treated chemically or physically to remove a single polymer phase, leveraging physical and chemical properties of the enriched sacrificial polymer to create nanopores. With this approach, high pore density ($>10^{11}$ pores/cm^2) can be achieved with uniform controllable pore sizes in the tens of nanometers (Vriezekolk et al., 2014). A higher density of open nanopores can be achieved in submicron thin films compared to thicker micron-scale films. To improve the mechanical integrity of films less than a micron thick, a nonrate limiting polymeric backing layer can be applied. Resulting pore characteristics and bulk film properties depend on the polymer and the processing parameters used to form nanopores. Material choice for block copolymers is limited by both on the need for thermodynamically incompatible homopolymers and a removal technique exclusive to one of the two polymers. A variety of block copolymers have successfully been used in creating nanoporous films, including polystyrene-block-poly2-vinylpyridine (PS-*b*-P2VP) (Yin et al., 2013), polystyrene-block-PMMA (PS-*b*-PMMA) (Park, 2007), and polyisoprene-block-polystyrene-block-poly4-vinylpyridine (PI-*b*-PS-*b*-P4VP) (Wang et al., 2013).The majority of work to date on block copolymers is in filtration and separation technologies and utilizes nonbiodegradable polymers, but the monodisperse and well-controlled pore structure in these polymers make them ideal materials for controlled drug delivery applications (Jackson and Hillmyer, 2010).

In this section, we have provided an overview of the basic techniques for creating nanoporous materials and their application to sustained drug delivery. The most prevalent approaches have been highlighted, and additional techniques are continuously developed and refined every year. We have focused on the use of nanoporous materials for loading and release of therapeutics, emphasizing sustained and controlled drug delivery. Another area of drug delivery from nanoporous materials is on-demand or pulsatile release from nano- or microparticles devices. On-demand or pulsatile devices often require stimuli-responsive nanoporous materials, and given existing high-quality resources in this area they are omitted from this chapter. Additional resources regarding such systems can be found elsewhere (James, 2014; Caldorera-Moore, 2009).

12.3 NANOTEMPLATING FOR ADHESION

There are many examples of the need for improved adhesion in drug delivery. First, adhesion can function in drug delivery systems to simply secure devices to the area of application. Adhesion to the skin is integral to transdermal drug delivery and wound healing. For oral, vaginal, pulmonary, and ocular therapies, mucoadhesive systems are beneficial to drug delivery: enhanced adhesion to the mucosal lining increases residence time and can improve the bioavailability of pharmaceuticals with poor permeability or solubility. There are many examples of mucoadhesive systems in ocular, buccal, intestinal, and pulmonary drug delivery utilizing polymer coatings or ligands. Recently, it has been demonstrated that micro- or nanostructures are another potential adhesive mechanism for drug delivery devices. In this section, we focus on materials that adhere through mechanical forces originating from micro- and nanotemplated features.

12.3.1 Gecko-Inspired Materials

Gecko-inspired materials are based on the strong adhesive properties inherent to gecko feet, which allow them to climb walls and hang from ceilings. Gecko feet have many micro- and nanoscale angled fibers (Figure 12.4). This micro- and nanotopography contributes to greater van der Waals forces through large numbers of surface contact points (Murphy et al., 2009). Depending on the design, these surfaces can improve both wet and dry adhesion, making them relevant to both transdermal drug delivery in the form of adhesive patches, mucosal delivery for improved mucoadhesion, and wound healing. Gecko-inspired materials employ micro- and nanopillars as fibrillar structures to mimic the "hairs" on a gecko's foot. With advanced fabrication techniques, these features can range from a single array to multilevel hierarchical structures (Murphy et al., 2009).

Synthetic gecko-inspired surfaces were made possible through nano- and microfabrication techniques such as nanomolding and electron-beam lithography. Since the understanding of this underlying adhesive surface topography in geckos, many studies have focused on the effects of structure, material, and geometry for optimizing synthetic gecko-inspired surfaces. When considering gecko-inspired adhesive

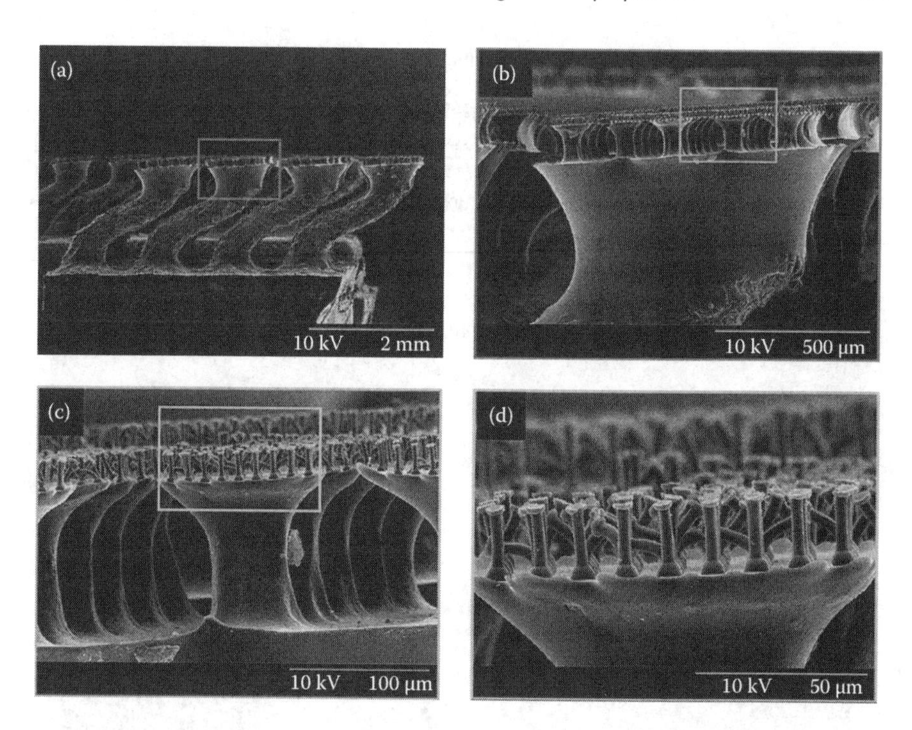

FIGURE 12.4 Hierarchical gecko-inspired topography. Scanning electron microscope image of three-level hierarchical polyurethane fibers created by Murphy et al. (a) Curved base-level fibers. (b, c) Mid-level fibers on top surface of base fibers. (d) Third-level fibers on top surface of mid-level fibers. (From Murphy, M. et al., *Small*, 5(2), 170, 2009.)

surfaces, both the geometry of the tips and the structure and organization of the fibers affect adhesive properties. Multilevel hierarchical structures increase adhesion relative to single-level structures. Pillars with larger tip-to-base diameter ratios and with larger tip diameter-to-pillar length ratios were shown to have higher adhesion (Mahdavi et al., 2008). Tip geometry is widely varied, including basic pillar or cone shapes and more exotic suction-cup or mushroom-shaped tips. As researchers look to incorporate gecko-inspired adhesive materials in drug delivery systems, one key focus is material compatibility. Gecko-inspired adhesives are often fabricated from polymers such as PDMS (Liu et al., 2012) and polyurethane (Murphy et al., 2009), but gecko adhesives have also been made from biodegradable and biocompatible polymer PGSA (polyglycerol-(co-sebacate acrylate)) (Mahdavi et al., 2008) and carbon nanotubes (Chen, 2012).

Gecko-inspired materials have many medical applications. Because of their adhesive properties, as well as the capacity to be biocompatible and biodegradable, they are suitable for wound sealing (Mahdavi et al., 2008). As a tissue adhesive, they can be made to work in wet conditions, which is particularly notable for use as a mucoadhesive (Greiner, 2012). These adhesives have the potential to be incorporated into drug delivery devices, possibly combined with additional controlled release and transmucosal delivery techniques.

12.3.2 Nanowire Coatings for Improved Mucoadhesion

While gecko-inspired surfaces can be applied to a range of drug delivery devices for improved adhesion, whether it be the gut wall or the skin, nanowires specifically leverage the features of mucosal layers to improve mucoadhesion. When tethered to surfaces of particles or microdevices, nanowires have been shown to improve mucoadhesion (Fischer et al., 2011). The mechanism for improved adhesion with nanowire coatings is twofold: nanofeatures result in a drastically increased surface area-to-volume ratio that leads to enhanced van der Waals interactions, and entanglement with proteins in the mucosal layer anchors the nanofeatured particle or device. Similar to gecko-inspired structures, these nanowire systems can be surface modified by chemical conjugation, such as pegylation, for the additional enhancement of mucosal adhesion or penetration (Uskokovic et al., 2012). Thus far, improved adhesion from nanowire-coated systems has been limited to inorganic particles, such as silica or stainless steel, but as these techniques mature and successful translation is demonstrated, polymeric

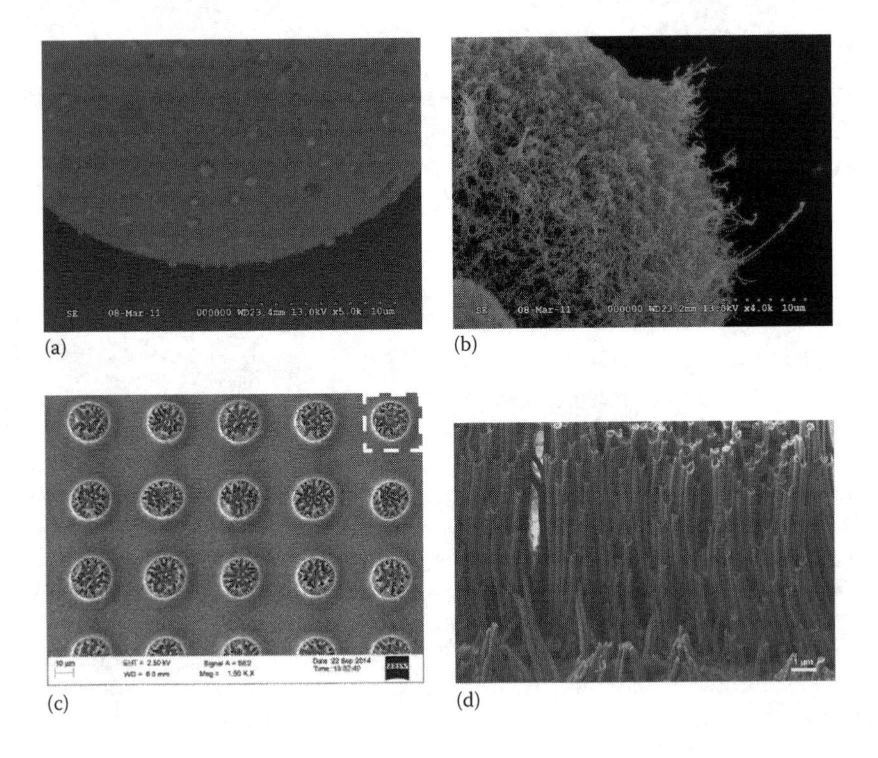

FIGURE 12.5 Nanowire coatings for enhanced mucoadhesion. Scanning electron microscope images of (a) Silica microparticle. (b) Silica microparticle with silicon nanowire coating for improved mucoadhesion. (c) Polycaprolactone microdevices with nanowire topography for improved mucoadhesion. (d) Nanowire topography on microdevices shown here. (From [a]: Reprinted with permission from Uskokovic, V., Lee, K., Lee, P., Fischer, K., Desai, T., *ACS Nano*, 6(9), 7832, 2012. Copyright 2009 American Chemical Society; [b]: Cade Fox, unpublished.)

examples are likely to follow as seen in other nanotechnologies. As discussed further in Section 12.4, nanofeatures can also be leveraged to disrupt tight junctions for improved permeability across epithelium. Drug delivery systems with nanofeatures could therefore be engineered to simultaneously improve adhesion and permeability, whether via a mucosal barrier or in a transdermal system (Figure 12.5).

Improved mucoadhesion in delivery systems should take advantage of delivering payload to the underlying epithelium locally to the vicinity of the adhered particle or device. Therefore, these systems are typically designed with a sustained release of drug tuned to the duration a device is adhered to the mucus. Devices with unidirectional release are being developed to further increase the percentage of drug available for absorption through the epithelium. In these planar devices, one surface is modified with nanotopography to improve mucoadhesion, and asymmetric layering causes drug release preferentially from only this side of the device. For oral delivery, this focuses drug release in the vicinity of epithelial cells where it can be absorbed and release of drug into the intestinal lumen can be avoided. Additionally, uptake of many drugs is limited by efflux transporter proteins and metabolizing enzymes, and the microdevice strategy of localizing drug may result in saturation of cellular transporters and enzymes, which is expected to improve the bioavailability of the released drug (International Transporter Consortium, 2010). Unidirectional drug release can be achieved with a multilayer structure composed of drug-impermeable and drug-permeable layers, where the drug-permeable layer may be contained within a reservoir in the impermeable layer to prevent drug release laterally from the device. Further, the drug-permeable layer may be the drug carrier as well, such as a drug-loaded hydrogel.

12.4 NANOTOPOGRAPHY IN DRUG DELIVERY SYSTEMS

Nanotopography has been integrated into drug delivery systems in several novel ways to improve drug transport, uptake, and loading. In some instances, nanotopography is used to enhance the physical properties of delivery vehicles to allow for higher loading capacity and surface area for drug release. In other approaches, nanotopography can alter tissue and cellular biology to impact drug uptake, drug permeability, or cellular proliferation and adhesion. Nanotopography has been demonstrated to influence endocytosis and free nanostructures can be internalized as drug carriers. For years researchers have demonstrated that microscale features interact with individual cells, influencing cell growth, alignment, and differentiation. Current research investigates how nanoscale features may interact on a subcellular or macromolecular level, influencing cell and tissue behavior through specific interactions with membrane proteins and receptors. Examples of nanostructures in a variety of materials have been shown to influence morphology, adhesion, and proliferation for a range of cell types, including fibroblasts, stem cells, and osteoblasts. In many cases, focal adhesions, clusters of transmembrane proteins that translate external mechanical cues to the cell interior, are influenced by nanostructures, which affects their localization and expression and elicits a response in the cytoskeleton and the activation or deactivation of signaling pathways. Cellular response to nanotopography is important to drug delivery for two principal reasons—improved uptake and permeability of therapeutics and reduced inflammatory or immunological response to delivery systems.

12.4.1 Nanotopography and Cellular Internalization

There are many drugs with intracellular sites of action that must pass through the cellular membrane to be effective. Drug transport through the cellular membrane may be active or passive and depends on drug properties and cell type. Nanotechnology can improve intracellular drug delivery by enhancing endocytosis, where a drug carrier is either internalized directly by a cell or exhibits enhanced properties for cell membrane permeation and uptake. For example, silicon nanowires (roughly 500 nm long and 100 nm in diameter) can be used as a carrier for the chemotherapeutic, doxorubicin (DOX). Nanoscale dimensions provide a high surface-to-volume ratio allowing high drug loading and a disposition for cellular internalization. Whereas typical nanocarrier systems have loading capacities less than 5,000 mg/g, up to 20,800 mg/g of DOX can be loaded onto these silicon nanowires. Exhibiting a pH-dependent release, silicon nanowires preferentially release the therapeutic after endocytosis in low-pH intracellular lysosomal compartments. In mouse tumor models, high loading capacity, sustained release, and lysosome-focused release result in prolonged accumulation of DOX in tumor cells and a reduction in tumor progression when compared to DOX alone (Peng et al., 2013).

In an alternative approach, rather than being fully internalized, hollow nanowires (termed *nanostraws*) penetrate the cell membrane and allow a direct path to the intracellular space. Inorganic (alumina, silicon, carbon, silica) and polymeric nanostraws (or nanotubes) are fabricated using several approaches, including vapor–liquid–solid (VLS) growth for silicon wires to template silica nanotubes (Fan et al., 2003), track etching, focused ion beam (FIB) etching (Han et al., 2005), and templating (Schilling et al., 2002). Nanostraws can have diameters ranging from tens of nanometers to several microns with lengths up to hundreds of microns. Drug payload may be integrated with nanostraws via direct surface adsorption onto nanostraws, a reservoir backing nanostraw arrays, or direct incorporation of drug into the nanostraw material. The use of nanostraws for intracellular delivery has been demonstrated in many *in vitro* systems to transfect plasmids and genetic material as well as for delivery of large and small molecules (Peer et al., 2012; Shalek et al., 2010). Variations among nanostraw approaches are vast and encompass a variety of materials and size scales, with a range of additional features, including porous structures, erodible or detachable needles, or integration with microfluidics and electroporation systems (Chiappini et al., 2010). A more detailed review of nanostraw technology can be found in *Semiconducting Silicon Nanowires for Biomedical Applications* (Chiappini and Almeida, 2014).

So far, we have discussed one scenario where nanowires are themselves internalized and another where nanostraws physically penetrate the cell membrane and open a direct path to the cytoplasm. A third example of nanostructure-enhanced cellular uptake used monolayers of nanoparticles or nanopillars embedded or patterned on surfaces that modulate endocytosis when brought into contact with cells. While enhanced uptake has been demonstrated with a range of materials, investigators continue to explore the impact of nanostructure size, distribution, and aspect ratio on endocytosis. In siRNA delivery to neural stem cells via silica nanoparticles, GFP knockdown was increased by a factor of 4 for 100 nm particles as a result of enhanced siRNA transfection compared to 700 nm particles (Solanki et al., 2013). In addition

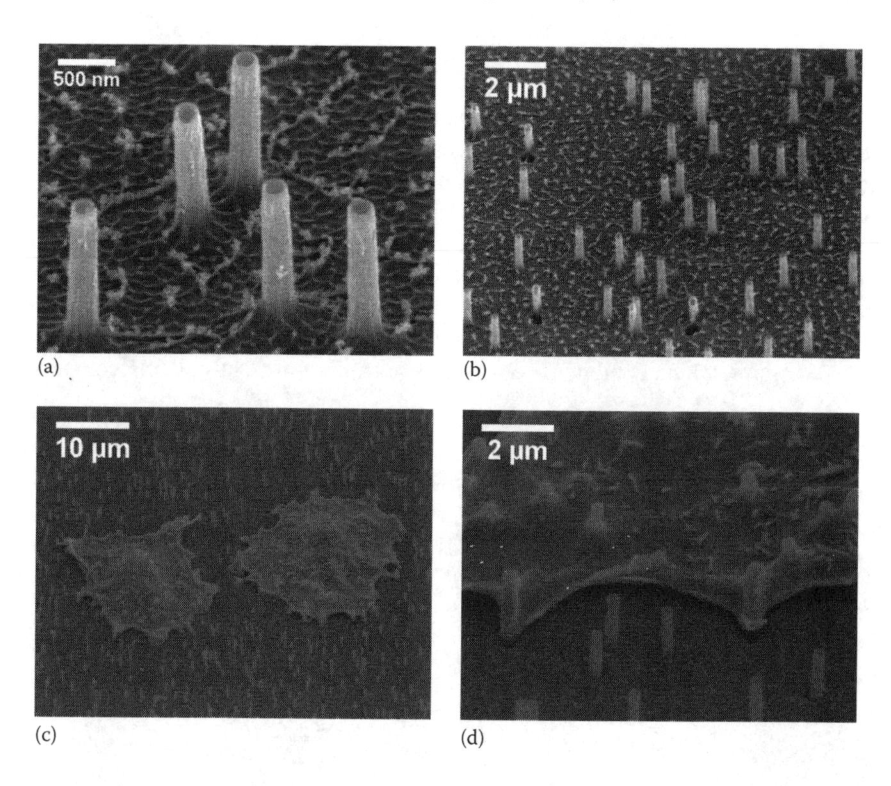

FIGURE 12.6 Scanning electron microscope images of nanostraws. Nanostraws (250 nm diameter, 0.2 straw/μm^2 density [a, b]), interface with cell membranes [c, d] to directly deliver drugs or genetic material to the intracellular space. (Reprinted with permission from Xie, X. et al., ACS Nano, 7(5), 4351, 2013. Copyright 2013 American Chemical Society.)

to size, nanostructure geometry and distribution may also affect endocytosis—in fact, the effect of size and shape may differ with each cell type! For example, in a study of cellular FITC-Dextran uptake, nano- and microtopography in PMMA increased uptake in COS7 kidney cells and human mesenchymal stem cells relative to unpatterned controls; the same topography lacked an uptake effect in MCF7 breast cancer cells (Teo et al., 2011). This approach is in development for *in vitro* transfection and is a potentially powerful tool to selectively affect cellular uptake from drug delivery systems. Nanotopography could be added to drug delivery systems for simultaneous cellular uptake enhancement and drug release; more complex nanostructured systems might serve first as a mechanical cue to increase endocytosis and subsequently detach their drug carrier to be internalized (Figure 12.6).

12.4.2 Nanotopography and Epithelial Permeability

At the cellular level, nanostructures are capable of improving intracellular delivery of drugs, genetic material, and drug carriers. On the tissue level, nanotopography can also increase epithelial drug permeability through mechanotransduction. Epithelia are interconnected cell layers that line organs, such as the intestine, cornea, and lung. As protective membranes optimized to exclude foreign bodies and toxic substances from the body, epithelia are a notable barrier to drug delivery. However, the most desirable administration routes for low-cost, patient-compliant therapeutics require delivery across an epithelial membrane, so tools that enhance drug permeability are essential (Figure 12.7).

As mentioned, cells have been shown to respond to nanotopography through the formation and localization of focal adhesions and associated downstream mechanotransduction and signaling pathways. In an intestinal epithelial layer, interactions with polypropylene nanopillars led to increased transport of macromolecules through the epithelium, which exhibited a characteristic ruffling of tight-junction complexes

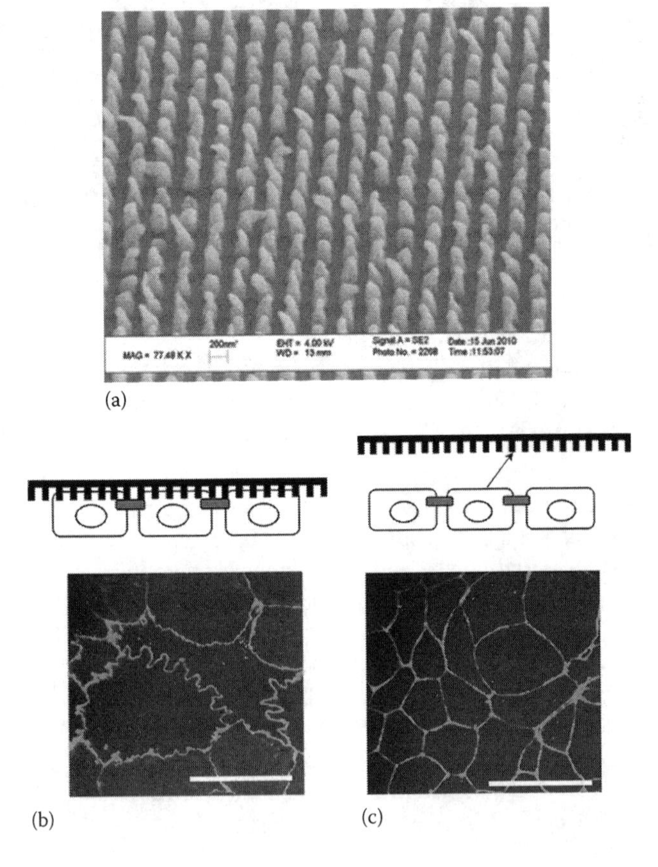

FIGURE 12.7 Polypropylene nanostructures disrupt cell–cell junctions. Contact with polypropylene nanostructures (a) causes a disruption to the tight junctions, causing a ruffling of the cell–cell boundary (b) compared to the normal smooth cell–cell interface that is reacquired with the removal of the structured surface (c). Cell–cell boundary is visualized with ZO-1 antibody stain, seen in white. (Reprinted with permission from Kam, K. et al., *Nanoletters*, 13(1), 164, 2013. Copyright 2012 American Chemical Society.)

between adjacent cells (Kam, 2013). Formation of focal adhesions between cells and nanostructures is hypothesized to weaken or disrupt intercellular adhesions, allowing for increased diffusion of molecules between adjacent cells. In addition to nanopatterned polypropylene thin films, this concept has also been demonstrated with silicon nanowire-coated silica particles. Interestingly, nanopatterned films with low aspect ratio pillars (300 nm tall and 200 nm wide) showed an effect that high aspect ratio pillars (16 μm high and 800 nm wide) lacked. Additional work is required to understand how these different nanostructure geometries and dimensions interact with cells to modulate adhesion complexes and permeability.

12.4.3 Nanotopography and Fibrosis

In addition to applications in drug delivery, nanotopography is also being integrated in drug delivery devices, implants, and tissue engineering scaffolds to improve long-term biocompatibility. Implanting a foreign object into the body commonly elicits an immune or fibrotic response. Implantation damage to the surrounding tissue can trigger fibrosis even for biocompatible materials, which may ultimately lead to fibrotic encapsulation. This response is not only harmful, resulting in irreversible damage and scar tissue, but fibrous encapsulation may impact release rates and device function for drug delivery systems. Surface modifications and coatings are a common approach to reduce fibrosis in medical implants and drug delivery devices. While not fully understood, several groups have recently observed that micro- and nanotopography can reduce fibroblast proliferation and fibrotic response (Baker, 2012; Kam, 2014; Smith et al., 2011). As discussed, nanotopographical features are on a size scale that allows interaction with individual cells as well as functional proteins. Nanotopography is hypothesized to affect adsorption and distribution of extracellular matrix proteins on implant surfaces and to interact directly with fibroblasts: this reduces proliferation and gene expression of key components of the fibrotic response (Kam, 2014).

Decreased proliferation of fibrotic phenotypes and reduced expression of marker proteins for fibrosis have been observed *in vitro*, and reduced fibrotic encapsulation has been observed *in vivo* for a range of geometries and materials. While still under investigation, reduction of fibrotic response by nanotopography may depend on the geometry and distribution of nanofeatures. For example, Kam et al. observed that an aspect ratio greater than five on polymeric thin films was necessary for nanopillars to reduce the fibrotic response (Kam, 2014); Von Recuum et al. observed that for silicone implants, uniform topography with 1–3 μm dimensions was ideal for biocompatibility (Von Recuum, 2012). Smith et al. demonstrated a reduced fibrotic encapsulation in a soft-tissue rat model for a titanium oxide nanotube–coated implant compared to a unstructured titania implant (Smith et al., 2011). As precision fabrication technologies expand to control nanofeatures in a wider range of materials, the properties of substrate material, topography dimensions, and their coupling will emerge. Translation of these recent studies on surface topography will improve safety profile and tolerability of novel biomedical and drug delivery devices, allowing for improved function and extended lifetimes for these systems (Figure 12.8).

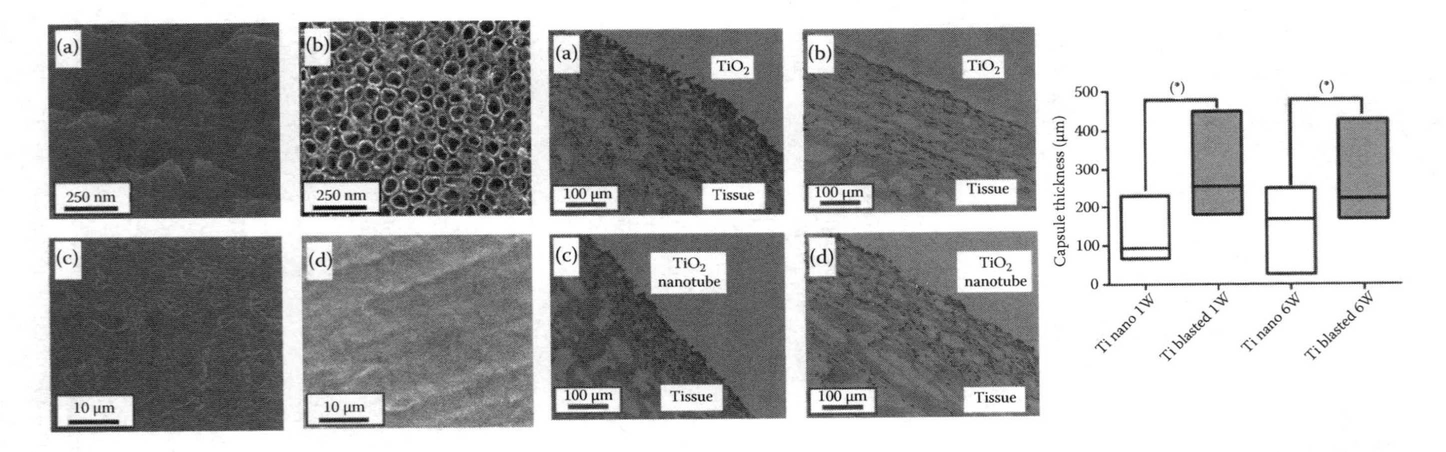

FIGURE 12.8 Nanotopography for reduced fibrosis. Titanium oxide nanotubes on implant surface (b, d—left) reduce fibrotic response in soft-tissue rat model as illustrated by histology images and quantified as fibrotic capsule thickness (c, d—center, "Ti nano" image right) compared to a flat titanium oxide control (a, c—left; a, b—center; "Ti blasted" image right). (From Smith, G. et al., *Acta Biomater.*, 7(8), 3209, 2011.)

12.5 CONCLUSION

In this chapter, we have discussed a variety of nanofeatured materials in development for improved drug delivery. Among these nanofeatured materials are inorganic and polymeric materials that are employed as nanoporous membranes, nanostructures as surface topography, and nanostraws. Such materials are being incorporated in drug delivery systems for drug loading, controlled drug release, improved tissue or mucosal adhesion, and enhanced intracellular drug uptake as well as through epithelium. Many of these materials and effects are limited to the dimensions and patterns possible with contemporary fabrication techniques. As the field of nano- and microfabrication advances, additional research may be able to elucidate the fundamental effects and underlying mechanisms for some of these nanofeatured systems.

A basic understanding of how nanotechnology and nanofeatured materials are integrated into drug delivery systems builds upon fundamental knowledge in cellular biology, mass transport, and material science and provides the necessary tools for creative design and critical evaluation of novel drug delivery systems based on nanotechnology. We have provided an introduction covering the basic concepts applied in each of these approaches with examples spanning a range of technologies in various developmental phases. While many of these technologies require further development to reach commercial products, as our understanding of and experience with nanotechnology and its utility in medicine expands, we expect to see the translation of such systems into viable products. The breadth of nanofeatured materials in drug delivery is vast, and the technologies highlighted here are not intended to comprise an exhaustive list of the concepts and approaches in development. In addition to alternative approaches, complex drug delivery systems also exist that combine multiple nanofeatured components for stimuli-responsive, targeted, and more effective drug delivery.

REFERENCES

Baker, D. (2012). Effect of microtopography on fibrocyte responses and fibrotic tissue reactions at the interface. In *Proteins at Interfaces III State of the Art*, Horbett, T., Brash, J. L., Norde, W., eds., Vol. 1120, pp. 339–353. Washington, DC: American Chemical Society. doi:10.1021/bk-2012-1120.ch015.

Bernards, D. and Desai, T. (2010). Nanoscale porosity in polymer films: Fabrication and therapeutic applications. *Soft Matter*, *6*, 1621–1631.

Bernards, D. A., Lance, K. D., Ciaccio, N. A., and Desai, T. A. (2012). Nanostructured thin film polymer devices for constant-rate protein delivery. *Nano Letters*, *12*(10), 5355–5361.

Briggs, E., Walpole, A., Wilshaw, P., Karlsson, M., and Palsgard, E. (2004). Formation of highly adherent nanoporous alumina on Ti-based substrates: A novel bone implant coating. *Journal of Materials Science: Materials in Medicine*, *15*(9), 1021–1029.

Caldorera-Moore, M. and Peppas, N. A. (2009). Micro- and nanotechnologies for intelligent and responsive biomaterial-based medical systems. *Advanced Drug Delivery Reviews*, *61*(15), 1391–1401.

Chen, B. (2012). Adhesive properties of gecko inspired mimetic via micropatterned carbon nanotube forests. *Journal of Physical Chemistry C*, *116*(37), 20047–20053.

Chiappini, C. and Almeida, C. (2014). Silicon nanoneedles for drug delivery. In *Semiconducting Silicon Nanowires for Biomedical Applications*, Coffer, J. L., ed., pp. 144–167. Waltham, MA: Woodhead Publishing Limited. doi:10.1533/9780857097712.2.144.

Chiappini, C., Liu, X., Fakhoury, J., and Ferrari, M. (2010). Biodegradable porous silicon barcode nanowires with defined geometry. *Advanced Functional Materials*, *20*(14), 2231–2239.

Desai, T., Hansford, D., and Ferrari, M. (2000). Micromachined interfaces: New approaches in cell immunoisolation and biomolecular separation. *Biomolecular Engineering*, *17*(1), 23–36.

Dewan, S. et al. (2014). *Global Markets and Technologies for Advanced Drug Delivery Systems—Focus on End Users*. Wellesley, MA: BCC Research.

Fan, R., Wu, Y., Li, D., Yue, M., Majumdar, A., and Yang, P. (2003). Fabrication of silica nanotube arrays from vertical silicon nanowire templates. *Journal of the American Chemical Society*, *128*(18), 5254–5255.

Fischer, K., Nagaraj, G., Daniels, R., Li, E., Cowles, V., Miller, J. et al. (2011). Hierarchical nanoengineered surfaces for enhanced cytoadhesion and drug delivery. *Biomaterials*, *32*, 3499–3506.

Gardner, P. (2006). Use of a nanopore membrane in a novel drug delivery device. *Future Drug Delivery*, 59–60.

Greiner, C. (2012). Gecko-inspired nanomaterials. *Nanotechnologies for the Life Sciences*, Wiley-VCH, Verlag GmbH & Co, KGaA, p. 1. 10.1002/9783527610419.ntls0203.

Gultepe, E., Nagesha, D., Sridhar, S., and Amiji, M. (2010). Nanoporous inorganic membranes or coatings for sustained drug delivery in implantable devices. *Advanced Drug Delivery Reviews*, *62*, 305–315.

Han, S., Nakamura, C., Obatav, I., Nakamura, N., and Miyake, J. (2005). A molecular delivery system by using AFM and nanoneedle. *Biosensors and Bioelectronics*, *20*(10), 2120–2125.

International Transporter Consortium. (2010). Membrane transporters in drug development. *Nature Reviews, Drug Discovery*, *9*(3), 215–236.

Jackson, E. and Hillmyer, M. (2010). Nanoporous membranes derived from block copolymers: From drug delivery to water filtration. *ACS Nano*, *4*(7), 3548–3553.

James, H. P., John, R., Anuj, A., and Annop, K. R. (2014). Smart polymers for the controlled delivery of drugs—A concise overview. *Acta Pharmaceutica Sinica B*, *4*(2), 120–127.

Kam, K. R., Walsh, L. A., Bock, S. M., Koval, M., Fischer, K. E., Ross, R. F., and Desai, T. A. (January 2013). Nanostructure-mediated transport of biologics across epithelial tissue: Enhancing permeability via nanotopography. *Nano Letters*, *13*(1), 164–171.

Lee, W. and Park, S.-J. (2014). Porous anodic aluminum oxide: Anodization and templated synthesis of functional nanostructures. *Chemical Reviews*, *114*(15), 7487–7556.

Liu, S., Zhang, P., Lu, H., Zhang, C., and Xia, Q. (2012). Fabrication of high aspect ratio microfiber arrays that mimic gecko foot hairs. *Bionic Engineering*, *57*(4), 404–408.

Mahdavi, A., Ferriera, L., Sundback, C., Nichol, J., Chan, E., Carter, D. et al. (2008). A biodegradable and biocompatible gecko-inspired tissue adhesive. *PNAS*, *105*(7), 2307–23012.

Martin, C. (1994). Nanomaterials: A membrane based synthetic approach. *Science*, *266*, 1961–1966.

Martin, F., Walczak, R., Bioarski, A., Cohen, M., West, T., Consetino, C., and Ferrari, M. (2005). Tailoring width of microfabricated nanochannels to solute size can be used to control diffusion kinetics. *Journal of Controlled Release*, *102*, 123–133.

Masuda, H., Yamada, H., Satoh, M., Asoh, H., Nakao, M., and Tamamura, T. (1997). Highly ordered nanochannel-array architecture in anodic alumina. *Applied Physics Letters*, *71*, 2770.

Moxon, K., Kalkhoran, N., Markert, M., Sambito, M., McKenzie, J., and Webster, J. (2004). Nanostructured surface modification of ceramic-based microelectrodes to enhance biocompatibility for a direct brain-machine interface. *IEEE Transactions on Biomedical Engineering*, *51*(6), 881–889.

Murphy, M., Aksak, B., and Sitti, M. (2009). Gecko-inspired directional and controllable adhesion. *Small*, *5*(2), 170–175.

Park, D. (2007). The fabrication of thin films with nanopores and nanogrooves from block copolymer thin films on the neutral surface of self-assembled monolayers. *Nanotechnology*, *18*(35), 355304. doi:10.1088/0957-4484/18/35/355304.

Peer, E., Artzy-Schnirman, A., Gepstein, L., and Sivan, U. (2012). Hollow nanoneedle array and its utilization for repeated administration of biomolecules to the same cells. *ACS Nano*, *6*(6), 4940–4946.

Peng, F., Su, Y., Wei, X., Lu, Y., Zhou, Y., Zhong, Y. et al. (2013). Silicon-nanowire-based nanocarriers with ultrahigh drug-loading capacity for in vitro and in vivo cancer therapy. *Angewandte Chemie International Edition*, *52*, 1457–1461.

Rao, V., Amar, J., Avasthi, D., and Narayana, R. (2003). Etched ion track polymer membranes for sustained drug delivery. *Radiation Measurements*, *36*(21), 585–589.

Rattan, S. and Sehgal, T. (2012). Stimuli responsive polymeric membranes through graft copolymerization of N-isopropylacrylamide onto polycarbonate track etched membranes for biomedical applications. *Procedia Chemistry*, *4*, 194–201.

Schilling, J., Steinhart, M., Wendorff, J., Greiner, A., Wehrspohn, R., Nielsch, K. et al. (2002). Polymer nanotubes by wetting of ordered porous templates. *Science*, *296*(5575), 1997–1997.

Shalek, A., Robinson, J., Karp, E., Lee, J., Ahn, D., Yoon, M. et al. (2010). Vertical silicon nanowires as a universal platform for delivery biomolecules into cells. *PNAS*, *107*(5), 1870–1875. doi:10.1073/pnas.0909350107.

Smith, G., Chamberlain, L., Faxius, L., Johnston, G., Jun, S., and Bjursten, L. (2011). Soft tissue response to titanium dioxide nanotube modified implants. *Acta Biomaterialia*, *7*(8), 3209–3215.

Solanki, A., Shah, S., Yin, P., and Lee, K. (2013). Nanotopography-mediated reverse uptake for siRNA delivery into neural stem cells to enhance neuronal differentiation. *Scientific Reports*, *3*, 1553.

Teo, B., Goh, S.-H., Kustandi, T., Loh, W., Low, H., and Yim, E. (2011). The effect of micro and nanotopography on endocytosis in drug and gene delivery systems. *Biomaterials*, *32*(36), 9866–9875. 10.1016/j.biomaterials.2011.08.038.

Tong, H., Jansen, H., Gadgil, V., BOstan, C., Berenschot, E., van Rijn, C., and Elwenspoek, M. (2004). Silicon nitride nanosieve membrane. *Nano Letters*, *4*(2), 283–287.

Ulbricht, M. (2006). Advanced functional polymer membranes. *Polymer*, *47*(7), 2217–2262. 10.1016/j.polymer.2006.01.084.

Uskokovic, V., Lee, K., Lee, P., Fischer, K., and Desai, T. (2012). Shape effect in the design of nanowire-coated microparticles as transepithelial drug delivery devices. *ACS Nano*, *6*(9), 7832–7841.

Vriezekolk, E., de Weerd, E., de Vos, W., and Nijmeijer, K. (2014). Control of pore size and pore uniformity in films based on self-assembling block copolymers. *Journal of Polymer Science Part B: Polymer Physics*, *52*(23), 568–1579. 10.1002/polb.23600.

Walczak, R. B., Cohen, M., West, T., Melnik, K., Shapiro, J., Sharma, S., and Ferrari, M. (2005). Long-term biocompatibility of NanoGATE drug delivery implant. *Nanobiotechnology*, *1*(1), 35.

Wang, C., Wang, T., and Wang, Q. (2013). Low dielectric, nanoporous polyimide thin films prepared from block copolymer templating. *eXPRESS Polymer Letters*, *7*(8), 667–672.

Xie, X., Xu, A., Leal-Oritz, S., Cao, Y., Garner, C., and Melosh, N. (2013). Nanostraw-electroporation system for highly efficient intracellular delivery and transfection. *ACS Nano*, *7*(5), 4351–4358.

Yin, J., Yao, X., Liou, J., Sun, W., Sun, Y., and Wang, Y. (2013). Membranes with highly ordered straight nanopores by selective swelling of fast perpendicularly aligned block copolymers. *ACS Nano*, *7*(11), 9961–9974.

Index

H

I

L

M

N

O

P

Q

R

S

T